CHEMOMETRICS
IN EXCEL

CHEMOMETRICS IN EXCEL

ALEXEY L. POMERANTSEV

WILEY

Published by John Wiley & Sons, Inc., Hoboken, New Jersey
Published simultaneously in Canada

Library of Congress Cataloging-in-Publication Data:

Pomerantsev, A. L. (Aleksei Leonidovich), 1954–
[Khemometrika v Excel. English]
Chemometrics in Excel / Alexey L. Pomerantsev.
pages cm
Includes index.
ISBN 978-1-118-60535-6 (cloth)
1. Chemometrics—Data processing. 2. Microsoft Excel (Computer file) I. Title.
QD75.4.C45P72 2014
543.0285'554—dc23

2013046049

Printed in the United States of America

ISBN: 9781118605356

10 9 8 7 6 5 4 3 2 1

To Vicky and Yanek

CONTENTS

PREFACE

Chemometrics is a very practical discipline. To learn it, one should not only understand the numerous chemometric methods but also adopt their practical application. This book can assist you in this difficult task. It is aimed primarily at those who are interested in the analysis of experimental data: chemists, physicists, biologists, and others. It can also serve as an introduction for beginners who have started to learn about multivariate data analysis.

The conventional way of chemometrics training utilizes either specialized programs (the Unscrambler, SIMCA, etc.) or the MATLAB®. We suggest our own method that employs the abilities of the world's most popular program, Microsoft Excel®. However, the main chemometric algorithms, for example, projection methods (principal component analysis (PCA), partial least squares (PLS)) are difficult to implement using basic Excel facilities. Therefore, we have developed a special supplement to the standard Excel version called Chemometrics Add-In, which can be used to perform such calculations.

This book is not only a manual but also a practical guide to multivariate data analysis techniques. Almost every chapter is accompanied by a special Excel workbook, which contains calculations related to the techniques discussed. For better understanding of the books's material, the reader should constantly refer to these workbooks and try to repeat the calculations. The workbooks with examples as well as Chemometrics Add-In are collected at the Wiley Book Support site (http://booksupport.wiley.com). The installation guide is presented in Chapter 3. We strongly recommend installing and using the software while studying this book.

My colleague Dr. Oxana Rodionova played a very active role in the development of the book. She has designed and developed the Chemometrics Add-In software. She has also assisted in writing many chapters of this book. Therefore, it can be stated without exaggeration that this book would have never been completed without her help. Chapter 15 was written together with my student, Yevgeny V. Mikhailov.

I am deeply grateful to the language editors Svetlana and Pieter Rol who contributed a lot to arrange this book.

We plan to revise and supplement the book continuously. You are welcome to send us your comments and suggestions at rcs@chph.ras.ru.

PART I

INTRODUCTION

1

WHAT IS CHEMOMETRICS?

Preceding chapters	
Dependent chapters	
Matrix skills	Low
Statistical skills	Low
Excel skills	Low
Chemometric skills	Low
Chemometrics Add-In	Not used
Accompanying workbook	None

Chemometrics is what chemometricians do

—a popular wisdom

Chemometricians are the people who drink beer and steal ideas from statisticians

—Svante Wold

1.1 SUBJECT OF CHEMOMETRICS

What is chemometrics and what it does can be explained in different ways using more or less sophisticated words. At present, we have no generally accepted definition and, apparently, there never will be. The most popular is designation by D. Massart that chemometrics is *the chemical discipline that uses mathematical, statistical and other methods employing formal logic to design or select optimal measurement procedures and experiments, and to provide maximum relevant chemical information by analyzing chemical data.* This definition gathered a lot of criticism; therefore, new delineations have been suggested. For instance, S. Wold suggested the following formula: *Chemometrics solves the following*

Chemometrics in Excel, First Edition. Alexey L. Pomerantsev.

tasks in the field of chemistry: how to get chemically relevant information out of measured chemical data, how to represent and display this information, and how to get such information into data.

This also was not adopted and everybody agreed that the best way to explain the chemometric essence is an old method, proven in other equally obscure areas. It was declared that *chemometrics is what chemometricians do*. Actually, it was a clear rip-off because a similar idea has already been published,[1] but nothing better has been thought up since.

Well, really, what are these chemometricians doing? Here is a small collection of subjects found in the papers published over the past 5 years. Substantially, chemometricians perform the following activities:

- manage the production of semiconductors, aspirin, beer, and vodka;
- investigate the causes of destruction of documents written by ancient Gallic ink;
- conduct doping control in sport;
- determine the composition of ancient Egyptian makeup;
- localize gold deposit in Sweden;
- control the state of forests in Canada;
- diagnose arthritis and cancer in the early stages;
- investigate the organics in comets;
- select pigs' diet;
- check how a diet affects the mental capacity;
- find traces of cocaine on banknotes collected in the British Parliament;
- detect counterfeit drugs;
- decide on the origin of wines, oils, and pigments.

As long as everybody has clearly apprehended *what* chemometricians do, it remains to explain *how* they do this. However, initially, we have to be acquainted with the basic principles of their activity. Those are not numerous, just three rules, and they are as simple as ABC. The first principle states that *no data are redundant*, that is, a lot is better than nothing. In practice, this means that if you record a spectrum, it would be stupid to throw out all readings except a few characteristic wavelengths. Scientifically speaking, this is the multivariate methodology in the experimental design and data analysis.

Any data contains undesirable component called *noise*. The nature of noise can be different and, in many cases, a noise is just a part of data that does not contain relevant information. Which data component should be considered as noise and which component as information is always decided considering the goals and the methods used to achieve them. This is the second chemometric principle: *noise is what we do not need*.

However, the noise and redundancy necessarily provoke nonsystematic (i.e., noncausal) but random (i.e., correlation) relations between the variables. The difference between causality and correlation has been illustrated humorously in a book by Box and Hunters.[2] There is an example of a high positive correlation between the number of inhabitants and the number of storks in Oldenburg (Germany) for the period from 1930 to 1936. This is good news for those who believe that the stork is a key factor in baby boom!

[1] P.W. Bridgman. On scientific method, in *Reflections of a Physicist*, Philosophical Library, NY, 1955.

[2] G.E.P. Box, W.G. Hunter, J.S. Hunter. *Statistics for Experimenters*, John Wiley & Sons Inc., NY, 1978.

However, the reason that these two variables are correlated is very simple. There was a somewhat hidden third variable, which had pair-wise causal relationship with both variables.[3] Therefore, the third principle of chemometrics states as follows: *seek for hidden variables*.

1.2 HISTORICAL DIGRESSION

Strangely enough, when can somebody claim beyond a doubt that a science was born on a certain date, in a certain place, and under specific circumstances? Who dares, for example, to specify where and when chemistry appeared first? It is only clear that it was long ago and very likely it did not happen under very nice circumstances, and it was related to an urgent necessity to quietly send a sixth priest of the Anubis temple to Kingdom Come or something. But we are quite certain that chemometrics was born on the evening, June 10, 1974, in a small Tex-Mex restaurant in Seattle, as the result of a pretty noisy revelry, arranged on the occasion of Swante's return home in Umea, after a long training under the renowned statistician George Box.

The American side was represented by the chemist Bruce Kowalski with his disciples, who had been intensively developing a software package for chemical data analysis. In fact, the word "chemometrics" has been used since the early 1970s by Svante Wold and his team from the University of Umea, in order to identify, briefly, what they were doing. Actually, they were engaged in an intensive implementation of the interesting ideas produced by Svante's father, Herman Wold, who was a famous statistician, developing many techniques (including the now famous method of the projection to latent structures – PLS) for data analysis in economic and social sciences.

Apparently, with an analogy to psychometrics, biometrics, and other related disciplines, the term *chemometrics* was coined. The circumstances and the place of birth have left an indelible life mark for the science itself and for the people who are involved in chemometrics. Much later, Svante Wold, being already a very venerable scholar, joked at a conference "chemometricians are the people who drink beer and steal ideas from statisticians." On the other hand, perhaps, the reason is that the first chemometrician, William Gosset, better known under the pseudonym Student, was employed as an analyst at the Guinness brewery. Anyway, beer and chemometric have gone together for many years. I remember a story told by another famous chemometrician, Agnar Höskuldsson at a Winter Symposium on Chemometrics held near Moscow, in February 2005. The atmosphere at the ceremony of awarding a young Russian chemometrician was semi-official, and Agnar, who presented the award, delivered a fitting address. This remarkable history, almost a parable, is worth presenting word by word. Thus spoke Agnar.

Chemometricians are the cheerful people who like to drink beer and sing merry songs. In the mid-1970s, when I (i.e., Agnar) first met Svante, he had only two students. And so, we, four men, were sitting in an Umea pub in the evening, drank beer, and sang merry songs. After 2 hours, I asked Svante:

"It may already be enough, and is it time to go home?"

"No," Svante said, "still not enough, we will drink beer and sing merry songs."

[3]The subsequent investigations have revealed the nature of a hidden variable, which was proved to be the area of the fields raiseing the cabbage in the Oldenburg neighborhood.

Another hour passed, and I asked Svante again

"It may already be enough, and is it time to go home?"
"No," he answered, "we still have to drink beer and sing more songs."

Meanwhile, a few students went down the street, near the pub where we were. They heard we had fun, and stopped and began to ask of the passersby:

"Who are these cheerful people who are singing these merry songs?"

No one could answer them, so the students came in the pub and asked Svante:

"Who are you guys, and why are you singing these merry songs?"

And Svante replied:

"We are chemometricians, the cheerful people, who like to drink beer and sing merry songs. Sit down, drink beer, and sing along with us."

That was the beginning of the famous Scandinavian school of chemometrics, which gave us many great scientists and cheerful people.

The appearance of chemometrics in Russia was also marked by a funny story. Soon after the establishment of the Russian Chemometrics Society in 1997, my colleagues decided that it was a proper time to announce our existence *urbi et orbi*. Just that time, we received the first call to the All-Russian conference "Mathematical Methods in Chemistry," which was to be held in the summer of 1998, in Vladimir city. This was a very popular scientific forum established in 1972 by the Karpov Institute of Physical Chemistry. The conference was a place where many scientists reported on their results; the hottest discussions of the latest mathematical methods applied in chemistry were conducted. We had a hope that using this high rostrum, we could tell fellows about this wonderful "chemometrics" developed by our friends around the world and how interesting and promising this science is. The abstract was written in the name of the Russian Chemometrics Society and timely delivered to the organizing committee.

Time had passed, but there was no response. The deadline was approaching, the program and the lists of speakers were already published, but our contribution was not included. Our bewildered requests went unanswered so we decided that there should be some misunderstanding. As Vladimir is rather close to Moscow, it was decided that we would take a chance and visit the conference without invitation in order to sort things out on the spot. So we did, appearing on the first conference day for the participant registration. This raised a big stink accompanied with an explicit fright, screaming, loud statements against provocation that finally resulted in the appearance of very responsible *tovarishchi*.

Things were heading toward us getting out of the university, facing the tough administrative sanctions, arrest, and finally imprisonment. Fortunately, in an hour, through the efforts of our old friends from previous conferences, the situation was resolved and we got the casus clarified. The reason was that the unsophisticated Vladimir scientists believed that chemometrics is something like Scientology and our society is a kind of a harmful sect that tries to use the science conference as a chance to embarrass the unstable minds of Vladimir's people and turn their comprehension away from true scientific values. At the end, we were saved by the fact that in the University library, a book titled "Chemometrics,"[4] published in 1987 was found. Final reconciliation occurred during the closing banquet, but our lecture, *de bene esse*, was not permitted.

It is possible that the committee members were guided by the sound judgment of Auguste Comte, who in 1825 warned[5] "Every attempt to employ mathematical methods in the study of chemical questions must be considered profoundly irrational and contrary to the spirit of chemistry. If mathematical analysis should ever hold a prominent place in chemistry – an aberration which is happily almost impossible – it would occasion a rapid and widespread degeneration of that science."

[4]M.A. Sharaf, D.L. Illman, B.R. Kowalski. *Chemometrics*, Wiley, New York, 1986 (Russian translation: Mir, 1987).

[5]A. Comte. *Cours de Philosophie Positive*, Bachelier, Paris, 1830.

2

WHAT THE BOOK IS ABOUT?

Preceding chapters	
Dependent chapters	
Matrix skills	Low
Statistical skills	Low
Excel skills	Low
Chemometric skills	Low
Chemometrics Add-In	Not used
Accompanying workbook	None

2.1 USEFUL HINTS

We remind the reader that this book is not only a text but also a practical guide. Almost every chapter is accompanied by a special Excel workbook, which contains the relevant calculations illustrating the discussed techniques. To understand the book's content better, we recommend the reader to refer to the corresponding workbook. The supporting materials can be found at the Wiley Book Support site[1]. These XLS files are intended for use with MS Excel versions 2003, 2007, and 2010.

In order to perform the calculations required for the projection models, you should install a special software called Chemometrics Add-In for Excel on your computer. To learn how to do this, please refer to Chapter 3. A full version of the add-in is also presented at the Wiley Book Support site.

Excel allows the user to set different levels of security (Section 7.3.1) when working with the macros. If it is set at the highest level of security, one is unable to use the special

[1] http://booksupport.wiley.com

Chemometrics in Excel, First Edition. Alexey L. Pomerantsev.

functions collected in our software. Therefore, the recommended level of security is *low*. Do not forget to restore the original level after finishing the work. All accompanying workbooks have the option **manually** preset for calculation. Therefore, after changing the XLS file's contents, it is necessary to press the **CTRL+ALT+F9** key combination to calculate the result. However, if one is using the file while Chemometrics Add-In has not been installed, this key combination removes the values in the cells that are associated with this add-in.

2.2 BOOK SYLLABUS

The book is divided into four parts. The first part is an **Introduction**, which contains most general information that can be useful when working with this book. The second part called **The Basics** collects the basic knowledge necessary to understand the main topics of the book. These chapters cannot serve as textbooks on the topics; they are rather synopses, that is, brief guides to the relevant mathematical areas. A more systematic and in-depth explanation can be found in the literature referenced in Chapter 4. The third part is the major book component. It gives an idea of the essence of **Chemometrics**, the main problems addressed, and the methods used to solve them. The last part contains **supplementary** materials that extend knowledge in specific areas.

Chapters	Workbook	Chemometrics Add In
Introduction		
1. What is chemometrics?	No	Not used
2. Whom is this book for?	No	Not used
3. Installation of Chemometrics Add-In	Installation.xls	Not used
4. Further reading on Chemometrics	No	Not used
The Basics		
5. Matrices and vectors	Matrix.xls	Not used
6. Statistics	Statistics.xls	Not used
7. Matrix calculations in Excel	Excel.xls	Not used
8. Projection methods in Excel	Projection.xls	Used
Chemometrics		
9. Principal component analysis (PCA)	People.xls	Used
10. Calibration	Calibration.xls	Used
11. Classification	Iris.xls	Used
12. Multivariate curve resolution	MCR.xls	Used
Supplements		
13. Extension of Chemometrics Add-In	Tricks.xls	Used
14. Kinetic modeling of spectral data	Grey.xls	Not used
15. MATLAB®. Beginner's Guide	No	Not used

Each chapter starts with a short list explaining the chapter's status. For example,

Preceding chapters	
Dependent chapters	
Matrix skills	Low
Statistical skills	Basic
Excel skills	Basic
Chemometric skills	Low
Chemometrics Add-In	Not used
Accompanying workbook	Statistics.xls

The "Preceding chapters" indicates the chapters that provide the basis for the current one; the "Dependent chapters" field refers to the chapters that use concepts from the current one. The list of skills (low, basic, or advanced) describes the levels of these skills required to understand the content. The last two items provide information on the necessity to use Chemometrics Add-In (used or not used) in the chapter and on the availability of an accompanying Excel workbook (name or none) that illustrates the chapter.

2.3 NOTATIONS

The following notation rules are used consistently throughout the whole book.

Small bold characters, for example, $\mathbf{x}$, stand for vectors, whereas capital bold characters, for example, $\mathbf{X}$, denote matrices. Nonbold italic characters indicate scalars and matrix elements, for example, $\mathbf{X} = (x_{ij})$. I and J denote the number of objects (samples) and variables, respectively; A denotes the number of latent variables (principal components). A corresponding small italic letter is used for the index, for example, $i = 1, \dots, I$.

There are three special styles used to emphasize particular parts of the text. The **menu** style points out the text associated with computer interface, such as **menu** items and **hot-key** combinations. The Excel style singles out Excel linked fragments such as Excel formulas and cells addresses. The `code` style emphasizes text elements that relate to programming, for example, `function names`, VBA `code fragments`, etc.

Figure 2.1 illustrates the usage of styles.

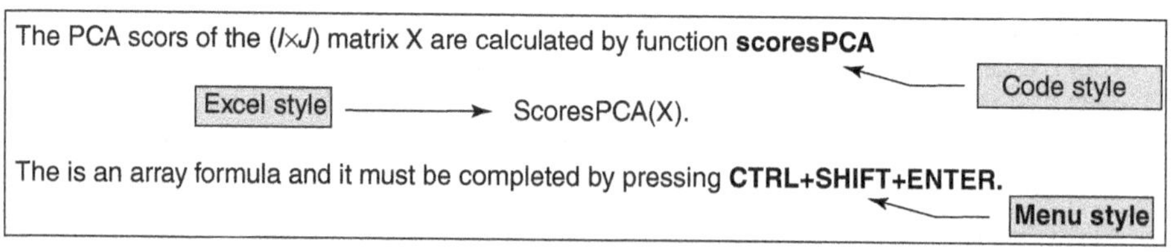

Figure 2.1 Styles usage.

3

INSTALLATION OF CHEMOMETRICS ADD-IN*

The chapter describes installation of Chemometrics Add-In software that is a specially designed add-in for Microsoft Excel. Chemometrics Add-In performs data analysis on the basis of the following projection methods: Principal Component Analysis (PCA) and Projection on Latent Structures (PLS1 and PLS2). The installation of Chemometrics Add-In can be done manually or automatically.

Preceding chapters	
Dependent chapters	7–13
Matrix skills	Low
Statistical skills	Low
Excel skills	Basic
Chemometric skills	Low
Chemometrics Add-In	Not used
Accompanying workbook	Installation.xls

3.1 INSTALLATION

In order to start working with Chemometrics Add-In, the software should be properly installed. The setup package consists of two files: *Chemometrics.dll* and *Chemometrics.xla*.[1] The setup package with a full version of Chemometrics Add-In at the Wiley Book Support site[2] in the encrypted archive. To open it, you should use the password: CMinXL2014.

*With contributions from Oxana Rodionova.

[1]The last minute updates can be found in file *Readme.txt*.

[2]http://booksupport.wiley.com

Chemometrics in Excel, First Edition. Alexey L. Pomerantsev.

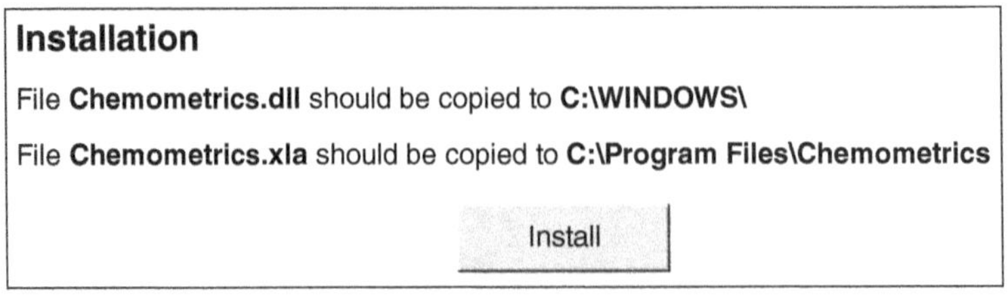

Figure 3.1 Automatic installation.

The setup files can be copied manually in accordance with the instructions provided in worksheet Install (see Fig. 3.1). Afterward Chemometrics Add-In can be installed similarly to any other Excel add-in. The sequence of actions is described in Section 7.3.6.

Chemometrics Add-In can also be installed automatically. First of all, the files *Chemometrics.dll* and *Chemometrics.xla* should be copied into a folder on your computer, for example, **G:\InstallAddIn**. After that, you may start installation with the help of the button **Install** located in worksheet Install.

If Chemometrics Add-In has already been installed, the procedure will be terminated. Otherwise, the following dialog window appears after pressing the Install button.

In the window shown in Fig. 3.2, one should select a folder where Chemometrics setup package is located. If you select a wrong folder, the message shown in Fig. 3.3 is displayed and you should locate the proper folder on your computer. During installation, various information boxes could appear, such as those shown in Fig. 3.4.

It is recommended to confirm all similar queries. If installation succeeds, one gets a message, which lists all completed operations as shown in Fig. 3.5.

If installation failed or was canceled, the message shown in Fig. 3.6 is displayed.

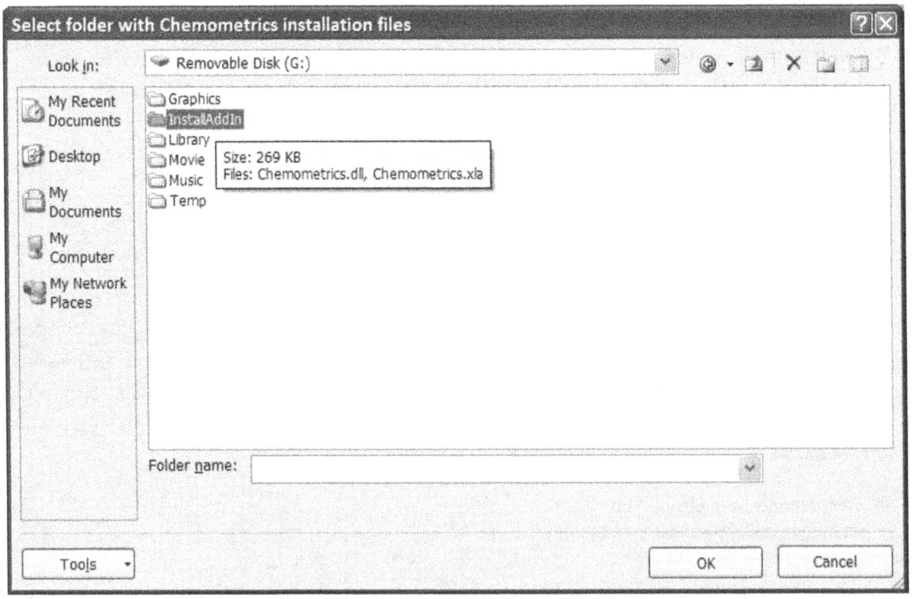

Figure 3.2 The selection of installation files folder.

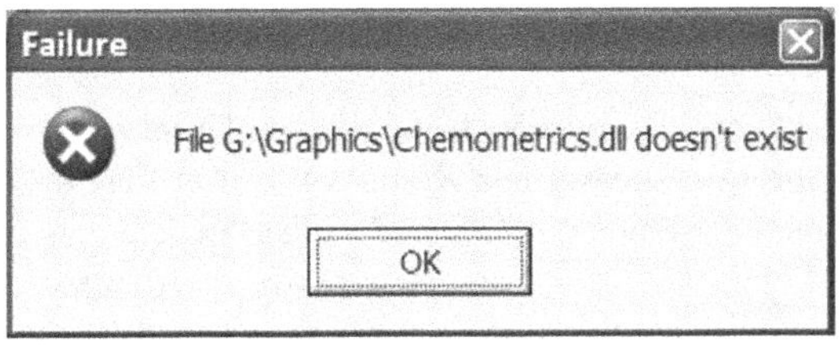

Figure 3.3 Message about a wrong folder selection.

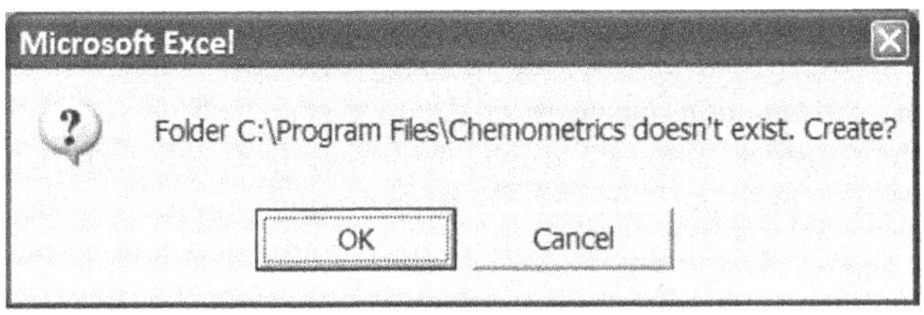

Figure 3.4 Folder creation dialog box.

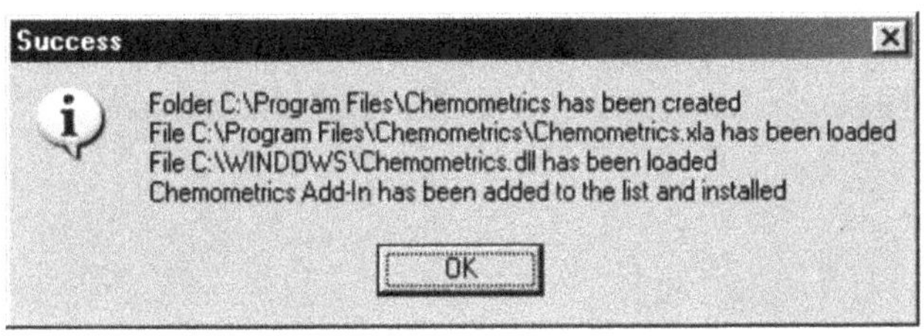

Figure 3.5 Information about a successful installation.

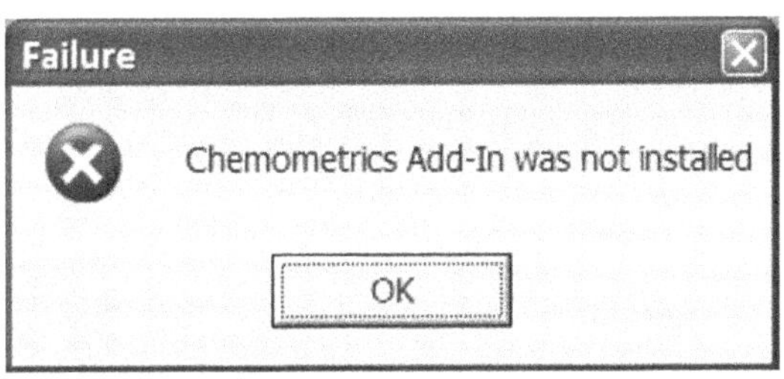

Figure 3.6 Unsuccessful installation message.

After the successful completion of installation, no other add-in-related operations are required, and one can immediately start using Chemometrics Add-In without restarting Excel.

In some cases, after the successful installation you will see "########" or "#VALUE!" in cells containing special chemometric functions. Press the **ALT-Ctrl-F9** key combination to recalculate the whole workbook.

3.2 GENERAL INFORMATION

Chemometrics Add-In uses all input data from an active Excel workbook. This data should be pasted directly in the worksheets (data **X** and **Y**). A user may organize one's working space in any way deemed convenient, for example, place all data onto one worksheet, share them between several worksheets, or even use worksheets from various workbooks. The results are displayed as arrays in the worksheets. The software does not limit input arrays' dimensionality, that is, the number of objects (*I*), the number of variables (*J*), and the number of responses (*K*). The input data size is limited by the supported computer memory, operation system in use, and by limitations of the Excel software.

Chemometrics Add-In includes special functions, which can be used as standard worksheet functions. To do so, one can use the **Insert-Function** dialog box and select **User Defined** category. All functions described in Chapter 8 can be found in the **Select a function** window. The returned values are arrays; therefore, the functions should be entered as array formulas, that is, pressing **Ctrl+Shift+Enter** at the end of the formula input. Function arguments can be numbers, names, arrays, or references containing numbers.

To insert an array formula, firstly, it is necessary to select the area on the worksheet, which corresponds to the dimensionality of the output array. If the selected area is larger than the output array, the excess cells will be filled with the #N/A symbols. On the contrary, when the selected area is smaller than the output array, a part of the output information will be lost.

Chemometrics Add-In provides a possibility to calculate scores and loadings values by projection methods. A detailed description is provided in Chapter 8.

4

FURTHER READING ON CHEMOMETRICS*

Studying any discipline (and chemometrics is not an exception) a student should follow more than just one book. It is always useful to look at things from different angles. In this chapter, we present a variety of information sources on chemometrics, which we consider useful.

Preceding chapters	
Dependent chapters	5, 7, 10
Matrix skills	Low
Statistical skills	Low
Excel skills	Low
Chemometric skills	Low
Chemometrics Add-In	Not used
Accompanying workbook	None

4.1 BOOKS

4.1.1 The Basics

1. I.M. Gelfand. *Lectures on Linear Algebra*, Dover Publications, NY, 1989.
2. F.R. Gantmacher, *Matrix Theory*, vol. 1, American Mathematical Society, Providence, Rhode Island, 1990.
3. F.R. Gantmacher, *Matrix Theory*, vol. 2, American Mathematical Society, Providence, Rhode Island, 2000.

*With contributions from Oxana Rodionova.

Chemometrics in Excel, First Edition. Alexey L. Pomerantsev.

4. G.E.P. Box, W.G. Hunter, J.S. Hunter. *Statistics for Experimenters*, John Wiley & Sons Inc., NY, 1978.
5. T.W. Anderson. *An Introduction to Multivariate Statistical Analysis*, 3rd Edition, John Wiley, NY, 2003.
6. N.R. Draper, H. Smith. *Applied Regression Analysis*, 3rd Edition, John Wiley, NY, 1998.
7. J. Walkenbach. *Excel 2007 Formulas*, John Wiley and Sons, NY, 2008.

4.1.2 Chemometrics

1. K.H. Esbensen. *Multivariate Data Analysis – In Practice*, 4th Edition, CAMO, 2000.
2. T. Næs, T. Isaksson, T. Fearn, T. Davies. *Multivariate Calibration and Classification*, Christerer, UK, 2002.
3. E.R. Malinowski. *Factor Analysis in Chemistry*, 2nd Edition, Wiley, NY, 1991.
4. A. Höskuldsson. *Prediction Methods in Science and Technology*, vol. 1, Thor Publishing, Copenhagen, 1996.
5. H. Martens, T. Næs. *Multivariate Calibration*, Wiley, New York, 1989.
6. P. Gemperline. *Practical Guide to Chemometrics*, Taylor & Francis, Boca Raton, USA, 2006.
7. M. Maeder. *Practical Data Analysis in Chemistry*, Elsevier, NY, 2007.
8. R.G. Brereton. *Chemometrics: Data Analysis for the Laboratory and Chemical Plant*, Wiley, Chichester, UK, 2003.

4.1.3 Supplements

1. D.L. Massart, B.G. Vandeginste, L.M.C. Buydens, S. De Jong, P.J. Lewi, J. Smeyers-Verbeke. *Handbook of Chemometrics and Qualimetrics Part A*, Elsevier, Amsterdam, 1997.
2. R. Pratap. *Getting Started with MATLAB®*, Oxford University Press, USA, 2009.
3. R. Kramer. *Chemometric Techniques for Quantitative Analysis*, Marcel-Dekker, NY, 1998.
4. F.T. Chau. *Chemometrics From Basics to Wavelet Transform*, John Wiley & Sons, Hoboken, New Jersey, 2004.
5. B.G. Vandeginste, D.L. Massart, L.M.C. Buydens, S. De Jong, P.J. Lewi, J. Smeyers-Verbeke. *Handbook of Chemometrics and Qualimetrics Part B*, Elsevier, Amsterdam, 1998.
6. L. Eriksson, E. Johansson, N. Kettaneh-Wold, S. Wold. *Multi- and Megavariate Data Analysis*, Umetrics, Umeå, 2001.
7. S.D. Brown, R. Tauler, B. Walczak. *Comprehensive Chemometrics. Chemical and Biochemical Data Analysis*, vol. 4 set, Elsevier, Amsterdam, 2009.
8. K.R. Beebe, R.J. Pell, M.B. Seasholtz. *Chemometrics: A Practical Guide*, Wiley, N.Y., 1998.

9. M.J. Adams. *Chemometrics in Analytical Spectroscopy*, RSC, Cambridge, UK, 1995.
10. J.W. Einax, H.W. Zweinziger, S. Geiß. *Chemometrics in Environmental Analysis*, Wiley-VCH, Weinheim, 1997.
11. H. Mark, J. Workman. *Chemometrics in Spectroscopy*, Elsevier, Amsterdam, 2007.
12. A. Smilde, R. Bro, P. Geladi. *Multi-Way Analysis with Applications in the Chemical Sciences*, Wiley, Chichester, UK, 2004.
13. B. Walczak. *Wavelets in Chemistry*, Elsevier, Amsterdam, 2000.
14. K. Varmuza, P. Filzmoser. *Introduction to Multivariate Statistical Analysis in Chemometrics*, Taylor & Francis, Boca Raton, FL, 2009.

4.2 THE INTERNET

4.2.1 Tutorials

1. Chemometrics[1]
2. Array Formulas in Excel[2]
3. Multivariate statistics: concepts, models, and applications[3]
4. Introductory statistics: concepts, models, and applications[4]
5. A short primer on chemometrics for spectroscopists[5]
6. Home page of chemometrics. Tutorials[6]
7. Reliability of univariate calibration[7]
8. Reliability of multivariate calibration[8]
9. PCR Tutorial[9]
10. Multivariate Curve Resolution Homepage[10]

4.3 JOURNALS

4.3.1 Chemometrics

1. Chemometrics and Intelligent Laboratory Systems[11]
2. Journal of Chemometrics[12]

[1] http://ull.chemistry.uakron.edu/chemometrics/
[2] http://www.databison.com/index.php/excel-array-formulas-excel-array-formula-syntax-array-constants/
[3] http://www.psychstat.missouristate.edu/MultiBook/mlt00.htm
[4] http://www.psychstat.missouristate.edu/sbk00.htm
[5] http://www.spectroscopynow.com/details/education/sepspec10349education/A-Short-Primer-on-Chemometrics-for-Spectroscopists.html
[6] http://chemometrics.se/category/tutorial/
[7] http://www.chemometry.com/Research/UVC.html
[8] http://www.chemometry.com/Research/MVC.html
[9] http://minf.vub.ac.be/~fabi/calibration/multi/pcr/index.html
[10] http://www.mcrals.info/
[11] http://www.sciencedirect.com/science/journal/01697439
[12] http://www.interscience.wiley.com/jpages/0886-9383/

4.3.2 Analytical

1. Trends in Analytical Chemistry[13]
2. Analytica Chimica Acta[14]
3. The Analyst[15]
4. Talanta[16]
5. Applied Spectroscopy[17]
6. Analytical Chemistry[18]
7. Analytical and Bioanalytical Chemistry[19]
8. Vibrational Spectroscopy[20]
9. Journal of Near Infrared Spectroscopy[21]

4.3.3 Mathematical

1. Computational Statistics and Data Analysis[22]
2. Computers & Chemical Engineering[23]
3. Technometrics[24]

4.4 SOFTWARE

4.4.1 Specialized Packages

1. Unscrambler – CAMO[25]
2. PLS_Toolbox – Eigenvector Research, Inc.[26]
3. SIMCA – UMETRICS[27]
4. GRAMS – Thermo[28]
5. Pirouette – Infometrix[29]

[13] http://www.elsevier.com/locate/inca/502695
[14] http://www.journals.elsevier.com/analytica-chimica-acta/
[15] http://www.rsc.org/publishing/journals/an/about.asp
[16] http://www.journals.elsevier.com/talanta
[17] http://www.opticsinfobase.org/as/home.cfm
[18] http://pubs.acs.org/page/ancham/about.html
[19] http://www.springer.com/chemistry/analytical+chemistry/journal/216
[20] http://www.journals.elsevier.com/vibrational-spectroscopy/
[21] http://www.impublications.com/content/journal-near-infrared-spectroscopy
[22] http://www.journals.elsevier.com/computational-statistics-and-data-analysis
[23] http://www.journals.elsevier.com/computers-and-chemical-engineering/
[24] http://www.tandfonline.com/loi/utch20#.UcRXkJxKSas
[25] http://www.camo.com/
[26] http://www.eigenvector.com/
[27] http://www.umetrics.com/
[28] http://www.thermoscientific.com/ecomm/servlet/productscatalog?categoryId=81850
[29] http://www.infometrix.com/software/pirouette.html

4.4.2 General Statistic Packages

1. Origin – Microcal[30]
2. Systat – SPSS[31]
3. Statistica – StatSoft[32]
4. Mathematica – Wolfram Research, Inc.[33]
5. MathCAD, S-PLUS, Axum – MathSoft[34]
6. MATLAB® – Mathworks[35]

4.4.3 Free Ware

1. Multivariate Analysis Add-in –Bristol University[36]
2. LIBRA – Leuven University[37]
3. TOMCAT – Silesian University[38]
4. MCR-ALS – Barcelona University[39]

[30] http://www.microcal.com/
[31] http://www.spss.com/
[32] http://www.statsoftinc.com/
[33] http://www.wri.com/
[34] http://www.mathsoft.com/
[35] http://www.mathworks.com/
[36] http://www.chm.bris.ac.uk/org/chemometrics/addins/index.html
[37] http://wis.kuleuven.be/stat/robust/LIBRA
[38] http://chemometria.us.edu.pl/RobustToolbox/
[39] http://www.mcrals.info/

PART II

THE BASICS

5

MATRICES AND VECTORS

This chapter summarizes the main facts of the matrix and vector algebra, which are used in chemometrics. The text cannot serve as a textbook on matrix algebra; it is rather a synopsis, a brief guide to the field. A deeper and more systematic exposition can be found in the literature presented in Chapter 4.

The text is divided into two parts named as *The basics* and *Advanced information*. The first part contains the provisions, which are necessary for the understanding of chemometrics and the second part presents the facts needed to know for a better comprehension of multivariate analysis techniques.

Each shaded figure in this chapter, for example, Fig. 5.1 has a corresponding worksheet in workbook Matrix.xls.

Preceding chapters	5
Dependent chapters	7, 9–12, 15
Matrix skills	Low
Statistical skills	Low
Excel skills	Basic
Chemometric skills	Low
Chemometrics Add-In	Not used
Accompanying workbook	Matrix.xls

5.1 THE BASICS

5.1.1 Matrix

A *matrix* is a rectangular table of numbers, as shown in Fig. 5.1.

Chemometrics in Excel, First Edition. Alexey L. Pomerantsev.

$$
A = \begin{vmatrix} 1.2 & -5.3 & 0.25 \\ 10.2 & 1.5 & -7.5 \\ 2.3 & -1.2 & 5.6 \\ 4.5 & -0.8 & 9.5 \end{vmatrix}
$$

Figure 5.1 Matrix.

Matrices are denoted by bold capital letters (**A**) and their elements by the corresponding lower case letters with indices, that is, a_{ij}. The first index stands for the rows and the second one stands for the columns. It is very common in chemometrics to denote the maximum index value by the same letter as the index itself, using capitals. Therefore, matrix **A** can also be presented as $\{a_{ij},\quad i = 1, \ldots, I;\ j = 1, \ldots, J\}$. In Fig. 5.1, $I = 4$, $J = 3$, and $a_{23} = -7.5$.

The pair of integers I and J is said to be the matrix *dimension*; this is denoted by $I \times J$. An example of a matrix in chemometrics is a set of spectra obtained for I samples and J wavelengths.

5.1.2 Simple Matrix Operations

A matrix can be *multiplied by a number*. To do so, each element of the matrix is multiplied by this number. An example is shown in Fig. 5.2.

Two matrices of the same dimension can be added and subtracted element-wise. See Fig. 5.3 for an example.

The result of the scalar-matrix multiplication and the matrix–matrix addition is a matrix of the same dimension. A zero matrix is the matrix consisting of zeros. It is denoted by **O**. Obviously, $\mathbf{A} + \mathbf{O} = \mathbf{A}$, $\mathbf{A} - \mathbf{A} = \mathbf{O}$, and $0\mathbf{A} = \mathbf{O}$.

A matrix can be *transposed*. During this operation, the matrix is flipped over, that is, the matrix rows and columns are swapped. Transposition is denoted by the apostrophe, $\mathbf{A}'$, or

$$
3 * \begin{vmatrix} 1.2 & -5.3 & 0.25 \\ 10.2 & 1.5 & -7.5 \\ 2.3 & -1.2 & 5.6 \\ 4.5 & -0.8 & 9.5 \end{vmatrix} = \begin{vmatrix} 3.6 & -16 & 0.75 \\ 30.6 & 4.5 & -23 \\ 6.9 & -3.6 & 16.8 \\ 13.5 & -2.4 & 28.5 \end{vmatrix}
$$

Figure 5.2 Multiplication by a number.

$$
\begin{vmatrix} 1.2 & -5.3 \\ 10.2 & 1.5 \\ 2.3 & -1.2 \\ 4.5 & -0.8 \end{vmatrix} + \begin{vmatrix} -2.5 & 13.8 \\ -4.2 & 11.2 \\ -3.5 & 4.8 \\ 1.4 & 6.5 \end{vmatrix} = \begin{vmatrix} -1.3 & 8.5 \\ 6 & 12.7 \\ -1.2 & 3.6 \\ 5.9 & 5.7 \end{vmatrix}
$$

Figure 5.3 Matrices addition.

$$\mathrm{A} = \begin{vmatrix} 1.2 & -5.3 & 0.25 \\ 10.2 & 1.5 & -7.5 \\ 2.3 & -1.2 & 5.6 \\ 4.5 & -0.8 & 9.5 \end{vmatrix} \qquad \mathrm{A}^{\mathrm{t}} = \begin{vmatrix} 1.2 & 10.2 & 2.3 & 4.5 \\ -5.3 & 1.5 & -1.2 & -0.8 \\ 0.25 & -7.5 & 5.6 & 9.5 \end{vmatrix}$$

Figure 5.4 Matrix transpose.

by the t index, $\mathbf{A}^{\mathrm{t}}$. Formally, if $\mathbf{A} = \{a_{ij}, \quad i = 1, \ldots, I; \; j = 1, \ldots, J\}$, then $\mathbf{A}^{\mathrm{t}} = \{a_{ji}, \quad j = 1, \ldots, J; \; i = 1, \ldots, I\}$. An example is shown in Fig. 5.4.

Obviously, $(\mathbf{A}^{\mathrm{t}})^{\mathrm{t}} = \mathbf{A}, \; (\mathbf{A} + \mathbf{B})^{\mathrm{t}} = \mathbf{A}^{\mathrm{t}} + \mathbf{B}^{\mathrm{t}}$.

In Excel, the standard worksheet function **TRANSPOSE** (see Section 7.2.5) is used to transpose a matrix.

5.1.3 Matrices Multiplication

Matrices can be *multiplied* but only when they have corresponding dimensions, that is, the number of columns of **A** matches the number of rows of **B**. The product of matrix **A** (dimension $I \times K$) and matrix **B** (dimension $K \times J$) is matrix **C** (dimension $I \times J$), whose elements are defined by

$$c_{ij} = \sum_{k=1}^{K} a_{ik} b_{kj}.$$

An example is shown in Fig. 5.5.

The rule of matrix multiplication can be formulated as follows. To calculate an element of matrix **C**, located on the intersection of the i-th row and the j-th column (c_{ij}), it is necessary to multiply each element of the i-th row of the first matrix **A** by the j-th column of the second matrix **B** and to sum up the results. In the example shown in Fig. 5.6, an element of **C** located in the third row and the second column is obtained as the sum of the element-wise products of the third row of **A** and the second column of **B**.

A matrix product depends on the sequence of multiplication, that is, $\mathbf{AB} \neq \mathbf{BA}$, at least because of the dimensionality. It is said that the matrix product is *not commutative*. However, this is an *associative* operation, that is, $\mathbf{ABC} = (\mathbf{AB})\mathbf{C} = \mathbf{A}(\mathbf{BC})$, and it is also *distributive* over matrix addition, that is, $\mathbf{A}(\mathbf{B} + \mathbf{C}) = \mathbf{AB} + \mathbf{AC}$. Obviously, $\mathbf{AO} = \mathbf{O}$.

In Excel, the standard worksheet function **MMULT** (see Section 7.2.6) is used for matrix multiplication.

$$\begin{vmatrix} 2.2 & 5.8 \\ 1.5 & 6.2 \\ -2.5 & 7.4 \\ 4.2 & -2.3 \end{vmatrix} * \begin{vmatrix} 5.2 & 3.2 & 3.8 \\ 6.8 & -7.5 & -1.9 \end{vmatrix} = \begin{vmatrix} 50.88 & -36.5 & -2.66 \\ 49.96 & -41.7 & -6.08 \\ 37.32 & -63.5 & -23.6 \\ 6.2 & 30.69 & 20.33 \end{vmatrix}$$

Figure 5.5 Matrices product.

$$\begin{vmatrix} \# & \# \\ \# & \# \\ -2.5 & 7.4 \\ \# & \# \end{vmatrix} * \begin{vmatrix} \# & 3.2 & \# \\ \# & -7.5 & \# \end{vmatrix} = \begin{vmatrix} \# & \# & \# \\ \# & \# & \# \\ \# & -63.5 & \# \\ \# & \# & \# \end{vmatrix}$$

Figure 5.6 The element of matrices product.

5.1.4 Square Matrix

If the number of columns is equal to the number of rows ($I = J = N$), then the matrix is said to be a *square matrix*. In this section, only square matrices are considered. Among them there are matrices with special properties.

An *identity* matrix (denoted by $\mathbf{I}$ and, sometimes, by $\mathbf{E}$) is a matrix whose elements equal zero, except for the diagonal elements, which are equal to 1, for example

$$\mathbf{I} = \begin{bmatrix} 1 & 0 & \dots & 0 \\ 0 & 1 & \dots & 0 \\ \dots & \dots & \dots & 0 \\ 0 & 0 & 0 & 1 \end{bmatrix}$$

It is obvious that $\mathbf{AI} = \mathbf{IA} = \mathbf{A}$.

Matrix $\mathbf{A}$ is said to be diagonal if all its elements except the diagonal (a_{ii}) are zeros, as shown in Fig. 5.7.

Matrix $\mathbf{A}$ is said to be *upper triangular* if all its elements below the diagonal are equal to zero, that is, $a_{ij} = 0$ for $i > j$. An example is presented in Fig. 5.8.

A *lower triangular* matrix is defined in a similar way.

Matrix $\mathbf{A}$ is said to be *symmetric*, if $\mathbf{A}^t = \mathbf{A}$. In other words, $a_{ij} = a_{ji}$. See example in Fig. 5.9.

Matrix $\mathbf{A}$ is said to be *orthogonal*, if $\mathbf{A}^t\mathbf{A} = \mathbf{A}\mathbf{A}^t = \mathbf{I}$.

Matrix $\mathbf{A}$ is said to be *normal*, if $\mathbf{A}^t\mathbf{A} = \mathbf{A}\mathbf{A}^t$.

$$\begin{vmatrix} 1 & 0 & 0 \\ 0 & 2 & 0 \\ 0 & 0 & -4 \end{vmatrix}$$

Figure 5.7 Diagonal matrix.

$$\begin{vmatrix} 1 & 5 & 3 \\ 0 & 2 & -1 \\ 0 & 0 & -4 \end{vmatrix}$$

Figure 5.8 Upper triangular matrix.

$$\begin{vmatrix} 1 & 5 & 3 \\ 5 & 2 & -1 \\ 3 & -1 & -4 \end{vmatrix}$$

Figure 5.9 Symmetric matrix.

5.1.5 Trace and Determinant

The *trace* of a square matrix (denoted by Tr(**A**) or Sp(**A**)) is defined as the sum of all diagonal elements

$$\mathrm{Sp}(\mathbf{A}) = \sum_{n=1}^{N} a_{nn}.$$

An example is shown in Fig. 5.10.

It is evident that

$$\mathrm{Sp}(\alpha\mathbf{A}) = \alpha\mathrm{Sp}(\mathbf{A})$$

$$\mathrm{Sp}(\mathbf{A} + \mathbf{B}) = \mathrm{Sp}(\mathbf{A}) + \mathrm{Sp}(\mathbf{B}).$$

It can be proved that

$$\mathrm{Sp}(\mathbf{A}) = \mathrm{Sp}(\mathbf{A}^{\mathrm{t}}), \quad \mathrm{Sp}(\mathbf{I}) = N,$$

and that

$$\mathrm{Sp}(\mathbf{AB}) = \mathrm{Sp}(\mathbf{BA}).$$

There is no standard worksheet function for this operation in Excel. Chemometric Add-In suggests using a worksheet function `MTrace`, which is explained in Section 8.6.4.

The matrix *determinant* (denoted by det(**A**)) is another important characteristic of a square matrix. A general definition of a determinant is rather complex; therefore, we demonstrate this in the simplest example for matrix **A** with dimension (2 × 2). Then

$$\det(\mathbf{A}) = a_{11}a_{22} - a_{12}a_{21}$$

For the (3 × 3) matrix, the determinant is equal to

$$\det(\mathbf{A}) = a_{11}a_{22}a_{33} - a_{13}a_{21}a_{31} + a_{12}a_{23}a_{31} - a_{12}a_{21}a_{33} + a_{13}a_{21}a_{31} - a_{11}a_{23}a_{32}$$

$$\mathrm{Sp}\begin{vmatrix} 1 & 5 & 3 \\ 5 & 2 & -1 \\ 3 & -1 & -4 \end{vmatrix} = -1$$

Figure 5.10 Trace of the matrix.

$$\det\begin{vmatrix} 1.2 & -5.3 & 0.25 & 2.2 & 13.8 \\ 10.2 & 1.5 & -7.5 & 1.5 & 11.2 \\ 2.3 & -1.2 & 5.6 & -2.5 & 4.8 \\ 4.5 & -0.8 & 9.5 & 4.2 & 6.5 \\ 2.5 & 8.2 & -3.5 & 5.9 & 1.1 \end{vmatrix} = 56044.308595$$

Figure 5.11 Matrix determinant.

In case of the $(N \times N)$ matrix, the determinant is calculated as the sum of $1{\cdot}2{\cdot}3{\cdot}\ \ldots\ {\cdot}N = N!$ terms, which is equal to

$$(-1)^r a_{1k_1} a_{2k_2} \ \cdots \ a_{Nk_N}.$$

Indices $k_1, k_2, \ldots, k_N$ are defined as all ordered permutations of r numbers in a set $(1, 2, \ldots, N)$. The determinant calculation is a complicated procedure that is performed in practice with the help of special programs. In Excel, the standard worksheet function **MDETERM** is used for this task (see Section 7.2.5). An example is given in Fig. 5.11.

The following properties are evident:

$$\det(\mathbf{I}) = 1, \quad \det(\mathbf{A}) = \det(\mathbf{A}^{\mathrm{t}}),$$
$$\det(\mathbf{AB}) = \det(\mathbf{A})\det(\mathbf{B})$$

5.1.6 Vectors

If a matrix has only one column ($J = 1$) it is called a *vector* or a column vector to be more precise. For example,

$$\mathbf{a} = \begin{pmatrix} 1.2 \\ 10.2 \\ 2.3 \\ 4.5 \end{pmatrix}.$$

One-row matrices are also in use. For example,

$$\mathbf{b} = (1.2 \quad -5.3 \quad 0.25).$$

b is also a vector but a row vector. Analyzing the data, it is important to understand whether we are dealing with columns or rows. For example, a sample spectrum can be considered as a row vector. On the other hand, the spectral intensities at a wavelength for all samples can be interpreted as a column vector.

Dimension of the vector is the number of its elements. Obviously, a column vector can be converted into a row vector using transposition, that is,

$$\begin{pmatrix} \mathrm{c} \\ \mathrm{o} \\ \mathrm{l} \\ \mathrm{u} \\ \mathrm{m} \\ \mathrm{n} \end{pmatrix}^{\mathrm{t}} = (\mathrm{r} \quad \mathrm{o} \quad \mathrm{w})$$

$$\begin{vmatrix} 1.2 \\ 10.2 \\ 2.3 \end{vmatrix} + \begin{vmatrix} 2.3 \\ -1.2 \\ 5.6 \end{vmatrix} = \begin{vmatrix} 3.5 \\ 9 \\ 7.9 \end{vmatrix} \qquad 3 \ * \begin{vmatrix} 2.3 \\ -1.2 \\ 5.6 \end{vmatrix} = \begin{vmatrix} 6.9 \\ -3.6 \\ 16.8 \end{vmatrix}$$

Figure 5.12 Vector operations.

When the type of a vector is not specified, a column vector is implied. We will also follow this rule. Vectors are denoted by bold lower case letters. A zero vector is a vector whose elements are zeros. It is denoted by $\mathbf{0}$.

5.1.7 Simple Vector Operations

Vectors can be added and multiplied by a number in the same way as it is done with matrices. An example is shown in Fig. 5.12.

Two vectors $\mathbf{x}$ and $\mathbf{y}$ are said to be *collinear* if there exists a number α such that

$$\alpha\mathbf{x} = \mathbf{y}.$$

5.1.8 Vector Products

Two vectors $\mathbf{x} = (x_1, x_2, \dots, x_N)^t$ and $\mathbf{y} = (y_1, y_2, \dots, y_N)^t$ of the same dimension N can be multiplied. Following the "row by column" multiplication rule, two products can be composed, namely, $\mathbf{x}^t\mathbf{y}$ and $\mathbf{x}\mathbf{y}^t$. The first product

$$\mathbf{x}^t\mathbf{y} = \begin{pmatrix} x_1 & x_2 & \dots & x_N \end{pmatrix} \cdot \begin{pmatrix} y_1 \\ y_2 \\ \vdots \\ y_N \end{pmatrix} = x_1y_1 + x_2y_2 + \dots + x_Ny_N$$

is named a *scalar* (or *dot* or *inner*) product. The result is a number. Notation $(\mathbf{x}, \mathbf{y}) = \mathbf{x}^t\mathbf{y}$ is also used. See Fig. 5.13 for an example.

The second product

$$\mathbf{x}\mathbf{y}^t = \begin{pmatrix} x_1 \\ x_2 \\ \vdots \\ x_N \end{pmatrix} \cdot \begin{pmatrix} y_1 & y_2 & \dots & y_N \end{pmatrix} = \begin{pmatrix} x_1y_1 & x_1y_2 & \cdots & x_1y_N \\ x_2y_1 & x_2y_2 & \cdots & x_2y_N \\ \cdots & \cdots & \cdots & \cdots \\ x_Ny_1 & x_Ny_2 & \cdots & x_Ny_N \end{pmatrix}$$

is called the *outer* product. The result is the $(N \times N)$ matrix. An example is given in Fig. 5.14.

Vectors are said to be *orthogonal* if their scalar product equals zero.

$$\begin{vmatrix} 1 & 5 & 2 \end{vmatrix} \ * \begin{vmatrix} 3 \\ -2 \\ 5 \end{vmatrix} = 3$$

Figure 5.13 Inner (scalar) product.

$$\begin{vmatrix} 3 \\ -2 \\ 5 \end{vmatrix} * \begin{vmatrix} 1 & 5 & 2 \end{vmatrix} = \begin{vmatrix} 3 & 15 & 6 \\ -2 & -10 & -4 \\ 5 & 25 & 10 \end{vmatrix}$$

Figure 5.14 Outer product.

$$\begin{Vmatrix} 1 \\ -2 \\ 2 \end{Vmatrix} = 3$$

Figure 5.15 Vector norm.

5.1.9 Vector Norm

A scalar product of a vector and itself is a scalar square. This value

$$(\mathbf{x}, \mathbf{x}) = \mathbf{x}^{\mathrm{t}}\mathbf{x} = \sum_{n=1}^{N} x_n^2$$

defines the squared length of vector **x**. The length itself is called the *vector norm*

$$\|\mathbf{x}\| = \sqrt{(\mathbf{x}, \mathbf{x})}$$

Figure 5.15 shows an example.

A unit length vector ($\|\mathbf{x}\| = 1$) is said to be *normalized.* A nonzero vector ($\mathbf{x} \neq \mathbf{0}$) can be normalized by dividing it by its length, that is, $\mathbf{x} = \|\mathbf{x}\|(\mathbf{x}/\|\mathbf{x}\|) = \|\mathbf{x}\|\mathbf{e}$. Here, $\mathbf{e} = \mathbf{x}/\|\mathbf{x}\|$ is the normalized vector.

Vectors are called *orthonormal* if they are normalized and mutually orthogonal.

5.1.10 Angle Between Vectors

The scalar product determines an angle φ between two vectors **x** and **y**

$$\cos\varphi = \frac{(\mathbf{x}, \mathbf{y})}{\|\mathbf{x}\| \cdot \|\mathbf{y}\|}$$

If vectors are orthogonal, then $\cos\varphi = 0$ and $\varphi = \pi/2$. If they are collinear, then $\cos\varphi = 1$ and $\varphi = 0$.

5.1.11 Vector Representation of a Matrix

Each matrix **A** of dimension $I \times J$ can be represented as a set of vectors

$$\mathbf{A} = \begin{pmatrix} \mathbf{a}_1 & \mathbf{a}_2 & \cdots & \mathbf{a}_J \end{pmatrix} = \begin{pmatrix} \mathbf{b}_1 \\ \mathbf{b}_2 \\ \vdots \\ \mathbf{b}_I \end{pmatrix}.$$

Here, vector $\mathbf{a}_j$ is the j-th column of the matrix $\mathbf{A}$, and a row-vector $\mathbf{b}_i$ is the i-th row of the matrix $\mathbf{A}$

$$\mathbf{a}_j = \begin{pmatrix} a_{1j} \\ a_{2j} \\ \vdots \\ a_{Ij} \end{pmatrix}, \qquad \mathbf{b}_i = \begin{pmatrix} a_{i1} & a_{i2} & \cdots & a_{iJ} \end{pmatrix}$$

5.1.12 Linearly Dependent Vectors

For vectors $\mathbf{x}_1, \mathbf{x}_2, \ldots, \mathbf{x}_K$ of the same dimension and numbers $\alpha_1, \alpha_2, \ldots, \alpha_K$, a vector

$$\mathbf{y} = \alpha_1 \mathbf{x}_1 + \alpha_2 \mathbf{x}_2 + \cdots + \alpha_K \mathbf{x}_K$$

is called the *linear combination* of vectors $\mathbf{x}_k$.

If there exist nonzero numbers $\alpha_k \neq 0, \quad k = 1, \ldots, K$, such that $\mathbf{y} = 0$, then the vector set $\{\mathbf{x}_k, \quad k = 1, \ldots, K\}$ is said to be *linearly dependent*. Otherwise, the vectors are called *linearly independent*. For example, vectors $\mathbf{x}_1 = (2, 2)^{\mathrm{t}}$ and $\mathbf{x}_2 = (-1, -1)^{\mathrm{t}}$ are linearly dependent because $\mathbf{x}_1 + 2\mathbf{x}_2 = \mathbf{0}$.

5.1.13 Matrix Rank

Consider a set of K vectors $\mathbf{x}_1, \mathbf{x}_2, \ldots, \mathbf{x}_K$ of dimension N. The maximum number of the linearly independent vectors in this set is called *rank*. For example, a set

$$\mathbf{x}_1 = \begin{pmatrix} 1 \\ 0 \\ 0 \end{pmatrix}, \quad \mathbf{x}_2 = \begin{pmatrix} 0 \\ 1 \\ 0 \end{pmatrix}, \quad \mathbf{x}_3 = \begin{pmatrix} 1 \\ 1 \\ 0 \end{pmatrix}, \quad \mathbf{x}_4 = \begin{pmatrix} 3 \\ 2 \\ 0 \end{pmatrix}$$

has only two linearly independent vectors, $\mathbf{x}_1$ and $\mathbf{x}_2$; therefore, the rank of the set is 2.

Obviously, if a set has more vectors than their dimension ($K > N$), then the vectors are necessarily linearly dependent.

A *matrix rank* (denoted as rank($\mathbf{A}$)) is the rank of a set of vectors that compose the matrix. Although any matrix can be represented in two ways (column-wise or row-wise), this does not affect the rank, because

$$\operatorname{rank}(\mathbf{A}) = \operatorname{rank}(\mathbf{A}^{\mathrm{t}}).$$

5.1.14 Inverse Matrix

A square matrix $\mathbf{A}$ is said to be *invertible* (*nondegenerate*) if it has a unique inverse matrix $\mathbf{A}^{-1}$, defined by

$$\mathbf{A}\mathbf{A}^{-1} = \mathbf{A}^{-1}\mathbf{A} = \mathbf{I}.$$

Some matrices are not invertible. The necessary and sufficient condition of invertibility is

$$\det(\mathbf{A}) \neq 0 \;\text{ or }\; \operatorname{rank}(\mathbf{A}) = N.$$

Matrix inversion is a complex procedure. In Excel, the standard worksheet function **`MINVERSE`** is used. It is explained in Section 7.2.5. For an example, see Fig. 5.16.

$$\begin{vmatrix} 1.2 & -5.3 & 0.25 & 2.2 & 13.8 \\ 10 & 1.5 & -7.5 & 1.5 & 11.2 \\ 2.3 & -1.2 & 5.6 & -2.5 & 4.8 \\ 4.5 & -0.8 & 9.5 & 4.2 & 6.5 \\ 2.5 & 8.2 & -3.5 & 5.9 & 1.1 \end{vmatrix}^{-1} = \begin{vmatrix} -0.099 & 0.104 & -0.065 & 0.0926 & -0.083 \\ -0.003 & -0.039 & 0.1782 & -0.082 & 0.1452 \\ -0.016 & -0.035 & 0.0633 & 0.0447 & 0.0097 \\ 0.0223 & -0.006 & -0.202 & 0.1126 & -0.003 \\ 0.0766 & -0.022 & 0.1052 & -0.058 & 0.0633 \end{vmatrix}$$

Figure 5.16 Matrix inverse.

The simplest case of the (2×2) matrix is presented by

$$\mathbf{A} = \begin{pmatrix} a_{11} & a_{12} \\ a_{21} & a_{22} \end{pmatrix}, \quad \mathbf{A}^{-1} = (1/\det(\mathbf{A})) \begin{pmatrix} a_{22} & -a_{12} \\ -a_{21} & a_{11} \end{pmatrix}.$$

If the matrices **A** and **B** are invertible, then

$$(\mathbf{AB})^{-1} = \mathbf{B}^{-1}\mathbf{A}^{-1}.$$

5.1.15 Pseudoinverse

If matrix **A** is degenerate and the inverse does not exist, it is sometimes possible to use a *pseudoinverse* matrix, which is defined as such a matrix $\mathbf{A}^+$ that

$$\mathbf{AA}^+\mathbf{A} = \mathbf{A}.$$

A pseudoinverse matrix is not unique and it depends on the method used. For example, for a rectangular matrix the Moore–Penrose method can be used. If the number of rows is less than the number of columns, then

$$\mathbf{A}^+ = (\mathbf{A}^t\mathbf{A})^{-1}\mathbf{A}^t.$$

An example is shown in Fig. 5.17.

If the number of columns exceeds the number of rows then,

$$\mathbf{A}^+ = \mathbf{A}^t(\mathbf{AA}^t)^{-1}.$$

$$\mathrm{A} = \begin{vmatrix} 1.2 & -1.1 & 0.3 \\ 1.1 & 1.5 & 0.8 \\ 2.3 & -1.2 & -0.5 \\ -1.2 & -1.8 & 2.1 \end{vmatrix} \quad \mathrm{A} = \begin{vmatrix} 1.2 & 1.1 & 2.3 & -1.2 \\ -1.1 & 1.5 & -1.2 & -0.8 \\ 0.3 & 0.8 & -0.5 & 2.1 \end{vmatrix} \quad \mathrm{A^tA} = \begin{vmatrix} 9.38 & -1.47 & -2.49 \\ -1.47 & 5.54 & -0.16 \\ -2.49 & -0.16 & 5.36 \end{vmatrix}$$

$$(\mathrm{A^tA})^{-1} = \begin{vmatrix} 0.13 & 0.04 & 0.06 \\ 0.14 & 0.19 & 0.02 \\ 0.06 & 0.02 & 0.22 \end{vmatrix} \quad \mathrm{A^+} = (\mathrm{A^tA})^{-1}\mathrm{A} = \begin{vmatrix} 0.13 & 0.24 & 0.22 & -0.06 \\ -0.16 & 0.34 & -0.16 & -0.15 \\ 0.10 & 0.27 & 0.01 & 0.36 \end{vmatrix} \quad \mathrm{AA^+A} = \begin{vmatrix} 1.2 & -1.1 & 0.25 \\ 1.1 & 1.5 & 0.8 \\ 2.3 & -1.2 & -0.5 \\ -1.2 & -0.8 & 2.1 \end{vmatrix}$$

Figure 5.17 Pseudoinverse matrix.

$$\begin{vmatrix} 1.2 & -3.5 & 1.8 & -2.2 & -5.3 \\ 10 & -2.8 & 3.2 & -2.8 & 1.5 \\ 2.3 & -6.5 & -9.2 & 6.5 & -1.2 \end{vmatrix} * \begin{vmatrix} 1.2 \\ 0.5 \\ 0.8 \\ 2.1 \\ -1.4 \end{vmatrix} = \begin{vmatrix} 3.93 \\ 5.42 \\ 7.48 \end{vmatrix}$$

Figure 5.18 Multiplication of matrix by vector.

5.1.16 Matrix–Vector Product

Matrix **A** can be multiplied by vector **x** of the appropriate dimension. A column-vector is right multiplied, that is, **Ax**, and a row-vector is left multiplied, that is, $\mathbf{x}^t\mathbf{A}$. The product of the J-dimensional vector and the $I \times J$ -dimensional matrix is a vector of dimension I. Figure 5.18 presents an example.

If **A** is a square matrix ($I \times I$), then vector **Ax** has the same dimension I. It is evident that

$$\mathbf{A}(\alpha_1\mathbf{x}_1 + \alpha_2\mathbf{x}_2) = \alpha_1\mathbf{A}\mathbf{x}_1 + \alpha_2\mathbf{A}\mathbf{x}_2.$$

Therefore, a matrix can be seen as a linear operator acting on vectors. In particular, $\mathbf{Ix} = \mathbf{x}$, $\mathbf{Ox} = \mathbf{0}$.

5.2 ADVANCED INFORMATION

5.2.1 Systems of Linear Equations

Let **A** be a matrix of dimension $I \times J$ and **b** be a vector of dimension J. Consider an equation

$$\mathbf{Ax} = \mathbf{b}$$

regarding the unknown vector **x** of dimension I. In fact, this is a system of I *linear equations* with J unknowns $\mathbf{x}_1, \ldots, \mathbf{x}_J$. The solution exists if and only if

$$\text{rank}(\mathbf{A}) = \text{rank}(\mathbf{B}) = R,$$

where **B** is an extended matrix of dimension $I \times (J + 1)$ consisting of matrix **A** augmented by column **b**, that is,

$$\mathbf{B} = (\mathbf{Ab}).$$

Otherwise, the system is inconsistent and has no solution.

If $R = I = J$, then the solution is unique

$$\mathbf{x} = \mathbf{A}^{-1}\mathbf{b}.$$

If $R < I$, there exists a set of different solutions, which can be represented as linear combinations of J–R vectors.

A homogeneous system $\mathbf{Ax} = \mathbf{0}$ with the square ($N \times N$) matrix **A** has a nontrivial solution if and only if $\det(\mathbf{A}) = 0$. If $R = \text{rank}(\mathbf{A}) < N$, then N–R linearly independent solutions exist.

5.2.2 Bilinear and Quadratic Forms

If $\mathbf{A}$ is a square matrix, and $\mathbf{x}$ and $\mathbf{y}$ are vectors of the appropriate dimensions, then the scalar product $\mathbf{x}^t\mathbf{A}\mathbf{y}$ is said to be the *bilinear form* defined by the matrix $\mathbf{A}$. When $\mathbf{x} = \mathbf{y}$, the product $\mathbf{x}^t\mathbf{A}\mathbf{x}$ is called the *quadratic* form.

5.2.3 Positive Definite Matrix

A square matrix $\mathbf{A}$ is said to be *positive definite*, if

$$\mathbf{x}^t\mathbf{A}\mathbf{x} > 0$$

for any nonzero vector $\mathbf{x} \neq 0$.

Similarly, a *negative* ($\mathbf{x}^t\mathbf{A}\mathbf{x} < 0$), *nonnegative* ($\mathbf{x}^t\mathbf{A}\mathbf{x} \geq 0$), and *nonpositive* ($\mathbf{x}^t\mathbf{A}\mathbf{x} \leq 0$) definite matrices are introduced.

5.2.4 Cholesky Decomposition

If a symmetric matrix $\mathbf{A}$ is positive definite, then there is a unique lower triangular matrix $\mathbf{U}$ with the positive diagonal elements, such that

$$\mathbf{A} = \mathbf{U}^t\mathbf{U}.$$

An example is presented in Fig. 5.19. This is called the *Cholesky decomposition*.

5.2.5 Polar Decomposition

Let $\mathbf{A}$ be a square invertible matrix of dimension $N \times N$. There exists unique *polar decomposition*

$$\mathbf{A} = \mathbf{S}\mathbf{R},$$

where $\mathbf{S}$ is a nonnegative symmetric matrix and $\mathbf{R}$ is an orthogonal matrix. Matrices $\mathbf{S}$ and $\mathbf{R}$ can be defined explicitly by

$$\mathbf{S}^2 = \mathbf{A}\mathbf{A}^t \quad \text{or} \quad \mathbf{S} = (\mathbf{A}\mathbf{A}^t)^{1/2}$$
$$\mathbf{R} = \mathbf{S}^{-1}\mathbf{A} = (\mathbf{A}\mathbf{A}^t)^{-1/2}\mathbf{A}.$$

An example is shown in Fig. 5.20.

If matrix $\mathbf{A}$ is degenerate, then the decomposition is not unique; namely, matrix $\mathbf{S}$ is always unique, but there are many matrices $\mathbf{R}$. The polar decomposition represents matrix $\mathbf{A}$ as a combination of stretching by $\mathbf{S}$ and rotation by $\mathbf{R}$.

$$\begin{vmatrix} 4 & 2 & 6 \\ 2 & 10 & 9 \\ 6 & 9 & 14 \end{vmatrix} = \begin{vmatrix} 2 & 0 & 0 \\ 1 & 3 & 0 \\ 3 & 2 & 1 \end{vmatrix} * \begin{vmatrix} 2 & 1 & 3 \\ 0 & 3 & 2 \\ 0 & 0 & 1 \end{vmatrix}$$

Figure 5.19 Cholesky decomposition.

$$\mathbf{A} = \begin{vmatrix} 4 & 2 & 6 \\ 2 & 10 & 9 \\ 3 & 8 & 5 \end{vmatrix} \quad \mathbf{AA}^t = \begin{vmatrix} 56 & 82 & 58 \\ 82 & 185 & 131 \\ 58 & 131 & 98 \end{vmatrix} \quad \mathbf{S} = \begin{vmatrix} -1.4 & 5.89 & 4.41 \\ 5.89 & 9.31 & 7.98 \\ 4.41 & 7.98 & 3.85 \end{vmatrix} \quad \mathbf{S}^2 = \begin{vmatrix} 56 & 82 & 58 \\ 82 & 185 & 131 \\ 58 & 131 & 98 \end{vmatrix}$$

$$\mathbf{R} = \begin{vmatrix} -0.5 & 0.85 & -0.2 \\ 0.81 & 0.53 & 0.27 \\ -0.3 & 0.01 & 0.95 \end{vmatrix} \quad \mathbf{RR}^t = \begin{vmatrix} 1 & 0 & 0 \\ 0 & 1 & 0 \\ 0 & 0 & 1 \end{vmatrix} \quad \mathbf{SR} = \begin{vmatrix} 4 & 2 & 6 \\ 2 & 10 & 9 \\ 3 & 8 & 5 \end{vmatrix}$$

Figure 5.20 Polar decomposition.

5.2.6 Eigenvalues and Eigenvectors

Let $\mathbf{A}$ be a square matrix. Vector $\mathbf{v}$ is said to be the *eigenvector* of matrix $\mathbf{A}$, if

$$\mathbf{Av} = \lambda\mathbf{v},$$

where number λ is said to be the *eigenvalue* of matrix $\mathbf{A}$. Thus, the multiplication of matrix $\mathbf{A}$ by eigenvector $\mathbf{v}$ can be replaced by a simple scaling of $\mathbf{v}$ by eigenvalue λ. An eigenvector is defined up to a factor $\alpha \neq 0$, that is, if $\mathbf{v}$ is an eigenvector, then $\alpha\mathbf{v}$ is also an eigenvector.

5.2.7 Eigenvalues

Matrix $\mathbf{A}$ of dimension $(N \times N)$ can have no more than N eigenvalues. They satisfy the *characteristic equation*

$$\det(\mathbf{A} - \lambda\mathbf{I}) = 0,$$

which is an algebraic equation of the N-th order. In particular, the characteristic equation for the 2×2 matrix has a form

$$\det(\mathbf{A} - \lambda\mathbf{I}) = \det\begin{pmatrix} a_{11} - \lambda & a_{12} \\ a_{21} & a_{22} - \lambda \end{pmatrix} = (a_{11} - \lambda)(a_{22} - \lambda) - a_{12}a_{21} = 0.$$

Figure 5.21 provides an example.

A set of eigenvalues $\lambda_1, \ldots, \lambda_N$ of the matrix $\mathbf{A}$ is called the *spectrum* of $\mathbf{A}$ and possesses certain properties

$$\det(\mathbf{A}) = \lambda_1 \times \ldots \times \lambda_N, \quad \mathrm{Sp}(\mathbf{A}) = \lambda_1 + \cdots + \lambda_N.$$

Eigenvalues of a matrix can be complex numbers. However, if a matrix is symmetric ($\mathbf{A}^t = \mathbf{A}$), then its eigenvalues are real numbers.

5.2.8 Eigenvectors

Matrix $\mathbf{A}$ of dimension $(N \times N)$ can have no more than N eigenvectors. Each eigenvector corresponds to its eigenvalue. Eigenvector $\mathbf{v}_n$ can be found as a solution of a homogeneous system of linear equations

$$(\mathbf{A} - \lambda_n\mathbf{I})\mathbf{v}_n = 0,$$

which has a nontrivial solution, because $\det(\mathbf{A} - \lambda_n\mathbf{I}) = 0$. An example is shown in Fig. 5.22.

A symmetric matrix has orthogonal eigenvectors.

$$\mathbf{A} = \left|\begin{array}{ccc} 11 & -6 & 2 \\ -6 & 10 & -4 \\ 2 & -4 & 6 \end{array}\right| \qquad \mathbf{I} = \left|\begin{array}{ccc} 1 & 0 & 0 \\ 0 & 1 & 0 \\ 0 & 0 & 1 \end{array}\right|$$

$$\lambda_1 = 18 \qquad \mathbf{A}-\lambda\mathbf{I} = \left|\begin{array}{ccc} -7 & -6 & 2 \\ -6 & -8 & -4 \\ 2 & -4 & -12 \end{array}\right| \qquad \mathbf{det(A}-\lambda\mathbf{I)} = 0$$

$$\lambda_2 = 6 \qquad \mathbf{A}-\lambda\mathbf{I} = \left|\begin{array}{ccc} 5 & -6 & 2 \\ -6 & 4 & -4 \\ 2 & -4 & 0 \end{array}\right| \qquad \mathbf{det(A}-\lambda\mathbf{I)} = 0$$

$$\lambda_3 = 3 \qquad \mathbf{A}-\lambda\mathbf{I} = \left|\begin{array}{ccc} 8 & -6 & 2 \\ -6 & 7 & -4 \\ 2 & -4 & 3 \end{array}\right| \qquad \mathbf{det(A}-\lambda\mathbf{I)} = 0$$

Figure 5.21 Eigenvalues.

$$\mathbf{A} = \left|\begin{array}{ccc} 11 & -6 & 2 \\ -6 & 10 & -4 \\ 2 & -4 & 6 \end{array}\right| \qquad \mathbf{I} = \left|\begin{array}{ccc} 1 & 0 & 0 \\ 0 & 1 & 0 \\ 0 & 0 & 1 \end{array}\right|$$

$$\lambda_1 = 18 \qquad \mathbf{v}_1 = \left|\begin{array}{c} 2 \\ -2 \\ 1 \end{array}\right| \quad \left|\begin{array}{ccc} -7 & -6 & 2 \\ -6 & -8 & -4 \\ 2 & -4 & -12 \end{array}\right| * \left|\begin{array}{c} 2 \\ -2 \\ 1 \end{array}\right| = \left|\begin{array}{c} 0 \\ 0 \\ 0 \end{array}\right|$$

$$\lambda_2 = 6 \qquad \mathbf{v}_2 = \left|\begin{array}{c} 2 \\ -1 \\ -2 \end{array}\right| \quad \left|\begin{array}{ccc} 5 & -6 & 2 \\ -6 & 4 & -4 \\ 2 & -4 & 0 \end{array}\right| * \left|\begin{array}{c} 2 \\ 1 \\ -2 \end{array}\right| = \left|\begin{array}{c} 0 \\ 0 \\ 0 \end{array}\right|$$

$$\lambda_3 = 3 \qquad \mathbf{v}_3 = \left|\begin{array}{c} 1 \\ 2 \\ 2 \end{array}\right| \quad \left|\begin{array}{ccc} 8 & -6 & 2 \\ -6 & 7 & -4 \\ 2 & -4 & 3 \end{array}\right| * \left|\begin{array}{c} 1 \\ 2 \\ 2 \end{array}\right| = \left|\begin{array}{c} 0 \\ 0 \\ 0 \end{array}\right|$$

Figure 5.22 Eigenvectors.

5.2.9 Equivalence and Similarity

Two rectangular $I \times J$ matrices **A** and **B** are called *equivalent*, if there exist two square matrices **S** of dimension $I \times I$ and **T** of dimension $J \times J$ such that

$$\mathbf{B} = \mathbf{SAT}.$$

Equivalent matrices have the same rank.

Two square $N \times N$ matrices **A** and **B** are said to be *similar*, if there exists an invertible matrix **T** such that

$$\mathbf{B} = \mathbf{T}^{-1}\mathbf{AT}.$$

$$\mathbf{A} = \begin{vmatrix} 11 & -6 & 2 \\ -6 & 10 & -4 \\ 2 & -4 & 6 \end{vmatrix} \quad \mathbf{T} = \begin{vmatrix} 2 & 2 & 1 \\ -2 & 1 & 2 \\ 1 & -2 & 2 \end{vmatrix} \quad \mathbf{T}^{-1} = \begin{vmatrix} 0.22 & -0.2 & 0.11 \\ 0.22 & 0.11 & -0.2 \\ 0.11 & 0.22 & 0.22 \end{vmatrix}$$

$$\Lambda = \begin{vmatrix} 18 & 0 & 0 \\ 0 & 6 & 0 \\ 0 & 0 & 3 \end{vmatrix} \quad \mathbf{T}\Lambda\mathbf{T}^{-1} = \begin{vmatrix} 11 & -6 & 2 \\ -6 & 10 & -4 \\ 2 & -4 & 6 \end{vmatrix}$$

Figure 5.23 Matrix diagonalization.

Matrix **T** is called a similarity transformation. Similar matrices have the same rank, trace, determinant, and spectrum.

5.2.10 Diagonalization

A normal (in particular, a symmetric) matrix **A** can be transformed into a *diagonal* matrix using similarity transformation

$$\mathbf{A} = \mathbf{T}\Lambda\mathbf{T}^{-1}$$

Here, $\Lambda = \mathrm{diag}(\lambda_1, \ldots, \lambda_N)$ is the diagonal matrix whose elements are the eigenvalues of **A**, and **T** is the matrix composed by the corresponding eigenvectors of **A**, that is, $\mathbf{T} = (\mathbf{v}_1, \ldots, \mathbf{v}_N)$. See Fig. 5.23 for an example.

5.2.11 Singular Value Decomposition (SVD)

Let us consider the rectangular $I \times J$ matrix **A** of rank $R\,(I \le J \le R)$. The matrix can be decomposed into the product of three matrices $\mathbf{P}_\mathrm{R}(I \times R)$, $\mathbf{D}_\mathrm{R}(R \times R)$, and $\mathbf{Q}_\mathrm{R}(J \times R)$

$$\mathbf{A} = \mathbf{P}_\mathrm{R}\mathbf{D}_\mathrm{R}\mathbf{Q}_\mathrm{R}^\mathrm{t}$$

in such a way that

$$\mathbf{P}_\mathrm{R}^\mathrm{t}\mathbf{P}_\mathrm{R} = \mathbf{Q}_\mathrm{R}^\mathrm{t}\mathbf{Q}_\mathrm{R} = \mathbf{I}_\mathrm{R}.$$

Here, $\mathbf{P}_\mathrm{R}$ is the matrix formed by R eigenvectors $\mathbf{p}_r$. These vectors correspond to the largest eigenvalues λ_r of matrix $\mathbf{AA}^\mathrm{t}$, that is,

$$\mathbf{AA}^\mathrm{t}\mathbf{p}_r = \lambda_r\mathbf{p}_r;$$

$\mathbf{Q}_\mathrm{R}$ is a matrix formed by R eigenvectors $\mathbf{q}_r$. These vectors correspond to the largest eigenvalues λ_r of matrix $\mathbf{A}^\mathrm{t}\mathbf{A}$, that is,

$$\mathbf{A}^\mathrm{t}\mathbf{A}\mathbf{q}_r = \lambda_r\mathbf{q}_r$$

$\mathbf{D}_R = \text{diag}(\sigma_1, \ldots, \sigma_R)$ is a nonnegative definite diagonal matrix with elements $\sigma_1 \geq \ldots \geq \sigma_R \geq 0$ that are called *singular* values of matrix $\mathbf{A}$. Singular values are equal to the square root of the eigenvalues of matrix $\mathbf{A}^t\mathbf{A}$

$$\sigma_r = \sqrt{\lambda_r}.$$

An example of singular value decomposition (SVD) is shown in Fig. 5.24.

By augmenting matrices $\mathbf{P}_R$ and $\mathbf{Q}_R$ by orthonormal columns and matrix $\mathbf{D}_R$ by zero entries, it is possible to construct matrices $\mathbf{P}$ $(I \times J)$, $\mathbf{D}$ $(J \times J)$, and $\mathbf{Q}$ $(J \times J)$ such that

$$\mathbf{A} = \mathbf{P}_R\mathbf{D}_R\mathbf{Q}_R^t = \mathbf{P}\mathbf{D}\mathbf{Q}^t.$$

Application of SVD is explained in Sections 9.2.3, 12.3.2, and 15.5.2.

5.2.12 Vector Space

Let us consider all possible vectors of dimension N. This set is called *vector space* of dimension N. It is denoted by R^N. As R^N spans all vectors (of dimension N), any linear combination of vectors from R^N also belongs to this space.

$$\mathrm{A} = \begin{vmatrix} -0.588 & 0.8317 & 0.7892 & 0.6727 \\ 0.1017 & 0.0217 & 0.1092 & 0.0647 \\ -0.118 & 0.2817 & 0.2192 & 0.2167 \\ 0.1817 & -0.138 & -0.041 & -0.067 \\ -0.738 & -0.868 & -0.671 & -0.776 \\ 0.0617 & 0.0917 & -0.041 & 0.0357 \\ -0.108 & 0.3417 & 0.2492 & 0.2597 \\ 0.1017 & 0.1217 & 0.1592 & 0.1347 \\ 0.4517 & -0.328 & -0.601 & -0.359 \\ 0.2717 & -0.158 & -0.071 & -0.08 \\ 0.2017 & -0.148 & -0.211 & -0.138 \\ 0.1817 & -0.048 & 0.1092 & 0.0377 \end{vmatrix} \qquad \mathrm{P_R} = \begin{vmatrix} 0.631 & -0.403 & -0.155 \\ 0.0464 & 0.0908 & 02616 \\ 0.1948 & -0.066 & -0.145 \\ -0.074 & 0.1395 & 0.2841 \\ -0.579 & -0.756 & 0.0623 \\ 0.021 & 0.0661 & -0.294 \\ 0.2297 & -0.049 & -0.205 \\ 0.1048 & 0.104 & 0.1551 \\ -0.358 & 0.3512 & -0.59 \\ -0.094 & 0.2152 & 0.2872 \\ -0.14 & 0.1563 & -0.115 \\ 0.0168 & 0.1509 & 0.4532 \end{vmatrix}$$

$$I = 12 \qquad J = 4 \qquad R = 3$$

$$\mathrm{Q_R} = \begin{vmatrix} -0.092 & 0.9865 & 0.1065 \\ 0.6136 & 0.093 & -0.662 \\ 0.5788 & -0.055 & 0.7412 \\ 0.5291 & 0.1233 & -0.024 \end{vmatrix} \qquad \mathrm{D_R} = \begin{vmatrix} 2.1825 & 0 & 0 \\ 0 & 1.149 & 0 \\ 0 & 0 & 0.2898 \end{vmatrix}$$

$$\mathrm{P_R D_R Q_R}^t = \begin{vmatrix} -0.588 & 0.8317 & 0.7892 & 0.6727 \\ 0.1017 & 0.0217 & 0.1092 & 0.0647 \\ -0.118 & 0.2817 & 0.2192 & 0.2167 \\ 0.1817 & -0.138 & -0.041 & -0.067 \\ -0.738 & -0.868 & -0.671 & -0.776 \\ 0.0617 & 0.0917 & -0.041 & 0.0357 \\ -0.108 & 0.3417 & 0.2492 & 0.2597 \\ 0.1017 & 0.1217 & 0.1592 & 0.1347 \\ 0.4517 & -0.328 & -0.601 & -0.359 \\ 0.2717 & -0.158 & -0.071 & -0.080 \\ 0.2017 & -0.148 & -0.211 & -0.138 \\ 0.1817 & -0.048 & 0.1092 & 0.0377 \end{vmatrix}$$

Figure 5.24 SVD.

5.2.13 Space Basis

Any set of N linearly independent vectors is said to be the *basis* in space R^N. The simplest basis is a set of coordinate vectors

$$\mathbf{e}_1 = \begin{pmatrix} 1 \\ 0 \\ \vdots \\ 0 \end{pmatrix}, \quad \mathbf{e}_2 = \begin{pmatrix} 0 \\ 1 \\ \vdots \\ 0 \end{pmatrix}, \ldots, \mathbf{e}_N = \begin{pmatrix} 0 \\ 0 \\ \vdots \\ 1 \end{pmatrix}.$$

Each coordinate vector has only one element that equals 1, the rest of the entries are zeros. Any vector $\mathbf{x} = (\mathbf{x}_1, \mathbf{x}_2, \ldots, \mathbf{x}_N)^t$ can be represented as a linear combination $\mathbf{x} = \mathbf{x}_1\mathbf{e}_1 + \mathbf{x}_2\mathbf{e}_2 + \cdots + \mathbf{x}_N\mathbf{e}_N$ of the coordinate vectors.

The basis consisting of mutually orthogonal vectors is called *orthogonal*. If the basis vectors are also normalized, then the basis is said to be *orthonormal*.

5.2.14 Geometric Interpretation

A geometrical interpretation can be given for a vector space. Let us imagine the N-dimensional space in which the basic vectors determine the direction of the coordinate axes. Then any vector $\mathbf{x} = (\mathbf{x}_1, \mathbf{x}_2, \ldots, \mathbf{x}_N)^t$ can be represented in this space by a point with the *coordinates* $(\mathbf{x}_1, \mathbf{x}_2, \ldots, \mathbf{x}_N)$. Figure 5.25 illustrates the coordinate space.

5.2.15 Nonuniqueness of Basis

A linear space can have an unlimited number of bases. For example, in an R^3 space besides the ordinary orthonormal basis

$$\mathbf{e}_1 = \begin{pmatrix} 1 \\ 0 \\ 0 \end{pmatrix}, \quad \mathbf{e}_2 = \begin{pmatrix} 0 \\ 1 \\ 0 \end{pmatrix}, \quad \mathbf{e}_3 = \begin{pmatrix} 0 \\ 0 \\ 1 \end{pmatrix}$$

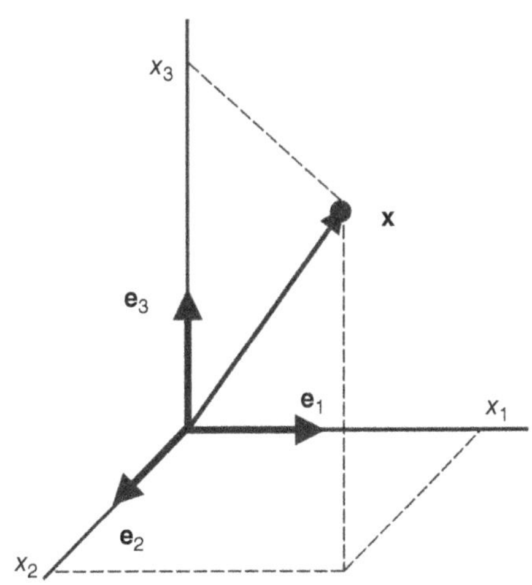

Figure 5.25 Coordinate space.

another orthonormal basis can be introduced, for example,

$$\mathbf{b}_1 = \begin{pmatrix} \sqrt{0.5} \\ 0.5 \\ 0.5 \end{pmatrix}, \quad \mathbf{b}_2 = \begin{pmatrix} -\sqrt{0.5} \\ 0.5 \\ 0.5 \end{pmatrix}, \quad \mathbf{b}_3 = \begin{pmatrix} 0 \\ -\sqrt{0.5} \\ \sqrt{0.5} \end{pmatrix}$$

Each basis can be represented by matrix $\mathbf{B} = (\mathbf{b}_1, \ldots, \mathbf{b}_N)$ consisting of basis vectors. The transition from one basis $\mathbf{B}_1$ to another basis $\mathbf{B}_2$ is carried out using a nondegenerate square matrix.

5.2.16 Subspace

Given a set of K linear independent vectors $\mathbf{x}_1, \mathbf{x}_2, \ldots, \mathbf{x}_K$ in space R^N, consider all possible linear combinations of these vectors

$$\mathbf{x} = \alpha_1\mathbf{x}_1 + \alpha_2\mathbf{x}_2 + \cdots + \alpha_K\mathbf{x}_K.$$

The obtained set Q is called a linear span, or it is said that it is spanned by vectors $\mathbf{x}_1, \mathbf{x}_2, \ldots, \mathbf{x}_K$. By definition of the vector space, the linear span Q is a vector space of dimension K. At the same time, Q belongs to space R^N; therefore, Q is called a *linear subspace* R^K in space R^N.

5.2.17 Projection

Let us consider subspace R^K spanned by vectors $\mathbf{X} = (\mathbf{x}_1, \mathbf{x}_2, \ldots, \mathbf{x}_K)$. The dimension of the basis matrix $\mathbf{X}$ is $(N \times K)$. Any vector $\mathbf{y}$ from R^N can be projected on subspace R^K, that is, represented in a form

$$\mathrm{y} = \mathbf{y}^{||} + \mathbf{y}^{\perp},$$

where vector $\mathbf{y}^{||}$ belongs to R^K and vector $\mathbf{y}^{\perp}$ is orthogonal to $\mathbf{y}^{||}$. See Fig. 5.26 for illustration.

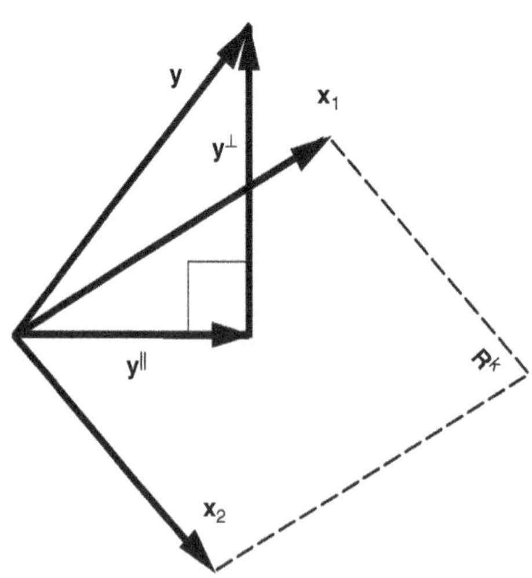

Figure 5.26 Projection on a subspace.

$$\mathbf{y} = \begin{vmatrix} 2 \\ 0 \\ 1 \end{vmatrix} \quad \mathbf{x}_1 = \begin{vmatrix} 1 \\ 1 \\ 0 \end{vmatrix} \quad \mathbf{x}_2 = \begin{vmatrix} 2 \\ 2 \\ 1 \end{vmatrix} \quad \mathbf{X} = \begin{vmatrix} 1 & 2 \\ 1 & 2 \\ 0 & 1 \end{vmatrix}$$

$$\mathbf{x}^t\mathbf{x} = \begin{vmatrix} 2 & 4 \\ 4 & 9 \end{vmatrix} \qquad (\mathbf{x}^t\mathbf{x})^{-1} = \begin{vmatrix} 4.5 & -2 \\ -2 & 1 \end{vmatrix}$$

$$\mathbf{P} = \begin{vmatrix} 0.5 & 0.5 & 0 \\ 0.5 & 0.5 & 0 \\ 0 & 0 & 1 \end{vmatrix} \quad \mathbf{y}^{||} = \mathbf{Py} = \begin{vmatrix} 1 \\ 1 \\ 1 \end{vmatrix} \quad \mathbf{y}^{\perp} = (\mathbf{I}\text{-}\mathbf{P})\mathbf{y} = \begin{vmatrix} 1 \\ -1 \\ 0 \end{vmatrix}$$

$$(\mathbf{y}^{||},\mathbf{y}^{\perp}) = 0 \quad \mathbf{y}^{||} = \mathbf{x}_2 - \mathbf{x}_1 = \begin{vmatrix} 1 \\ 1 \\ 1 \end{vmatrix} \quad \mathbf{y}^{||} + \mathbf{y}^{\perp} = \begin{vmatrix} 2 \\ 0 \\ 1 \end{vmatrix}$$

Figure 5.27 Projection.

The projected vector $\mathbf{y}^{||}$ can be represented as the result of application of the projection matrix $\mathbf{P}$

$$\mathbf{y}^{||} = \mathbf{Py}.$$

The projection matrix is defined by

$$\mathbf{P} = \mathbf{X}(\mathbf{X}^t\mathbf{X})^{-1}\mathbf{X}^t.$$

An example is shown in Fig. 5.27.

6

STATISTICS

This chapter presents mathematical statistics, namely the main statistical methods that are used in chemometrics. The text below cannot serve as a textbook on statistics; it is rather a synopsis, a brief guide to mathematical statistics. A more systematic and in-depth explanation can be found in the literature referenced in Chapter 4.

Preceding chapters	5
Dependent chapters	7, 9–11, 15
Matrix skills	Basic
Statistical skills	Basic
Excel skills	Basic
Chemometric skills	Low
Chemometrics Add-In	No
Accompanying workbook	Statistics.xls

6.1 THE BASICS

6.1.1 Probability

The world is full of events whose outcome is not known in advance. Everyone knows the classic example of tossing a coin, which ends with a random draw of heads or tails. The probability of such random events is a number between zero and one. However, not all events have a probability. Repeatability is the key condition for this. For instance, it is meaningless to ask about the probability of rain tomorrow as "tomorrow" has no repetitions. This unique event cannot be repeated. Nevertheless, we can talk about the probability that

Chemometrics in Excel, First Edition. Alexey L. Pomerantsev.

it will rain on July 7. The event "July 7" is repeated every year, and the rain on that day can be predicted with some probability.

The concept of probability is only applicable to the events that have not happened or the outcomes of which are not yet known. For example, one can calculate the probability of winning a lottery but as soon as everybody knows the result of the draw, that is, the event has occurred, the calculated probability of winning becomes meaningless.

Another important concept is sample space, that is, the complete set of all possible outcomes. In the example with the coin, the experiment has only two outcomes: heads or tails. Consider another experiment such as measuring the height of a randomly selected person. If the measurement accuracy is 1 cm, the sample space is a set of numbers between 30 cm (newborn) and 272 cm (Guinness record), 243 possible outcomes in total. However, if the height is measured with an accuracy of 1 m, the sample space has only three events: less than 1 m, between 1 and 2 m, and greater than 2 m.

6.1.2 Random Value

Random variable is a variable whose value is not known before the experiment. Every random variable is characterized by the following properties:

- a range of possible values (the sample space);
- an unlimited number of repeated realizations;
- a probability to hit a given area in the sample space.

The sample space can be discrete, continuous, and discrete–continuous. The random variables are named respectively.

6.1.3 Distribution Function

Let X be a random variable, the possible values of which are real numbers and let P be the probability of the event that realization of X is not greater than a given value x. Considering this probability P as a function of x, we obtain a function $F(x)$, which is called the *(cumulative) distribution function* of a random variable,

$$F(x) = \Pr\{X \le x\}. \tag{6.1}$$

The distribution function is a nondecreasing function, which tends to 0 for a small x and tends to 1 for the large values of the argument. The fact that the random variable X has the distribution function F is denoted by

$$X \sim F.$$

A distribution is said to be *symmetric* (relative to a), if $F(a + x) = 1 - F(a - x)$.

For a discrete random variable, the distribution function is a piece-wise constant with jumps at $x = x_i$.

The derivative of the distribution function $F(x)$ is called the *probability density function* $f(x)$

$$F(x) = \int_{-\infty}^{x} f(t)dt$$

An illustration is given in Fig. 6.1.

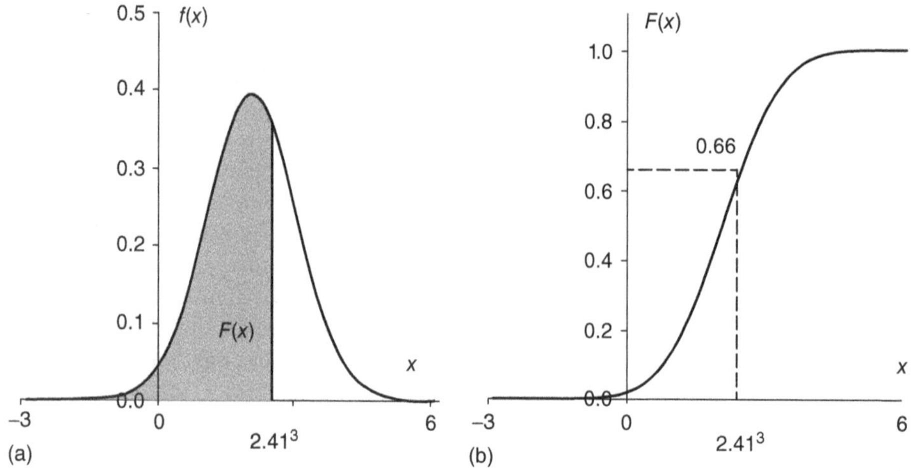

Figure 6.1 (a) Probability density function $f(x)$ and (b) cumulative distribution function $F(x)$.

6.1.4 Mathematical Expectation

Let X be a random variable with the probability density function $f(x)$. An *expected value* of X is defined by

$$E(X) = \int_{-\infty}^{+\infty} xf(x)dx.$$

6.1.5 Variance and Standard Deviation

Let X be a random variable with the probability density function $f(x)$. The *variance* of X is equal to

$$V(X) = \int_{-\infty}^{+\infty} (x - E(X))^2 f(x)dx = E((X - E(X))^2).$$

The square root of the variance is said to be the *standard deviation.*

6.1.6 Moments

Let X be a random variable with the probability density function $f(x)$. The *n-th moment* of X is defined by

$$\mu_n = \int_{-\infty}^{+\infty} x^n f(x)dx.$$

By definition, $\mu_1 = E(X)$.
The *n-th central moment* is

$$m_n = \int_{-\infty}^{+\infty} (x - \mu_1)^n f(x)dx = E((X - \mu_1)^n). \tag{6.2}$$

By definition, $m_2 = V(X)$.

6.1.7 Quantiles

Let $F(x)$ be a (cumulative) distribution function of a random variable

$$F(x) = \int_{-\infty}^{x} f(t)dt.$$

Consider a function $F^{-1}(P)$, $\quad 0 \le P \le 1$, which is the inverse of $F(x)$, that is,

$$F^{-1}(F(x)) = x, \quad F(F^{-1}(P)) = P.$$

Function $F^{-1}(P)$ is said to be the *P-quantile* of distribution F.

The quantile for $P = 0.5$ is called the *median* of the distribution. The quantiles for $P = 0.25, 0.75$ are said to be *quartiles* and for $P = 0.01, 0.02, \ldots, 0.99$ are said to be *percentiles*.

6.1.8 Multivariate Distributions

Two (or more) random variables can be considered together. The joint cumulative distribution function of two random variables X and Y is defined by

$$F(x, y) = \Pr\{(X \le x) \wedge (Y \le y)\} = \int_{-\infty}^{x} \int_{-\infty}^{y} f(\xi, \eta)d\xi d\eta.$$

Similar to the univariate case, the function $f(x, y)$ is called the *probability density function*.

Random variables X and Y are said to be *independent* if their joint probability density function is a product of partial densities, that is,

$$f(x, y) = f(x)f(y). \tag{6.3}$$

6.1.9 Covariance and Correlation

A *covariance* of random variables X and Y is a (deterministic) value defined by

$$\mathrm{cov}(X, Y) = \int_{-\infty}^{+\infty} \int_{-\infty}^{+\infty} (\xi - E(X))(\eta - E(Y))f(\xi, \eta)d\xi d\eta,$$

where $f(x, y)$ is the joint probability density function.

Value

$$\mathrm{cor}(X, Y) = \mathrm{cov}(X, Y)/\sqrt{V(X)V(Y)}$$

is said to be the *correlation* between random variables X and Y. If random variables X and Y are independent, then their covariance and correlation are equal to zero. The converse is not true.

For a joint distribution of multivariate random variables $X_1, \ldots, X_n$, the *covariance matrix* $\mathbf{C}$ is defined by

$$c_{ij} = \mathrm{cov}(X_i, X_j), \quad i, j = 1, \ldots, n$$

This matrix plays the same role as the variance of a univariate distribution.

6.1.10 Function

A function of a random variable is also a random variable. Let the random variable X have a distribution function $F_X(x)$ and the random variables X and Y be connected via the biunique relations

$$y = \varphi(x), \quad x = \psi(y).$$

If $\varphi(x)$ is an increasing function, then the distribution and quantiles of the random variable Y are defined by

$$F_Y(y) = F_X(\psi(y)), \quad y(P) = \varphi(x(P)).$$

If $\psi(y)$ is a differentiable function, then the probability density function of the random variable Y is calculated by

$$f_Y(y) = f_X(\psi(y))|d\psi/dy|$$

For a linear transformation $y = ax + b$, the following relations hold

$$f_Y(y) = (1/|a|)f_X((y-b)/a),$$

$$E(aX + b) = aE(X) + b, \qquad V(aX + b) = a^2V(X)$$

$$E(X + Y) = E(X) + E(Y), \qquad V(X + Y) = V(X) + V(Y) + \text{cov}(X, Y). \tag{6.4}$$

6.1.11 Standardization

If a random variable X has an expectation m and variance s^2:

$$E(X) = m, \quad V(X) = s^2,$$

then a random variable

$$Y = (X-m)/s$$

is the standardized (normalized) variable, for which

$$E(Y) = 0, \quad V(Y) = 1.$$

6.2 MAIN DISTRIBUTIONS

6.2.1 Binomial Distribution

A discrete random variable X has a *binomial distribution* if its probability density function is given by

$$f(k|p,n) \equiv \Pr(X = k) = \binom{n}{k} p^k(1-p)^{n-k},$$

where $\binom{n}{k} = n!/(n-k)!k!$ is the binomial coefficient. Figure 6.2 illustrates the binomial distribution.

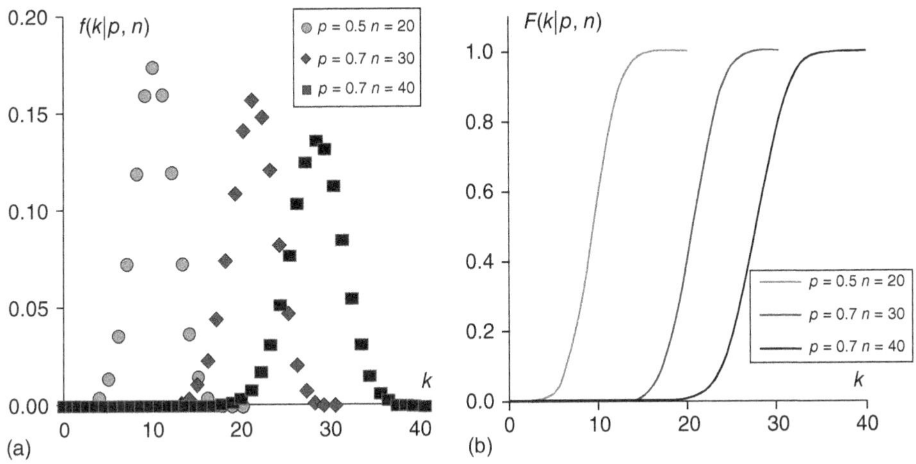

Figure 6.2 (a,b) Probability density function and distribution function of binomial distribution.

Binomial distribution is a distribution of k successes in a series of n independent trials, provided that the probability of success in each trial is p. The expectation and variance are, respectively, equal to

$$E(X) = np, \quad V(X) = np(1 - p). \tag{6.5}$$

For a large n, the binomial distribution is well approximated by the normal distribution with parameters given in Eq. (6.5).

To calculate the binomial distribution, the standard Excel function **BINOMDIST** is used.

Syntax

BINOMDIST (number_s= k, **trials=** n, **probability_s=** p, `cumulative = TRUE | FALSE)`

If `cumulative` = `TRUE`, the function returns the cumulative distribution function, and if `cumulative` = `FALSE`, it returns the probability density

An example is given in worksheet Binomial and shown in Fig. 6.3.

6.2.2 Uniform Distribution

A random variable X is *uniformly* distributed over interval $[a, b]$, if its distribution function $U(x|a, b)$ and, therefore, the probability density function $u(x|a, b)$ are given by

$$U(x|a,b) = \begin{cases} 0, & x \le a \\ (x-a)/(b-a), & a < x \le b \\ 1, & x > b \end{cases}, \qquad u(x|a,b) = \begin{cases} 0, & x \le a \\ 1/(b-a), & a < x \le b \\ 0, & x > b \end{cases}$$

The expectation and variance are, respectively, equal to

$$E(X) = 0.5(a + b), \qquad V(X) = (b - a)^2/12.$$

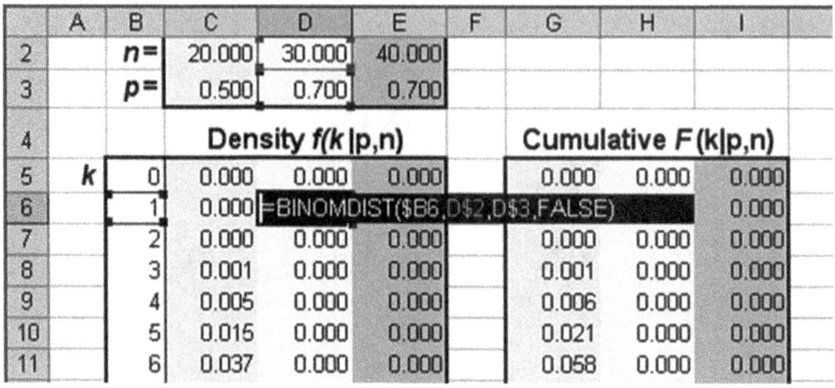

	A	B	C	D	E	F	G	H	I
2		n=	20.000	30.000	40.000				
3		p=	0.500	0.700	0.700				
4			Density f(k|p,n)				Cumulative F(k|p,n)		
5	k	0	0.000	0.000	0.000		0.000	0.000	0.000
6		1	0.000	=BINOMDIST($B6,D$2,D$3,FALSE)					0.000
7		2	0.000	0.000	0.000		0.000	0.000	0.000
8		3	0.001	0.000	0.000		0.001	0.000	0.000
9		4	0.005	0.000	0.000		0.006	0.000	0.000
10		5	0.015	0.000	0.000		0.021	0.000	0.000
11		6	0.037	0.000	0.000		0.058	0.000	0.000

Figure 6.3 Example of calculation of the binomial distribution.

The fact that random variable X is uniformly distributed over interval $[a, b]$, is denoted by

$$X \sim U(a, b).$$

Figure 6.4 illustrates the binomial distribution.

6.2.3 Normal Distribution

The normal (or Gaussian) distribution is probably the most important distribution in statistics. A probability density function of this distribution is defined by

$$f(x|m, \sigma) = (1/\sigma\sqrt{2\pi})\exp(-(x-m)^2/\sigma^2). \tag{6.6}$$

The normal distribution depends on two parameters: m and σ^2. It is usually denoted by $N(m, \sigma^2)$, that is,

$$X \sim N(m, \sigma^2).$$

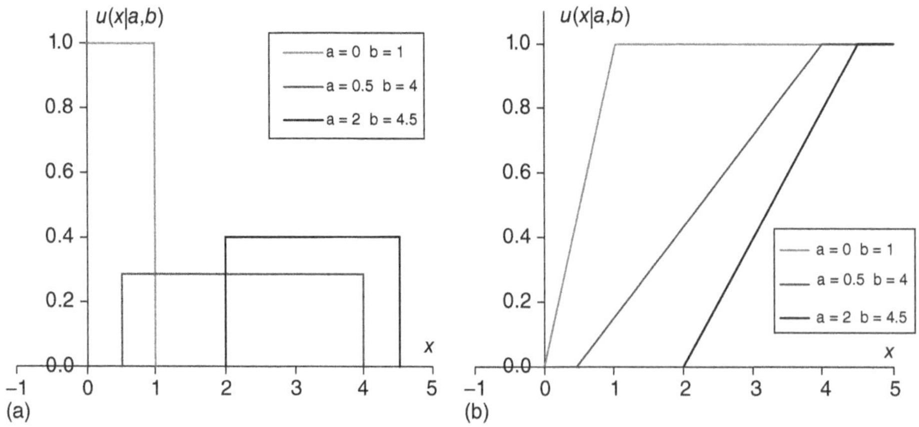

Figure 6.4 (a,b) Probability density and distribution of the uniform distribution.

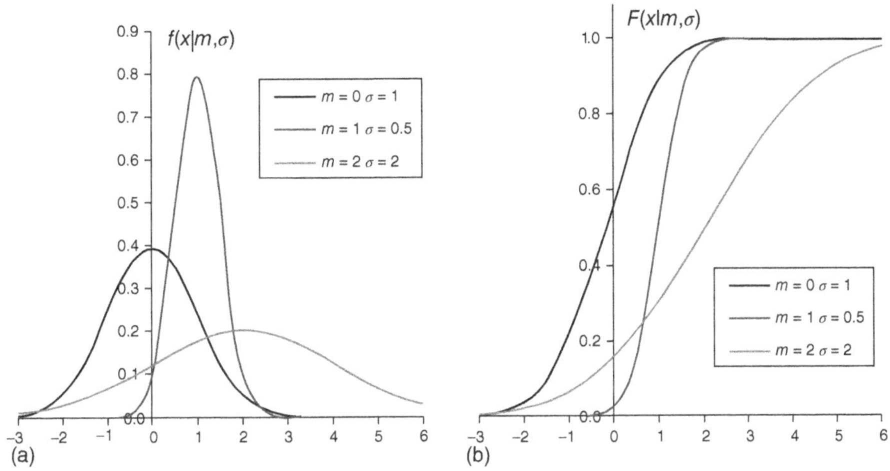

Figure 6.5 (a,b) Probability density function and distribution function of the normal distribution.

The expectation and variance of the normal distribution are, respectively, equal to

$$E(X) = m, \quad V(X) = \sigma^2.$$

The normal distribution is said to be *standard*, if $m = 0$ and $\sigma^2 = 1$. If $X_0 \sim N(0, 1)$, then $X = m + \sigma X_0 \sim N(m, \sigma^2)$.

The cumulative distribution function of the standard normal distribution, that is,

$$\Phi(x) = \int_{-\infty}^{x} f(t)dt$$

is a special function because it cannot be expressed in the terms of elementary functions. The quantile of the standard normal distribution is denoted by $\Phi^{-1}(P)$. The standard normal distribution is symmetric; therefore, the following relations hold

$$\Phi(-x) = 1-\Phi(x), \quad \Phi^{-1}(1-P) = -\Phi^{-1}(P).$$

Examples of the normal distribution are shown in Fig. 6.5.

One can calculate normal distribution in Excel by means of several standard worksheet functions, such as **NORMDIST** and **NORMSDIST**, as well as **NORMINV** and **NORMSINV**.

Syntax

NORMDIST(x, mean = m, **standard_dev** = σ, `cumulative = TRUE|FALSE)`

If `cumulative` = `TRUE`, the function returns the cumulative distribution function $\Phi(x|m, \sigma^2)$ and if `cumulative` = `FALSE`, it returns the probability density.

NORMSDIST(x)

Returns the cumulative standard normal distribution function at point x.

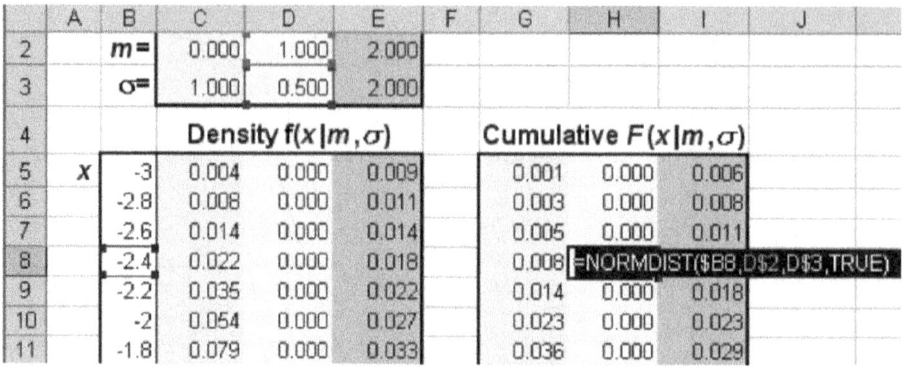

Figure 6.6 Example of calculation of the normal distribution.

NORMINV(probability= P, **mean**= m, **standard_dev**= σ)
Returns the quantile $\Phi^{-1}(P|m, \sigma^2)$ of the normal distribution for probability P.

NORMSINV(probability= P)
Returns the quantile $\Phi^{-1}(P|0, 1)$ of the standard normal distribution for probability P.

An example of the Excel calculations is given in worksheet Normal and shown in Fig. 6.6.

6.2.4 Chi-Squared Distribution

Consider N independent and identically distributed standard normal random variables $X_1, \ldots, X_n, \ldots, X_N$ with zero expectation and unit variance, that is,

$$X_n \sim N(0, 1).$$

A variable

$$\chi^2(N) = X_1^2 + \cdots + X_N^2$$

is random. Its distribution is said to be *chi-squared*. This distribution depends on a single parameter N, which is called the *number of degrees of freedom*. The chi-squared probability density is given by

$$f(x|N) = ((1/2)^{N/2}/\Gamma(N/2))x^{(N/2)-1}e^{-x/2}.$$

An illustration of the chi-squared distributions is shown in Fig. 6.7.

The chi-squared distribution is widely used in statistics, for example, in hypothesis testing. The expectation and variance of the $\chi^2(N)$ distribution are respectively, equal to

$$E(\chi^2(N)) = N, \quad V(\chi^2(N)) = 2N. \tag{6.7}$$

For a large N, the chi-squared distribution is well approximated by the normal distribution with the parameters given in Eq. (6.7). A quantile of the $\chi^2(N)$ distribution is denoted by $\chi^{-2}(P|N)$.

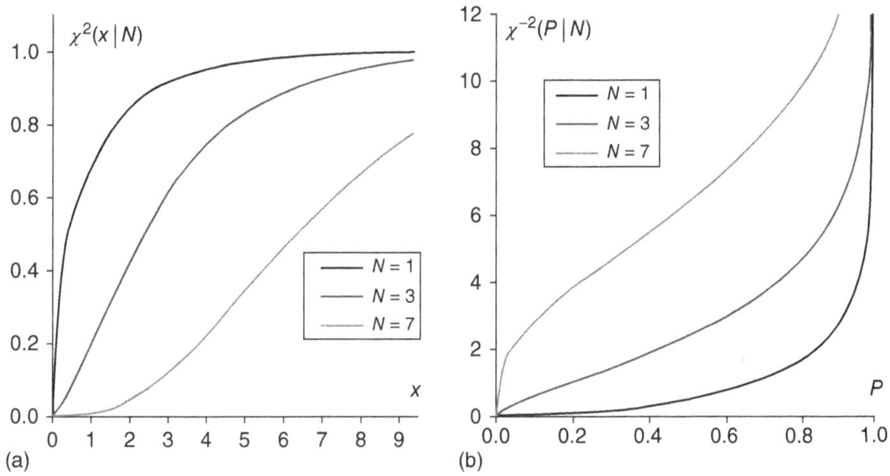

Figure 6.7 (a,b) Probability distribution function and quantile of the chi-squared distribution.

To calculate the chi-squared distribution in Excel, two standard worksheet functions **CHIDIST** and **CHIINV** are used.

Syntax

CHIDIST(x, degrees_freedom $= N$)

Returns $1 - \chi^2(x|N)$, where $\chi^2(x|N)$ is the cumulative chi-squared distribution function.

CHIINV(probability $= 1-P$, **degrees_freedom** $= N$)

Returns quantile $\chi^{-2}(1-P|N)$ of the chi-squared distribution for probability $1 - P$.

An example of the Excel calculations is given in worksheet Chi-square and shown in Fig. 6.8.

	A	B	C	D	E	F	G	H	I	J
2										
3		N=	1	3	7					
4			Cumulative $\chi^2(x\|N)$					Quantile $\chi^{-2}(P\|N)$		
5	x	0	0.000	0.000	0.000	P	0	0.000	0.000	0.011
6		0.2	0.345	0.022	0.000		0.02	0.001	0.185	1.564
7		0.4	0.473	0.060	0.000		0.04	0.003	0.300	1.997
8		0.6	0.561	=1-CHIDIST($B8,D$3)			0.06	0.006	0.401	2.320
9		0.8	0.629	0.151	0.003		0.08	0.010	0.495	2.592
10		1	0.683	0.199	0.005		0.1	0.016	0.584	2.833

Figure 6.8 Example of calculating the chi-squared distribution.

6.2.5 Student's Distribution

Consider two random variables X and Y. X has the standard normal distribution $X \sim N(0, 1)$ and Y has the chi-squared distribution with N degrees of freedom $Y \sim \chi^2(N)$. A random variable

$$T(N) = \sqrt{N}(X/\sqrt{Y})$$

has a distribution, which bears the name *Student*. This distribution depends on parameter N, which is also called *the number of degrees of freedom*. Student's distribution is used to test hypotheses and to develop confidence intervals. Examples of Student's distribution are shown in Fig. 6.9.

The expectation of $T(N)$ is zero and the variance is

$$V(T(N)) = N/(N-2), \quad N > 2.$$

T-distribution is symmetric and at $N > 20$ it is indistinguishable from the normal distribution. The formula for the Student's probability density function is given in many textbooks. The quantiles of Student's distribution are denoted by $T^{-1}(P|N)$.

To calculate Student's distribution in Excel, two standard worksheet functions **TDIST** and **TINV** are used.

Syntax

TDIST(x, degrees_freedom = *N*, **tails** = 1|2)

If **tails** = 1, the function returns the probability $\Pr\{T(N) > x\}$ and if **tails** = 2, it returns the probability $\Pr\{|T(N)| > x\}$. The values for $x < 0$ are not returned. Therefore, in order to calculate the cumulative distribution function of Student's $T(x|N)$ in Excel, it is necessary to use the following worksheet formula

IF(x > 0, 1−TDIST(x, N, 1), −TDIST(-x, N, 1)).

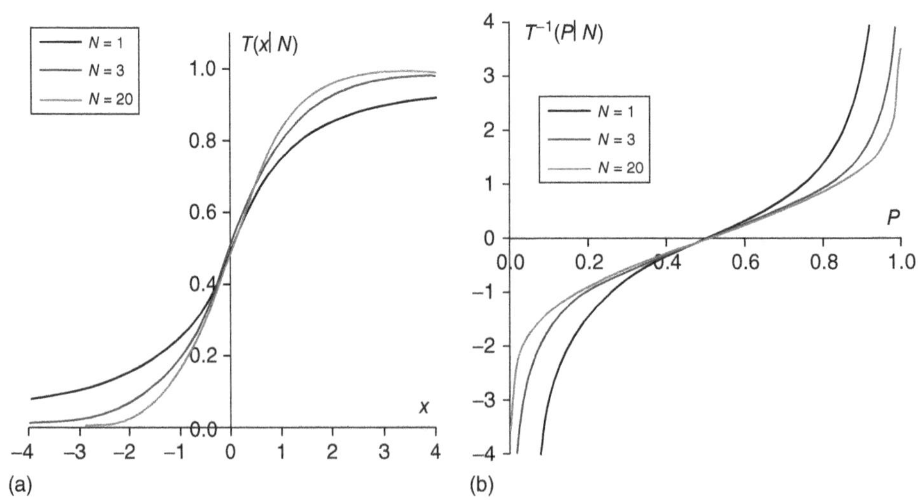

Figure 6.9 (a,b) Probability distribution function and quantile of Student's distribution.

TINV(probability, degrees_freedom $= N$)

Returns the value of x, for which the probability $\Pr\{|T(N)| > x\} =$ **probability**.

For the calculation of the quantile of Student's distribution $T^{-1}(P|N)$ in Excel, the following worksheet formula should be used

$$\mathrm{IF}(\mathrm{P} < 0.5, \mathrm{TINV}(2 * \mathrm{P}, \mathrm{N}), -\mathrm{TINV}(2-2 * \mathrm{P}, \mathrm{N})).$$

An example of the Excel calculations is given in worksheet Student and shown in Fig. 6.10.

6.2.6 *F*-Distribution

Consider two independent random variables X_1 and X_2, each of which has a chi-squared distribution with N_1 and N_2 degrees of freedom correspondingly, that is,

$$X_1 \sim \chi^2(N_1) \text{ and } X_2 \sim \chi^2(N_2).$$

A random variable

$$F(N_1, N_2) = N_2 X_1 / N_1 X_2$$

has a distribution, which is called the *F-distribution*. This distribution depends on two parameters N_1 and N_2, which are also called *the degrees of freedom*. The expectation and variance of the distribution $F(N_1, N_2)$ are, respectively,

$$E(F(N_1, N_2)) = N_2/(N_2-2), \quad N_2 > 2$$

$$V(F(N_1, N_2)) = (2N_2^2(N_1 + N_2 - 2)/N_1(N_2 - 2)^2(N_2 - 4)), \quad N_2 > 4.$$

The formula for the probability density function of the F-distribution is given in many textbooks. Examples of the F-distribution are shown in Fig. 6.11.

If $X \sim F(N_1, N_2)$, then $1/X \sim F(N_2, N_1)$.

The quantile of the F-distribution $F(N_1, N_2)$ is denoted by $F^{-1}(P|N_1, N_2)$.

To calculate the F-distribution in Excel, two standard worksheet functions, **FDIST** and **FINV**, are used.

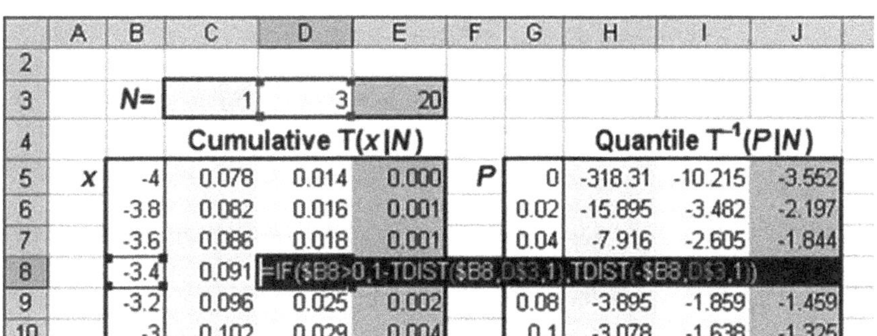

	A	B	C	D	E	F	G	H	I	J
2										
3		N=	1	3	20					
4			Cumulative T(x|N)					Quantile T⁻¹(P|N)		
5	X	-4	0.078	0.014	0.000	P	0	-318.31	-10.215	-3.552
6		-3.8	0.082	0.016	0.001		0.02	-15.895	-3.482	-2.197
7		-3.6	0.086	0.018	0.001		0.04	-7.916	-2.605	-1.844
8		-3.4	0.091	=IF($B8>0,1-TDIST($B8,D$3,1),TDIST(-$B8,D$3,1))						
9		-3.2	0.096	0.025	0.002		0.08	-3.895	-1.859	-1.459
10		-3	0.102	0.029	0.004		0.1	-3.078	-1.638	-1.325

Figure 6.10 Example of calculating Student's distribution.

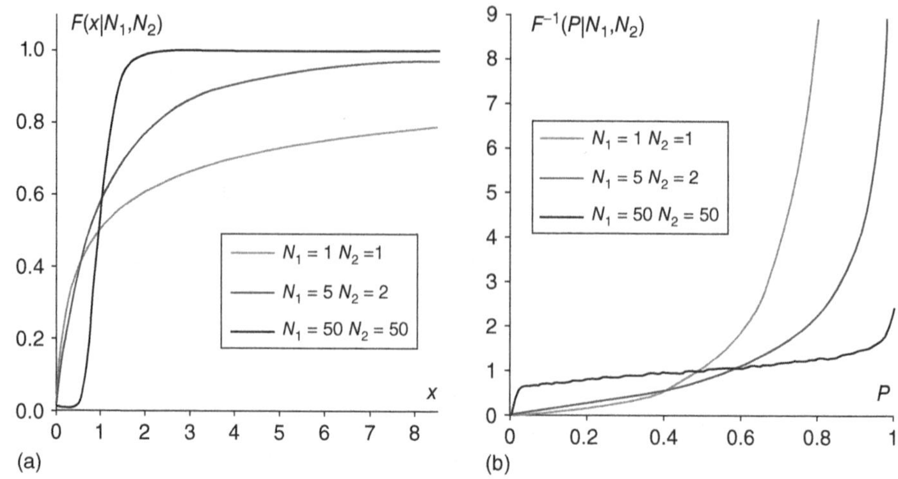

Figure 6.11 (a,b) Probability distribution function and quantile function of F-distribution.

Syntax

FDIST(x, degrees_freedom1 $= N_1$, **degrees_freedom2** $= N_2$)

The function returns $1-F(x|N_1,N_2)$, where $F(x|N_1,N_2)$ is the probability distribution function of the F-distribution.

FINV(probability= 1-*P*, **degrees_freedom1** $= N_1$, **degrees_freedom2** $= N_2$)

Returns the quantile $F^{-1}(1-P|N_1,N_2)$ of the F-distribution for the probability $1-P$.

An example of the Excel calculations is given in worksheet Fisher and shown in Fig. 6.12.

6.2.7 Multivariate Normal Distribution

This distribution is a generalization of a one-dimensional (univariate) normal distribution for the case of a multivariate random variable, that is, a random n-dimensional vector $\mathbf{x}$. The probability density function is given by

$$f(\mathbf{x}|\mathbf{m},\boldsymbol{\Sigma}) = (1/\sqrt{\det(\boldsymbol{\Sigma})(2\pi)^n})\exp(-(1/2)(\mathbf{x}-\mathbf{m})^t\boldsymbol{\Sigma}^{-1}(\mathbf{x}-\mathbf{m})),$$

where $\boldsymbol{\Sigma}$ is the $(n \times n)$ symmetric positive definite matrix.

The multivariate normal distribution depends on two groups of parameters:

$$\mathbf{x} \sim N(\mathbf{m},\boldsymbol{\Sigma}).$$

The expectation of $\mathbf{x}$ is equal to vector $\mathbf{m}$ and the covariance matrix is equal to matrix $\boldsymbol{\Sigma}$.

Let $\mathbf{x}_1$ and $\mathbf{x}_2$ be two random vectors drawn from the multivariate normal distribution, that is, $\mathbf{x}_i \sim N(\mathbf{m},\boldsymbol{\Sigma})$, $\quad i = 1,2$. Then a value

$$d = (\mathbf{x}_1-\mathbf{x}_2)^t\boldsymbol{\Sigma}^{-1}(\mathbf{x}_1-\mathbf{x}_2)$$

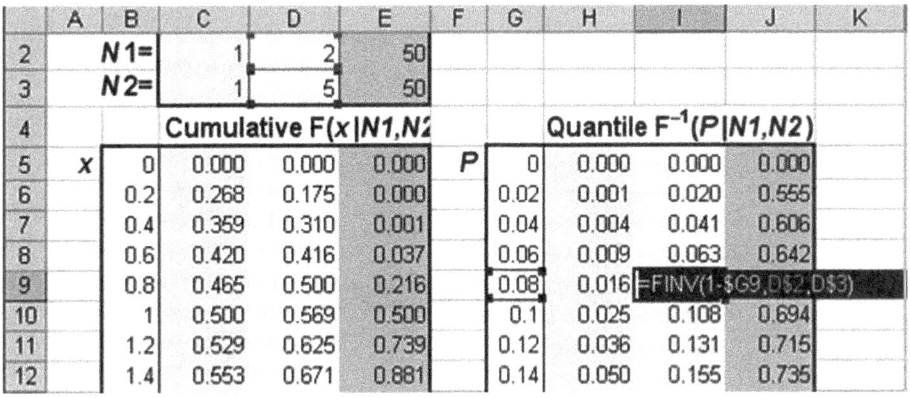

	A	B	C	D	E	F	G	H	I	J	K
2		N1=	1	2	50						
3		N2=	1	5	50						
4			Cumulative F(x|N1,N2					Quantile F-1(P|N1,N2)			
5	x	0	0.000	0.000	0.000	P	0	0.000	0.000	0.000	
6		0.2	0.268	0.175	0.000		0.02	0.001	0.020	0.555	
7		0.4	0.359	0.310	0.001		0.04	0.004	0.041	0.606	
8		0.6	0.420	0.416	0.037		0.06	0.009	0.063	0.642	
9		0.8	0.465	0.500	0.216		0.08	0.016	=FINV(1-$G9,D$2,D$3)		
10		1	0.500	0.569	0.500		0.1	0.025	0.108	0.694	
11		1.2	0.529	0.625	0.739		0.12	0.036	0.131	0.715	
12		1.4	0.553	0.671	0.881		0.14	0.050	0.155	0.735	

Figure 6.12 Example of *F*-distribution calculation.

is called the *square of the Mahalanobis distance* between these points. The Mahalanobis metrics (as opposed to Euclidean) accounts for the internal structure of the distribution.

6.2.8 Pseudorandom Numbers

Sometimes, it is useful to create an artificial sample of random numbers, which have a given distribution. This can be done by using the following simple proposition.

Let $F(x)$ and $F^{-1}(P)$ be a probability distribution function and a quantile function, respectively. If a random variable X has a uniform distribution over interval [0, 1], that is,

$$X \sim U(0, 1),$$

then a random variable

$$Y = F^{-1}(X)$$

is distributed according to F.

Therefore, if a set of uniformly distributed random values is obtained, these random values can be transformed into new values, which have a given distribution.

To generate uniformly distributed pseudorandom values in Excel, the standard worksheet function **RAND** is used.

Syntax

RAND ( )

The function returns a random value uniformly distributed over interval [0,1]. A new random value is returned every time a worksheet is recalculated.

Worksheet Random and Fig. 6.13 present an example of random numbers generation for various distributions.

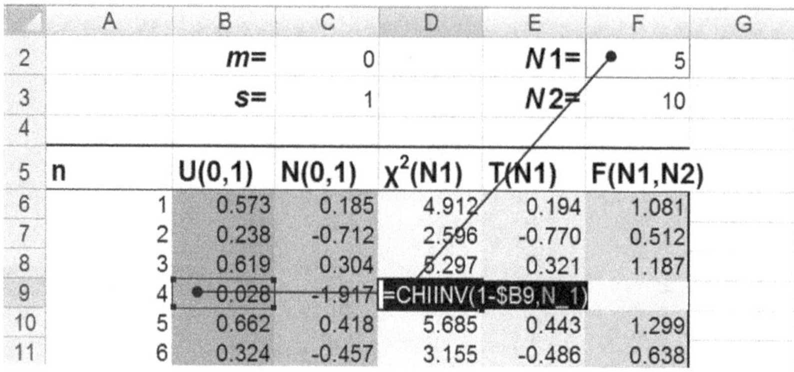

A B C D E F G
m= 0
s= 1
N1= 5
N2= 10
n U(0,1) N(0,1) χ²(N1) T(N1) F(N1,N2)
1 0.573 0.185 4.912 0.194 1.081
2 0.238 -0.712 2.596 -0.770 0.512
3 0.619 0.304 5.297 0.321 1.187
4 0.028 -1.917 =CHIINV(1-$B9,N_1)
5 0.662 0.418 5.685 0.443 1.299
6 0.324 -0.457 3.155 -0.486 0.638

Figure 6.13 Example of pseudorandom numbers generation.

6.3 PARAMETER ESTIMATION

6.3.1 Sample

Suppose there is a set of values $\mathbf{x} = (x_1, \ldots, x_I)$ and each x_i is a single realization of a random variable, that is, all x_i have the same distribution. This set is said to be a *sample* and the number I is called the *sample size*. In case of a univariate distribution, the sample is vector $\mathbf{x}$ and in the multivariate case the sample is matrix $\mathbf{X}$ of dimension $I \times J$. Each row of matrix $\mathbf{X}$ represents a realization (measurement) of the same multivariate random variable of dimension J.

It is usually assumed that all the elements of a sample are statistically independent. In practical applications, the word "sample" is often replaced by the word "data."

6.3.2 Outliers and Extremes

Some elements of a sample may be quite different from other elements. Suppose there is a sample drawn from the standard normal distribution $N(0,1)$ and this sample has an element $x_{\text{out}} = 3.2$. For the $N(0, 1)$ distribution, the probability α of the event $x_{\text{out}} \geq 3.2$ is small; it is equal to $\alpha = 0.0007$. However, value x_{out} is a member of an independent sample of size I. Therefore, it is necessary to calculate the probability of the event of "at least one time among I trials"

$$P_{\text{out}} = 1 - (1 - \alpha)^I \approx 1 - \exp(-I\alpha).$$

For $I = 10$, $P_{\text{out}} = 0.007$, for $I = 100$, $P_{\text{out}} = 0.07$, and for $I = 1000$, $P_{\text{out}} = 0.50$. Naturally, the larger the sample the higher the probability to get such extreme value.

Thus, the interpretation of an out-of-order sample element significantly depends on the sample size; at small I, it should be considered as an *outlier* (bundle), which has to be removed from the sample. For a large I, such objects are the acceptable *extremes*, which must be kept in the sample.

6.3.3 Statistical Population

In statistics, a sample creation is called *drawing*. This term emphasizes that a given sample $\mathbf{x}_1$ is not unique and that it is possible (often only in theory) to obtain other similar samples $\mathbf{x}_2, \mathbf{x}_3, \ldots, \mathbf{x}_n$. The word *similar* means that all these samples are arranged in the same way, that is, they have the same distribution, the same size I, etc. The entire infinite set of such samples forms the *statistical population*.

6.3.4 Statistics

In mathematics, the word *statistics* has two meanings.

First of all, it is a branch of mathematics pertaining to data (sample) analysis. Using the data (experimental results), statistics determines the type of distribution from which this sample was drawn, estimates the parameters of this distribution, and tests hypotheses about this distribution.

The second meaning of the word *statistics* is a (measurable) function of a sample. Because sample elements are random values, a statistics is a random variable. In this sense, a statistics is an estimator (measure) of some attribute (parameter) of the distribution from which the sample was drawn. Examples of such estimators are shown below.

6.3.5 Sample Mean and Variance

The *sample mean* is a statistics defined by

$$\bar{x} = (1/I)\sum_{i=1}^{I} x_i. \tag{6.8}$$

The *sample variance* can be calculated using two statistics.

In case of a sample with unknown expectation, the sample variance is given by

$$s^2 = (1/(I-1))\sum_{i=1}^{I} (x_i - \bar{x})^2, \tag{6.9}$$

In case of the known expectation m, the sample variance is calculated by

$$s_m^2 = (1/I)\sum_{i=1}^{I} (x_i - m)^2. \tag{6.10}$$

The sample moments are defined in a similar way. For example, a statistics

$$\overline{m}_k = (1/(I-1))\sum_{i=1}^{I} (x_i - \bar{x})^k, \tag{6.11}$$

is an estimator of the *k-th central moment*.

To calculate the sample statistics in Excel, the following standard worksheet functions are used. They are illustrated in Section 7.1.7.

Syntax

`AVERAGE(x)`

Returns the sample mean of sample **x**, calculated by Eq. (6.8).

VAR(x)
Returns the sample variance of sample **x**, calculated by Eq. (6.9).

VARP(x)
Returns the sample variance of sample **x**, calculated by Eq. (6.10).

STDEV(x)
Returns the standard deviation, that is, the square root of the sample variance calculated by Eq (6.9).

STDEVP(x)
Returns the standard deviation, that is, the square root of the sample variance calculated by Eq. (6.10).

6.3.6 Sample Covariance and Correlation

For two samples $\mathbf{x} = (x_1, \ldots, x_I)$ and $\mathbf{y} = (y_1, \ldots, y_I)$, the *sample covariance* and *correlation* can be calculated. Covariance c is obtained by the formula

$$c = (1/I)\sum_{i=1}^{I}(x_i - \bar{x})(y_i - \bar{y}),$$

and the correlation coefficient r is calculated by the formula

$$r = \sqrt{c/s_y^2/s_y^2}.$$

For a general case of data matrix $\mathbf{X}$ with the dimension of I objects by J variables, two sample covariance matrices $\mathbf{C}$ can be calculated. The object-wise covariance matrix $\mathbf{C}_I$ is computed as follows

$$\mathbf{C}_I = \mathbf{X}\mathbf{X}^{\mathrm{t}}.$$

The variable-wise covariance matrix $\mathbf{C}_J$ is calculated by the formula

$$\mathbf{C}_J = \mathbf{X}^{\mathrm{t}}\mathbf{X}.$$

To calculate the pair-wise sample covariances in Excel, the standard worksheet functions **COVAR** and **CORREL** are used.

Syntax

COVAR(x, y)
Returns the sample covariance between the samples **x** and **y**.

CORREL(x, y)
Returns the sample correlation coefficient between the samples **x** and **y**.

6.3.7 Order Statistics

An original sample $(x_1, \ldots, x_I)$ can be sorted in the nondescending order:

$$x(1) \le x(2) \le \ldots \le x(i) \le \ldots \le x(I).$$

The elements of this series are the *order statistics*. The central element of the series (and if I is an even number, the half-sum of two central elements) is the *sample median*

$$\text{median}(\mathbf{x}) = \begin{cases} x(k+1), & I = 2k+1 \\ 0.5(x(k) + x(k+1)), & I = 2k \end{cases} \tag{6.12}$$

The sample estimators of the quartiles $\hat{x}(P)$ and percentiles are developed in a similar way.

The *range* of a sample is the difference between the largest and the smallest elements, that is, the statistics

$$x(I) - x(1).$$

The *interquartile range* of the sample $\mathbf{x}$ is a statistics

$$\text{IQR}(\mathbf{x}) = \hat{x}(0.75) - \hat{x}(0.25),$$

that is, it is the difference between the sample quartiles for $P = 0.75$ and $P = 0.25$

To calculate the order statistics in Excel, the following standard worksheet functions are used **MEDIAN**, **QUARTILE**, **PERCENTILE**.

Syntax

MEDIAN(**x**)

Returns the sample median for sample **x**.

QUARTILE(**x**, **quart** = 0|1|2|3|4)

Returns the sample quartile of sample **x** depending on the value of argument **quart**

0 The minimum value in **x**

1 The first quartile (the 25th percentile)

2 The median (the 50th percentile)

3 The third quartile (the 75th percentile)

4 The maximum value in **x**

PERCENTILE(**x**, **k**)

Returns the *k-th* sample percentile of sample **x**. Valid **k** values are $0 \le \mathbf{k} \le 1$.

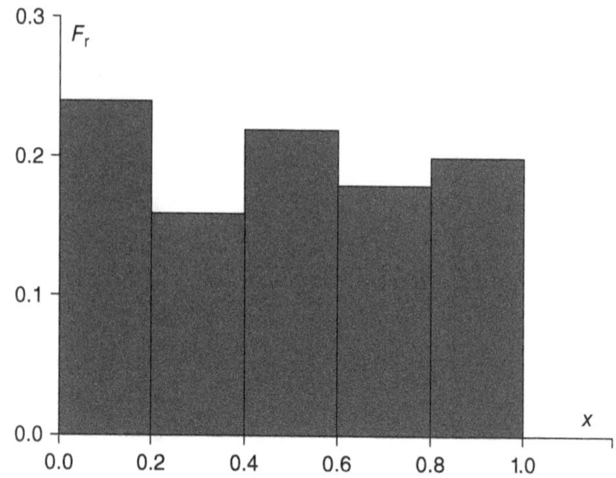

Figure 6.14 Histogram.

6.3.8 Empirical Distribution and Histogram

Empirical (cumulative) distribution function is a nondecreasing function $F_I(x)$, which is equal to zero for $x < x(1)$, and equals 1, if $x \geq x(I)$. Between these two points, $F_I(x)$ is a step function that increases by $1/I$ at each point $x(i)$, that is,

$$F_I(x) = (\text{number}\{x(i) \leq x\})/I$$

An empirical distribution function has a theoretical value as it converges to the true distribution function as the sample size I is increased. However, in practical applications *histogram* is often used as an alternative. The histogram is an estimator of the probability density function.

To construct a histogram, the sample range $[x(1), x(I)]$ is divided into R intervals of equal size. Then, one counts the number of elements grouped within each of the intervals: $I_1 + I_2 + \cdots + I_R = I$. After that, the frequencies $F_r = I_r/I$ are presented in a column-wise plot similar to that shown in Fig. 6.14.

In Excel, the standard worksheet function **FREQUENCY** is employed to construct a histogram.

Syntax

FREQUENCY(**data_array, bins_array**)

Returns a vertical array with the numbers of the grouped values $I_1, I_2, \ldots, I_R$.

data_array is a worksheet range with the sample values;

bins_array is a worksheet range with intervals in which the sample values are grouped.

This function must be entered as an *array formula*, that is, using the keyboard shortcut **CTRL+SHIFT+ENTER** (see Section 7.2.1). The size of the returned array is one cell

	A	B	C	D	E	F	G	H
2		X			Bins	I_r	F_r	P_r
3	1	0.823		1	0.0	=FREQUENCY(X,Bins)		
4	2	0.583		2	0.2	12	0.24	0.2
5	3	0.886		3	0.4	8	0.16	0.2
6	4	0.934		4	0.6	11	0.22	0.2
7	5	0.847		5	0.8	9	0.18	0.2
8	6	0.179		6	1.0	10	0.2	0.2
9	7	0.329				0	0	
10	8	0.485						

Figure 6.15 Example of **FREQUENCY** function.

larger than the number of elements in the array **bins_array**. The additional cell contains the number of elements in **data_array** that are greater than the maximum value in **bins_array**.

An example of **FREQUENCY** function application is given in worksheet Chi-Test and shown in Fig. 6.15.

6.3.9 Method of Moments

None of the sample analysis methods above accounts for a specific distribution from which this sample is drawn. Such methods are called *nonparametric estimators*. Now let us take a look at a typical *parametric* method, which is called the *method of moments* (MM). Let a sample $\mathbf{x} = (x_1, \ldots, x_I)$ has a distribution function

$$x_i \sim F(x|\mathbf{p}),$$

which is known up to the values of parameters $\mathbf{p} = (p_1, \ldots, p_M)$. In order to estimate the parameters, M sample moments $\overline{m}_m$ (6.11) are calculated and equated to their corresponding theoretical values determined by Eq. (6.2). The MM-estimator is defined by a solution (if there is one) of the following set of equations.

$$\begin{cases} m_1(p_1, \ldots, p_M) = \overline{m}_1 \\ m_2(p_1, \ldots, p_M) = \overline{m}_2 \\ \vdots \\ m_M(p_1, \ldots, p_M) = \overline{m}_M \end{cases}.$$

For example, consider a random variable $X = aY$, such that Y has the chi-squared distribution

$$Y \sim \chi^2(N).$$

Sample $\mathbf{x} = (x_1, \ldots, x_I)$ is employed for the estimation of two unknown parameters a and N. From Eqs (6.4) and (6.7), it follows that

$$E(X) = aE(Y) = aN, \quad V(X) = a^2V(Y) = 2a^2N.$$

Therefore,

$$\hat{a}_{MM} = V(X)/(2E(X)) = s^2/(2\overline{m}), \quad \hat{N}_{MM} = E(X)/a = 2\overline{m}^2/s^2.$$

6.3.10 The Maximum Likelihood Method

The most popular method of parametric estimation is the *maximum likelihood* (ML) method. Given that each element of the sample $\mathbf{x} = (x_1, \ldots, x_I)$ has the same probability density $f(x_i|\mathbf{p})$, the joint density of the whole sample is defined by

$$L(\mathbf{x}|\mathbf{p}) = f(x_1|\mathbf{p}) \times f(x_2|\mathbf{p}) \times \ldots \times f(x_I|\mathbf{p}) = \prod_{i=1}^{I} f(x_i|\mathbf{p}). \tag{6.13}$$

Function $L(\mathbf{x}|\mathbf{p})$ is said to be the *likelihood* of a sample. The likelihood function depends on two groups of variables: the sample values $\mathbf{x} = (x_1, \ldots, x_I)$ measured during the experiment and the unknown parameters $\mathbf{p} = (p_1, \ldots, p_M)$, which are to be estimated.

ML-estimators are the $\mathbf{p}$ values, for which the likelihood function (or its logarithm) has a maximum

$$\hat{\mathbf{p}} = \underset{\mathbf{p}}{\operatorname{argmax}}(\ln L(\mathbf{x}|\mathbf{p})).$$

Consider, for example, estimating the parameters of the normal distribution $N(m,\sigma^2)$. From Eqs. (6.3) and (6.6) it follows that

$$L(\mathbf{x}|m, \sigma) = (\sigma\sqrt{2\pi})^{-I} \exp\left(-\sigma^{-2}\sum_{i=1}^{I}(x_i - m)^2\right).$$

The maximum of this function is attained for the following values of m and σ^2

$$\hat{m}_{ML} = \frac{1}{I}\sum_{i=1}^{I} x_i, \quad \hat{\sigma}^2_{ML} = \frac{1}{I}\sum_{i=1}^{I}(x_i - \hat{m}_{ML})^2. \tag{6.14}$$

Thus, for the normal distribution the ML-estimators coincide with the sample estimates presented in Eqs (6.8) and (6.10).

6.4 PROPERTIES OF THE ESTIMATORS

6.4.1 Consistency

Any estimator $p(\mathbf{x})$ of a parameter p is a statistics, that is, a random variable. Like any random variable, it has its own distribution function, expectation, variance, etc. These characteristics can be used to compare different estimators, assessing their properties and qualities. Below is a brief overview of the main properties of the estimators.

An estimator $p(\mathbf{x})$ is said to be *consistent* if it converges in probability to the value of the estimated parameter p as the sample size I increases. More precisely, statistics $p(\mathbf{x})$ is a consistent estimator of parameter p if and only if the following holds for any positive value ε

$$\lim_{I\to\infty} \Pr(|p(\mathbf{x}) - p| > \varepsilon) = 0.$$

Most of the estimators used in practical applications are consistent.

6.4.2 Bias

An estimator $p(\mathbf{x})$ is said to be *unbiased* if

$$E[p(\mathbf{x})] = p.$$

The biased estimators are often encountered in applications. For example, the ML-estimator of the normal distribution variance given in Eq. (6.14) is biased

$$E[\hat{\sigma}^2_{\mathrm{ML}}(\mathbf{x})] = (1 - 1/I)\sigma^2.$$

Variance $V[p(\mathbf{x})]$ is a measure of the accuracy of unbiased estimators; the smaller the better. For biased estimators the expected mean square error

$$d(\mathbf{x}) = E[(p(\mathbf{x}) - p)^2]$$

should be used. The following formula holds

$$d(\mathbf{x}) = V[p(\mathbf{x})] + \{E([p(\mathbf{x})] - p)\}^2. \tag{6.15}$$

6.4.3 Effectiveness

An unbiased estimator is said to be *efficient* if it has the smallest possible variance. Estimators (6.14) of the normal distribution are efficient but the sample median (see Section 6.3.7) is not; it is less efficient than the sample mean.

A biased estimator could be more accurate than an unbiased one. It means that sometimes it is possible to construct a biased estimator for which the mean squared error is less than the lowest effective variance. Several methods of estimation such as principal component regression (PCR), partial least squares (PLS) (see Chapter 8) employ this principle.

6.4.4 Robustness

The *estimator robustness* is an important characteristic, which, however, is poorly formalized.

An estimator $p(\mathbf{x})$ is said to be *robust* if it is resistant to the sample outliers. As a rule, efficient estimators are less robust than ineffective. Therefore, efficiency is a price for a more robust estimation. For the normal distribution, the sample median, given in Eq. (6.12), is the robust estimate and the median absolute deviation *(MAD) estimator*

$$s_{\mathrm{MAD}} = 1.4826 \ \mathrm{median}(|\mathbf{x} - \mathrm{median}(\mathbf{x})|),$$

can be used for a robust estimation of the standard deviation. Here, the **median** is calculated by Eq. (6.12).

Worksheet Robust and Fig. 6.16 demonstrate the comparison of the classical and robust estimators of the standard normal distribution $N(0,1)$. To simulate an outlier, the first element of the sample is replaced by a random value drawn from the $N(0,100)$ distribution.

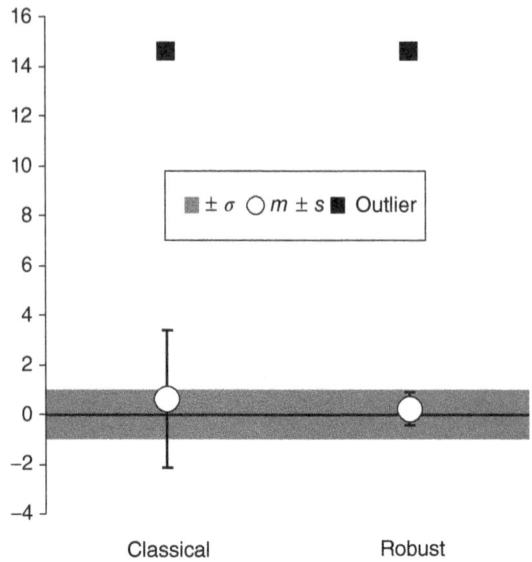

Figure 6.16 Conventional and robust estimators.

6.4.5 Normal Sample

If a sample $\mathbf{x} = (x_1, \ldots, x_I)$ is drawn from the normal distribution

$$x_i \sim N(m, \sigma^2),$$

and the estimators $\bar{x}, s^2, s_m^2$ are defined by Eqs (6.8)–(6.10), the following holds:

$$\sqrt{I}(\bar{x} - m)/\sigma \sim N(0, 1), \text{ i.e., that is, it has the standard normal distribution;} \tag{6.16}$$

$I(s_m^2/\sigma^2) \sim \chi^2(I)$, i.e., that is, it has the chi-squared distribution with I degrees of freedom;

$(I-1)(s^2/\sigma^2) \sim \chi^2(I-1)$, i.e., that is, it has the chi-squared distribution with $I-1$ degrees of freedom;

$\sqrt{I-1}((\bar{x} - m)/s) \sim T(I-1)$, i.e., that is, it has Student′s distribution with $I-1$ degrees of freedom.

6.5 CONFIDENCE ESTIMATION

6.5.1 Confidence Region

Oftentimes, in addition to point estimators of the unknown parameters, it is desirable to specify a random area in which the true values of these parameters are located with a given probability. This area is called the *confidence region*.

Let a sample $\mathbf{x} = (x_1, \dots, x_I)$ has a probability distribution function $F(x|\mathbf{p})$,

$$x_i \sim F(x|\mathbf{p}),$$

all parameters $\mathbf{p} = (p_1, \dots, p_M)$ of which are known. Statistics $P(\mathbf{x}) \in \mathrm{R}^M$ is said to be the confidence region that corresponds to the confidence probability γ, if

$$\Pr\{\mathbf{p} \in P(\mathbf{x})\} \geq \gamma.$$

6.5.2 Confidence Interval

Often a one-dimensional region, that is, the *confidence interval*, is developed for each parameter p. The confidence limits are two statistics $p^-(\mathbf{x})$ and $p^+(\mathbf{x})$, such that

$$\Pr\{p^-(\mathbf{x}) \leq p \leq p^+(\mathbf{x})\} \geq \gamma.$$

For a one-sided confidence interval, the corresponding boundary is replaced by $-\infty$, 0, or $+\infty$. In practice, the confidence intervals are usually developed for the (asymptotically) normal samples using the relations given in Section 6.4.5.

6.5.3 Example of a Confidence Interval

Below we show how to construct a confidence interval. Assume we have sample $\mathbf{x} = (x_1, \dots, x_I)$ drawn from the normal distribution $N(m, \sigma^2)$ with the known variance σ^2. We want to build a confidence interval for the expectation parameter m. From Eq. (6.16), it follows that

$$\Pr\{\Phi^{-1}(1-\alpha_1) \leq ((\bar{x} - m)/\sigma)\sqrt{I} \leq \Phi^{-1}(1-\alpha_2)\} = \alpha_1 + \alpha_2 - 1,$$

where Φ^{-1} is the quantile of the standard normal distribution. Then,

$$\Pr\{\bar{x} - (\sigma/\sqrt{I})\Phi^{-1}(\alpha_2) \leq m \leq \bar{x} + (\sigma/\sqrt{I})\Phi^{-1}(\alpha_1)\} = \alpha_1 + \alpha_2 - 1.$$

In order to construct a symmetric confidence interval with the confidence probability γ, we set $\alpha_1 = \alpha_2 = 0.5(1+\gamma)$. To construct one-sided confidence intervals, we set $\alpha_1 = 1$, $\alpha_2 = \gamma$ or $\alpha_1 = \gamma$, $\alpha_2 = 1$.

6.5.4 Confidence Intervals for the Normal Distribution

We can develop confidence intervals for the parameters of the normal distribution $N(m,\sigma^2)$ using the relations given in Section 6.4.5. Let estimators $\bar{x}, s^2, s_m^2$ be defined by Eqs (6.8)–(6.10), then the following relations hold.

The confidence interval for the mean value m with an *unknown variance* σ^2 is

$$\Pr\{\bar{x} - (s/\sqrt{I})T^{-1}(\alpha_2|I-1) \leq m \leq \bar{x} + (s/\sqrt{I})T^{-1}(\alpha_1|I-1)\} = \alpha_1 + \alpha_2 - 1,$$

where $T^{-1}(\alpha|I-1)$ is the quantile of Student's distribution with $I-1$ degrees of freedom.

The confidence interval for the variance σ^2 with the *known mean* value m is

$$\Pr\{I(s_m^2/(\chi^{-2}(\alpha_2|I))) \le \sigma^2 \le I(s_m^2/(\chi^{-2}(\alpha_1|I)))\} = \alpha_1 + \alpha_2 - 1,$$

where $\chi^{-2}(\alpha|I)$ is the quantile of the chi-squared distribution with I degrees of freedom.

The confidence interval for the variance σ^2 with *unknown mean* m is

$$\Pr\{(I-1)(s^2/(\chi^2(\alpha_2|I-1))) \le \sigma^2 \le (I-1)(s^2/(\chi^2(\alpha_1|I-1)))\} = \alpha_1 + \alpha_2 - 1,$$

where $\chi^{-2}(\alpha|I-1)$ is the quantile of the chi-squared distribution with $I-1$ degrees of freedom.

6.6 HYPOTHESIS TESTING

6.6.1 Hypothesis

Statistical hypothesis is a consistent statement about the form of a sample distribution. The *null hypothesis* is the hypothesis to be tested. Any other hypothesis is the *alternative hypothesis*. For example, assume a sample drawn from the chi-squared distribution with N degrees of freedom. The null hypothesis is

$$\mathrm{H}_0 : N = 2,$$

The alternative hypothesis is

$$\mathrm{H}_1 : N > 2.$$

In practice, an alternative is often omitted and only the null hypothesis is stated.

The hypothesis is *simple* if it uniquely determines sample distribution. Otherwise, the hypothesis is said to be composite. In the example above, H_0 is a simple hypothesis and H_1 is a composite alternative.

Hypotheses are *parametric* if the distribution is known in advance and defined up to the values of its parameters, as in the example above. Otherwise, the hypothesis is called *nonparametric*. Suppose we have a sample with an unknown distribution F. The null hypothesis is

$$\mathrm{H}_0 : F \text{ is the uniform distribution.}$$

6.6.2 Hypothesis Testing

After formulating a hypothesis, one has to find an appropriate method to test it. The hypothesis testing is based on a *test statistics* $S(\mathbf{x})$, which is a function of sample $\mathbf{x} = (x_1, \ldots, x_I)$. The *critical* area C is associated with the statistics S in such a way that when $S(\mathbf{x}) \in C$, the hypothesis is rejected, otherwise it is accepted.

Statistics $S(\mathbf{x})$ must be designed in such a way that its distribution does not depend on the unknown parameters of the sample distribution. Moreover, the distribution function of S should be tabulated in advance. In most practical applications, statistics S is developed under the assumptions of normality.

6.6.3 Type I and Type II Errors

Testing a statistical hypothesis does not mean its "logical" confirmation or rejection. The test is a statistical procedure, which concludes that "the data are (not) contradicting the proposed statement." We can reject a false null hypothesis or accept a true null hypothesis. Also, there is a possibility to make an error, which can be of two types.

A Type I error occurs when a true null hypothesis is rejected according to the criteria.
A Type II error occurs when a false null hypothesis is accepted according to the criteria.

The probability of making Type I error is called the *significance level* and is denoted by α. Usually, the significance level is chosen to be 0.01, 0.05, or 0.1. This value determines the critical region C_α.

6.6.4 Example

Let us consider a sample $\mathbf{x} = (x_1, \dots, x_I)$ drawn from the normal distribution, that is,

$$x_i \sim N(m, \sigma^2)$$

with the *known variance* σ^2 and an *unknown mean* m. Let us test a simple null hypothesis

$$\mathrm{H}_0 : m = 0$$

An alternative will be formulated later. The function

$$S(\mathbf{x}) = \sqrt{I}(\overline{x}/\sigma),$$

is used as the test statistics. For $m = 0$ it has the standard normal distribution $S \sim N(0, 1)$. For a given significance level α, the critical region is defined by $\Pr\{|S| > C_\alpha\} = \alpha$. Hence, $C_\alpha = \Phi^{-1}(1-\alpha/2)$.

Now, let us formulate an alternative hypothesis.

$$\mathrm{H}_1 : \quad m = a$$

The probability of Type II error is given by the formula $\beta = \Pr\{|S| < C_\alpha | m = a\}$, which is calculated under the condition $S \sim N(a, 1)$. Therefore,

$$\beta = \Phi(\mathrm{C}_\alpha - a) - \Phi(-\mathrm{C}_\alpha - a).$$

Worksheet Hypothesis and Fig. 6.17 present the calculations for this example.

6.6.5 Pearson's Chi-Squared Test

Pearson's goodness-of-fit test compares the empirical frequency estimators $I_1/I, I_2/I, \dots$ with the corresponding theoretical probabilities $P_1, P_2, \dots$. For example, consider sample $\mathbf{x} = (x_1, \dots, x_I)$ drawn from an unknown distribution

$$x_i \sim F(x).$$

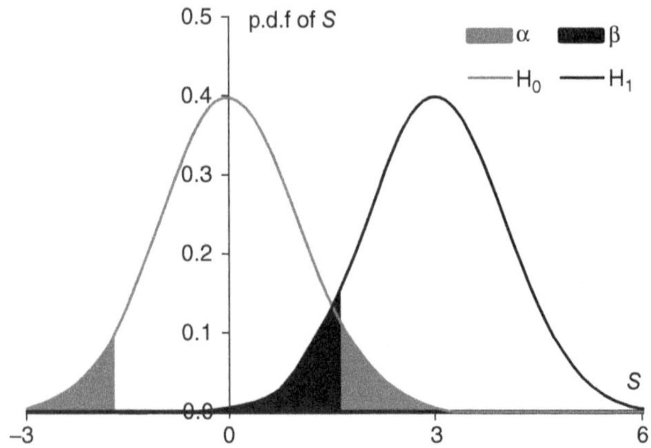

Figure 6.17 Type I and Type II errors in hypothesis testing.

The null hypothesis is a statement about the specification of this distribution, for example, "F is the normal distribution with the zero mean and variance of 2."

For the hypothetical distribution, the range of random variable X is divided into R classes (bins) for which the theoretical probabilities $P_1, P_2, \ldots, P_R$ are calculated. At the same time, the empirical frequencies $F_r = I_r/I$ are computed using the numbers of the samples $I_1, I_2, \ldots, I_R$ grouped in R classes.

The goodness-of-fit statistics is defined by

$$S = \sum_{r=1}^{R} (I_r - IP_r)^2/I/P_r = I\sum_{r=1}^{R} (F_r - P_r)^2/P_r. \tag{6.17}$$

When $I \to \infty$, the statistics tends to the chi-squared distribution with $R-1$ degrees of freedom. The number and size of classes should be selected in such a way that $IP_r > 6$.

The critical area at the significance level α is defined by

$$S > \chi^{-2}(1-\alpha|R-1).$$

Pearson's chi-squared test can be applied when theoretical distribution $F(x|\mathbf{p})$ is known up to the values of the unknown parameters $\mathbf{p} = (p_1, \ldots, p_M)$. The parameters are estimated using the same sample $\mathbf{x}$ and substituted into function $F(x|\mathbf{p})$. In this case, the number of the degrees of freedom is changed to $R-M-1$

Pearson's chi-squared test can be performed in Excel using the standard worksheet function **CHITEST**.

Syntax

CHITEST(actual_range, expected_range)

This function computes statistics S given by Eq. (6.17) using **actual_range** $= (I_1, I_2, \ldots, I_R)$ and **expected_range** $= (IP_1, IP_2, \ldots, IP_R)$, and returns probability $P = 1-\chi^2(S|R-1)$.

To accept the hypothesis at the significance level α, it is necessary that $P > 1-\alpha$.

An example of the Excel calculations is given in worksheet Chi-Test and shown in Fig. 6.18.

6.6.6 *F*-Test

This criterion is used to test the null hypothesis that two normally distributed samples $\mathbf{x} = (x_1, \ldots, x_I)$ and $\mathbf{y} = (y_1, \ldots, y_J)$ have the same variance. Let

$$s_x^2, s_y^2$$

be the sample variance estimators calculated by Eq. (6.9).

If $s_x^2 > s_y^2$, then we denote

$$s_1^2 = s_x^2, \quad N_1 = I - 1$$
$$s_2^2 = s_y^2, \quad N_2 = J - 1.$$

Otherwise,

$$s_1^2 = s_y^2, \quad N_1 = J - 1$$
$$s_2^2 = s_x^2, \quad N_2 = I - 1.$$

The F-test statistics is defined by

$$S = s_1^2/s_2^2 \sim F(N_1, N_2).$$

It has the F-distribution with N_1, N_2 degrees of freedom. The critical region at the significance level α is given by

$$S > F^{-1}(1-\alpha|N_1, N_2).$$

The F-test is very sensitive to nonnormality of sample distributions, hence it is not recommended for practical application.

	D	E	F	G	H	I	J	K	L	M	N
2		Bins	I_r	F_r	P_r	I_rP_r	$(I_r-IP_r)^2/IP_r$				
3	1	0.0	0	0					$R=$	5	
4	2	0.2	12	0.24	0.2	10	0.4		$I=$	50	
5	3	0.4	8	0.16	0.2	10	0.4		$\alpha=$	0.1	
6	4	0.6	11	0.22	0.2	10	0.1		$S=$	1	
7	5	0.8	9	0.18	0.2	10	0.1		$C_\alpha=$	7.779	
8	6	1.0	10	0.2	0.2	10	0		CHITEST=	=CHITEST(ActN,ExpN)	
9			0	0					$1-\chi^2(S\|R-1)=$	0.910	
10											

Figure 6.18 An example of Pearson's chi-squared test.

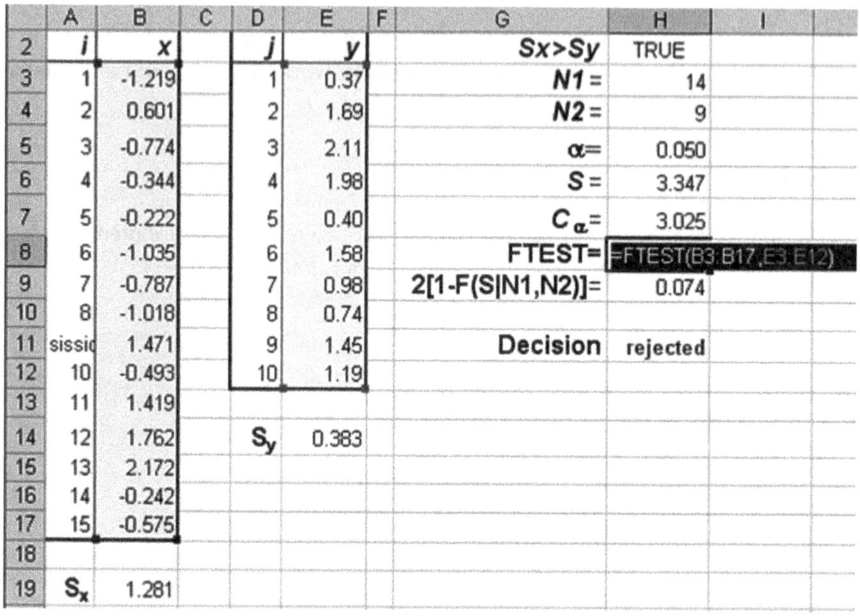

	A	B	C	D	E	F	G	H	I
2	*i*	x		*j*	y		Sx>Sy	TRUE	
3	1	-1.219		1	0.37		N1=	14	
4	2	0.601		2	1.69		N2=	9	
5	3	-0.774		3	2.11		α=	0.050	
6	4	-0.344		4	1.98		S=	3.347	
7	5	-0.222		5	0.40		C_α=	3.025	
8	6	-1.035		6	1.58		FTEST=	=FTEST(B3:B17,E3:E12)	
9	7	-0.787		7	0.98		2[1-F(S|N1,N2)]=	0.074	
10	8	-1.018		8	0.74				
11	sissic	1.471		9	1.45		Decision	rejected	
12	10	-0.493		10	1.19				
13	11	1.419							
14	12	1.762		S_y	0.383				
15	13	2.172							
16	14	-0.242							
17	15	-0.575							
18									
19	S_x	1.281							

Figure 6.19 The *F*-test example.

To implement the *F*-test in Excel the standard worksheet function **FTEST** is used.

Syntax

FTEST(**x**, **y**)

Returns probability $P = 2[1-F(S|N_1, N_2)]$.

To accept the hypothesis at the significance level α, it is necessary that $P > 2\alpha$.

An example is given in worksheet F-test and shown in Fig. 6.19.

6.7 REGRESSION

6.7.1 Simple Regression

The most common form of the regression analysis is the linear regression model. Let us consider two samples: a set of deterministic values $\mathbf{x} = (x_1, \ldots, x_I)$ and a set of random values $\mathbf{y} = (y_1, \ldots, y_I)$. Set **x** is the set of *predictors* and set **y** is called the *responses*. We assume that there is a linear relationship between these variables, such as

$$y_i = ax_i + b + \varepsilon_i,$$

where a and b are unknown parameters and ε_i are the *errors*, that is, independent random values with zero expectation and unknown variance σ^2.

If the errors are normally distributed

$$\varepsilon_i \sim N(0, \sigma^2), \tag{6.18}$$

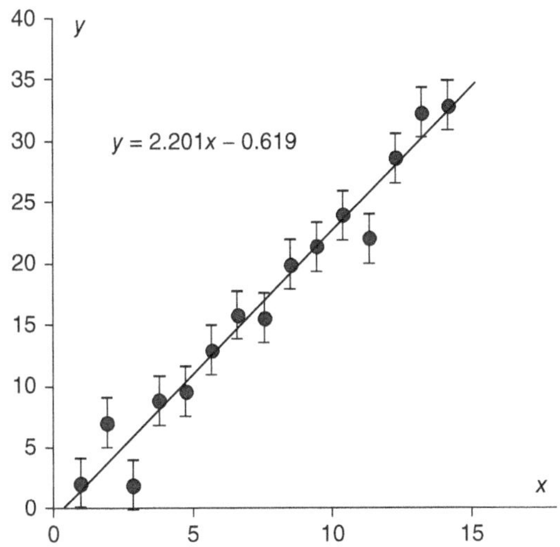

Figure 6.20 Simple linear regression.

then the ML-estimators (see Section 6.3.10) of parameters a and b are calculated as a point that attains the minimum of the sum of squares

$$Q(a,b) = \sum_{i=1}^{I} (y_i - ax_i - b)^2.$$

See Fig. 6.20 for illustration.

6.7.2 The Least Squares Method

In practice, the assumption of normality is redundant and the least squares method is used for any type of errors. The minimum of Q is attained at point

$$\hat{a} = \bar{y} - \hat{b}\bar{x}, \qquad \hat{b} = \sum_{i=1}^{I}(y_i - \bar{y})(x_i - \bar{x}) \Big/ \sum_{i=1}^{I} (x_i - \bar{x})^2$$

$$\bar{x} = \frac{1}{I}\sum_{i=1}^{I} x_i, \qquad \bar{y} = \frac{1}{I}\sum_{i=1}^{I} y_i$$

Variances of the parameter estimates are

$$V(\hat{a}) = (\sigma^2/I)\left(\sum_{i=1}^{I} x_i^2 \Big/ \sum_{i=1}^{I} (x_i - \bar{x})^2\right), \quad V(\hat{b}) = \left(\sigma^2 \Big/ I\sum_{i=1}^{I} (x_i - \bar{x})^2\right)$$

An estimator for σ^2 is found by applying the formula

$$s^2 = Q(\hat{a},\hat{b})/(I-2) = (1/I-2)\sum_{i=1}^{I} (y_i - \hat{a}x_i - \hat{b})^2$$

If statement in Eq. (6.18) is true, the following holds

$$(I-2)(s^2/\sigma^2) \sim \chi^2_{I-2}, \quad \hat{a} \sim N(a, V(\hat{a})), \quad \hat{b} \sim N(b, V(\hat{b})).$$

The estimators of parameters a and b are independent from the estimator of σ^2. This allows calculating confidence intervals for parameters a and b using Student's statistics.

In Excel, the regression parameters are calculated using two standard worksheet functions: **SLOPE** and **INTERCEPT**, which are discussed in Section 7.2.7.

Syntax

SLOPE(known_y's = y, known_x's = x)

Returns an estimator of parameter a.

INTERCEPT(known_y's = y, known_x's = x)

Returns an estimator of parameter b.

An example of regression is given in worksheet Regression and shown in Fig. 6.21.

6.7.3 Multiple Regression

A multiple regression

$$\mathbf{y} = \mathbf{Xa} + \boldsymbol{\varepsilon},$$

is a natural generalization of a simple univariate regression. This equation relates the response vector $\mathbf{y}$ and the matrix of predictors $\mathbf{X}$. The usual assumption about error $\boldsymbol{\varepsilon}$ is that

$$E(\varepsilon_i) = 0, \quad \mathrm{cov}(\boldsymbol{\varepsilon}, \boldsymbol{\varepsilon}) = \sigma^2\mathbf{I},$$

where $\mathbf{I}$ is the $(I \times I)$ identity matrix.

Assume we have I observations and J variables, then matrix $\mathbf{X}$ is of dimension $I \times J$. The goal of the regression analysis is to find estimators of unknown parameters $\mathbf{a} = (a_1, \ldots, a_J)^t$. The least squares method suggests choosing such values of the parameters that minimize the sum of the squared residuals

$$Q(\mathbf{a}) = \|\mathbf{y} - \mathbf{Xa}\|^2 = \sum_{i=1}^{I} \left| y_i - \sum_{j=1}^{J} x_{ij}a_j \right|^2.$$

The ordinary least squares (OLS) method assumes that matrix $\mathbf{X}^t\mathbf{X}$ is invertible. In this case, the minimum of $Q(\mathbf{a})$ is attained at the point

$$\mathbf{a}_{\text{hat}} = (\mathbf{X}^t\mathbf{X})^{-1}\mathbf{X}^t\mathbf{y}.$$

The estimator of σ^2 is

$$s^2 = Q(\mathbf{a}_{\text{hat}})/(I-J).$$

The covariance matrix estimator is

$$\mathbf{C} = s^2(\mathbf{X}^t\mathbf{X})^{-1}.$$

	A	B	C	D	E	F	G	H
2	a=	2.000			a hat=	2.039		
3	b=	1.000			b hat=	=INTERCEPT(B7:B21,A7:A21)		
4	σ=	2			b hat=	1.751		
5								
6	x	y	y hat	y–y hat				
7	1	2.514	2.969	-0.455				
8	2	6.016	5.009	1.007				
9	3	6.682	7.048	-0.365				
10	4	7.660	9.087	-1.427				
11	5	13.508	11.126	2.381				
12	6	13.819	13.166	0.654				
13	7	15.518	15.205	0.313				
14	8	16.445	17.244	-0.799				
15	9	19.074	19.284	-0.209				
16	10	20.381	21.323	-0.942				
17	11	22.319	23.362	-1.044				
18	12	21.893	25.402	-3.508				
19	13	29.583	27.441	2.143				
20	14	32.628	29.480	3.148				
21	15	30.625	31.519	-0.894				
22								

Figure 6.21 Linear regression parameters calculation.

A multiple regression can be constructed in Excel using two standard worksheet functions: **`TREND`** and **`LINEST`**, which are discussed in Section 7.2.7.

CONCLUSION

Statistical methods are widely used in data analysis, including chemometric methods.

7

MATRIX CALCULATIONS IN EXCEL

This chapter is devoted to Excel, which is a popular environment for calculations, analysis, and graphic representation of data. The goal is to present the main Excel features used for data processing. The main focus is on operations with the multivariate data and matrix calculations. A special section is dedicated to Add-In programs, which extend the Excel possibilities. This text is not a comprehensive Excel manual but only a short introduction. Detailed Excel features are explained elsewhere in numerous books and articles listed in Chapter 4.

Information presented here refers to the basic properties of Excel 2003 and Excel 2007. This chapter is written mainly for Excel 2003 users. Features essential for Excel 2007 are discussed separately in the text.

Preceding chapters	4, 6
Dependent chapters	8–14
Matrix skills	Basics
Statistical skills	Low
Excel skills	Low
Chemometric skills	Low
Chemometrics Add-In	Not used
Accompanying workbook	Excel.xls

7.1 BASIC INFORMATION

7.1.1 Region and Language

Excel software can be used in different countries and languages. Regional and Language settings change the software appearance, names, and syntax of the standard worksheet

Chemometrics in Excel, First Edition. Alexey L. Pomerantsev.

functions. For example, recently, in the Russian or French versions, the comma (,) was used as a delimiter for the decimal part of a number; therefore, the semicolon (;) played the role of a list separator (particularly for the function arguments). Luckily, this is becoming a thing of the past, and in scientific calculations the full stop or period (.) is widely used as a fraction separator.

One can use **Control Panel** for changing **Regional and Language Options** and **Customize** button for changing the data appearance (see Fig. 7.1). The names of the standard Excel functions depend on the Language settings chosen for the computer. For example, in an English version, the summation function is called SUM. The same function in a French version is SOMME. In German, it is SUMME, in Italian SOMMA, in Polish SUMA, etc. The list of English and local function names is presented in file FUNCS.XLS that is usually located in folder **C:\Program Files\Microsoft Office\OfficeVer\Local**. Here, **OfficeVer** is the name of an Excel version, for example, Office 11 and **Local** is the number of the local version, for example, **1049** for Russian and **1033** for English. The file Excel_Functions.xls presented at the Wiley Book Support site[1] contains the names of all Excel functions in 16 languages.

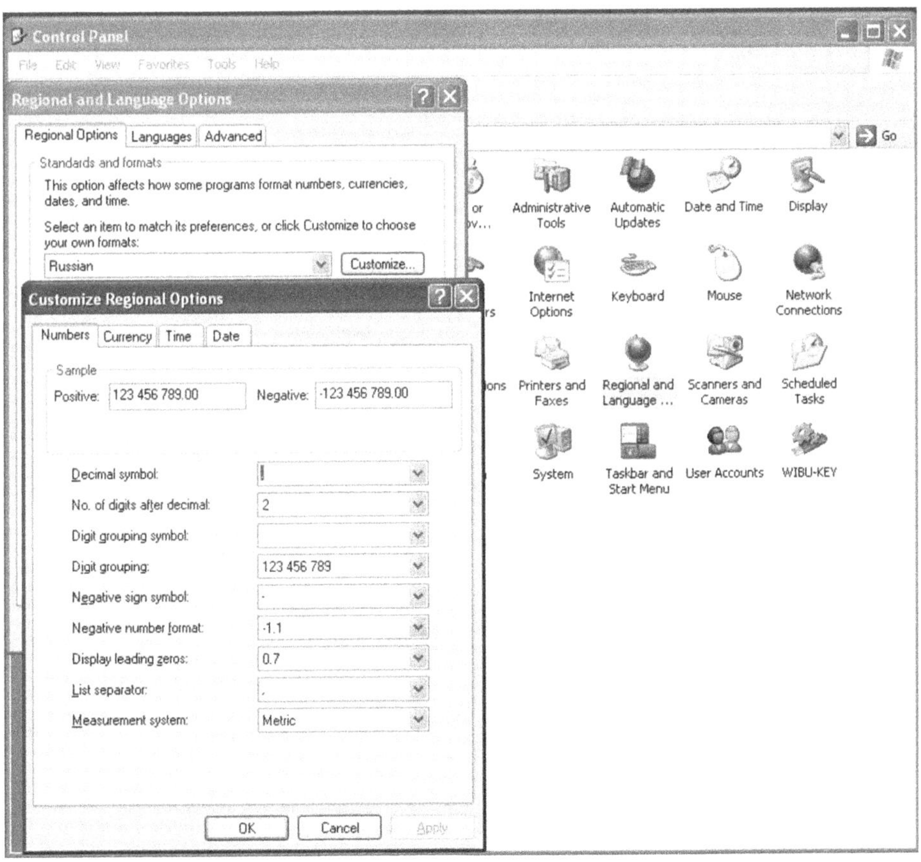

Figure 7.1 Regional options.

[1]http://booksupport.wiley.com

In this book, we use an English version of Excel 2003 with a period as a fraction separator and a comma as a list delimiter.

7.1.2 Workbook, Worksheet, and Cell

An Excel file with extension XLS (XLSX in Excel 2007) is called a *workbook*. When starting the Excel program, for example, by clicking on an icon, a new workbook opens (see Fig. 7.2).

If a workbook already exists, you can open it using **Explorer** by clicking on the file icon as shown in Fig. 7.3.

A workbook consists of several *worksheets*. Worksheet names are shown at the bottom of the open window. You can delete, add, and rename the worksheets. For this purpose,

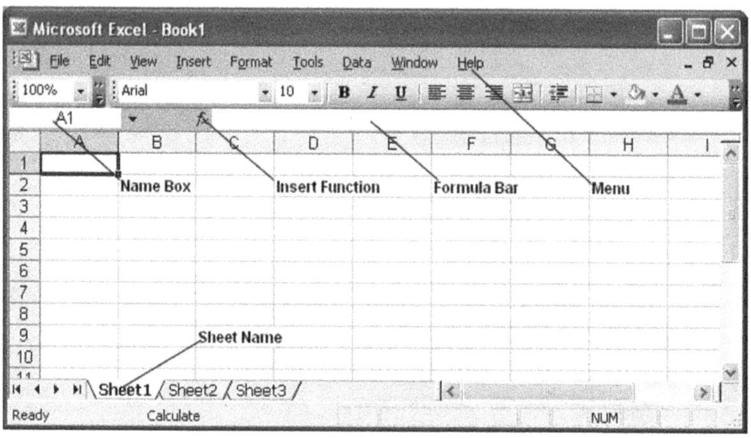

Figure 7.2 New workbook.

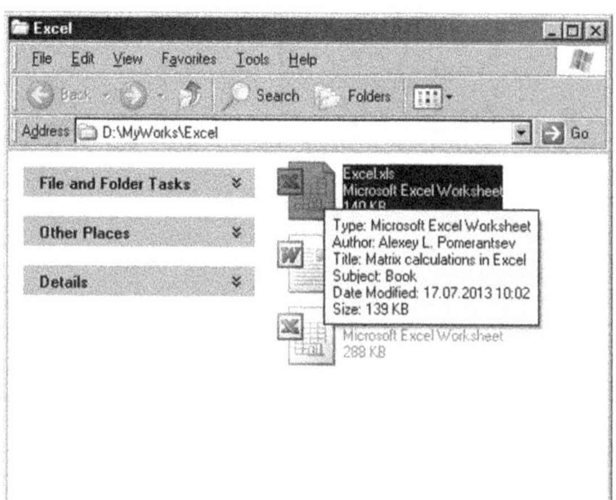

Figure 7.3 Opening workbook with **Explorer**.

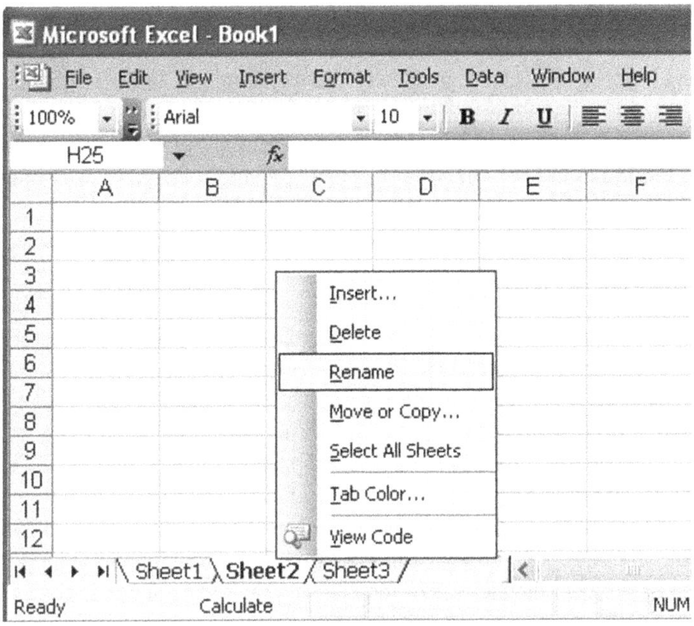

Figure 7.4 Worksheets manipulations.

right-click on the Sheet name and select an operation in the pop-up menu as shown in Fig. 7.4.

A standard worksheet name is Sheet1, but you can give it any name you like, for example, *Data*. Various objects can be inserted in the sheet, for example, charts and pictures. Each worksheet consists of *cells*. The cells form a spreadsheet with 256 columns and 65,536 rows (in Excel 2007, there are 16,384 columns and 1,048,576 rows). The rows are labeled numerically, 1, 2, 3 ..., and columns are labeled alphabetically, A, B, ..., Z, AA, AB ..., etc. till column IV (in Excel 2007 till XFD). This reference style is called A1. Another reference style is called R1C1 where the columns are also labeled numerically. The latter style is used rarely and we will not employ it below.

The columns and rows can be deleted, added, and hidden. It is also possible to change the height and width of the columns and rows. A cell can contain different content such as a number, a text, or a formula. For visualization purposes, the cell can be formatted by changing font, color, border, etc. All Excel operations are executed with the help of the Menu items located at the top of the window (Fig. 7.2). The Menu of Excel 2007 substantially differs from the previous versions. A special ribbon substitutes the habitual icons. We will not explore these differences in detail and redirect an interested reader to the numerous Excel 2007 manuals.

7.1.3 Addressing

In Excel, each cell has its own address consisting of the corresponding row and column headings. For example, the address of the first cell in a worksheet is A1. The address of the cell located at the intersection of the third column and the fifth row is C5. An active cell is indicated by the thick frame around it and its address is displayed in the **Name Box** window (see Fig. 7.2). To copy the cell content (e.g., cell A1) into another cell (e.g., cell F1), a simple formula = address should be used in the destination cell (e.g., = A1).

Addressing (referencing) can be absolute, relative, or mixed. For example, the first worksheet cell has an absolute address $A$1, a relative address A1, and two mixed addresses $A1 and A$1. The differences in addressing types are manifested when the cell formula is copied or moved into another cell. An example is given in worksheet Address. Figure 7.5 illustrates this example also.

A worksheet fragment is presented in Fig. 7.5a. Different types of references to cell A1 (highlighted) are presented nearby (column F and row 6). The reference type is indicated next to each cell. Let us copy (one by one) the cells from this range and paste them in columns G and H and rows 7 and 8 (Fig. 7.5b). One can see that the result depends on the referencing method. The reference to cell A1 is not changed for the absolute reference. When relative addressing is used, the reference shifts right and down preserving the relative distance between the cells. For mixed addressing, the results depend on the location of the cell with the reference and on the fixed (invariant) part of the reference indicated by symbol $. Figure 7.5c shows the results after replication of the formula.

For addressing the cell located in another worksheet of the same workbook, the reference should include a worksheet name, for example, Data!B2. An exclamation mark (!) separates the worksheet name from the address of the cell. If a worksheet name includes a space, the name should be enclosed within single quotation marks, for example, 'Raw Spectra'!C6. For referencing a cell from another workbook, the workbook name is enclosed in brackets, for example, [Other.xls]Results!P24.

7.1.4 Range

A matrix occupies a range (i.e., a set of cells) in a worksheet. Figure 7.6 shows a matrix with 9 rows (from row 2 till row 10) and 3 columns (from B till D).

The addresses of the upper left and the lower right cells connected by a semicolon are used for range referencing, for example, B2:D10 or $B$2:$D$10. Often, it is useful to *name* a range while working with the matrices. There are two ways to do this. The most straightforward way is to select the area in a worksheet, then click on the **Name Box** window (see Fig. 7.2), delete the current address, and input a name, for example, Data (Fig. 7.6). Another way is to use the **Insert-Name-Define** menu. The range name may be global, that is, accessible from any worksheet in a book or local. The local name is accessible only in its worksheet. In the latter case, the name should be defined as SheetName!RangeName.

7.1.5 Simple Calculations

Excel uses different formulas to carry out calculations. A formula starts with an equal sign (=) and may include references, operators, constants, and functions. Operators for simple arithmetic calculations are presented in Fig. 7.7.

7.1.6 Functions

A function is a standard formula that operates using given values called *arguments*. Examples of some useful functions are presented in Fig. 7.8.

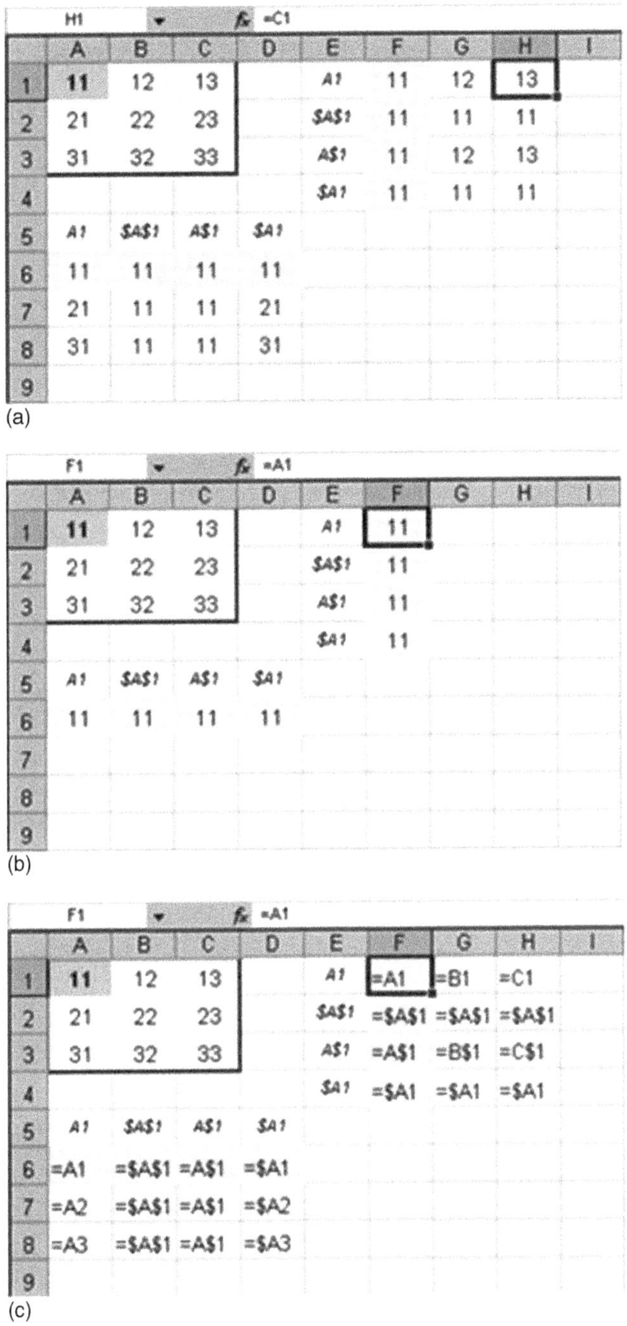

Figure 7.5 (a–c) Absolute and relative addressing.

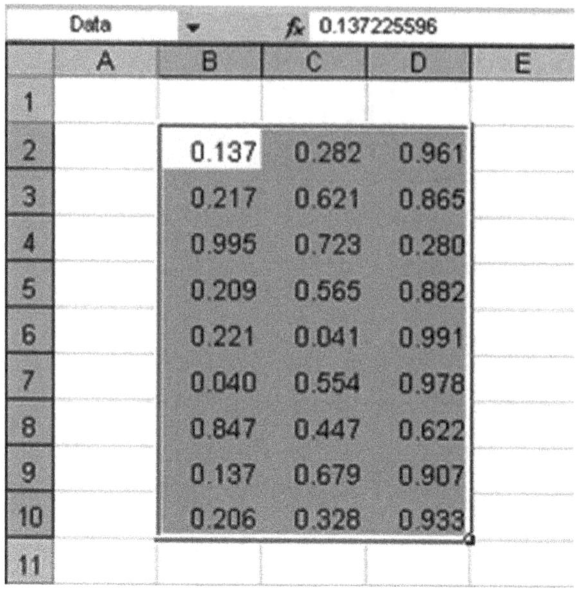

Figure 7.6 Range.

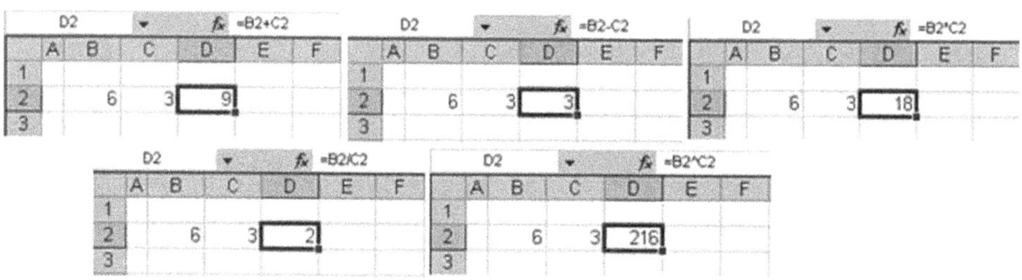

Figure 7.7 Simple calculations.

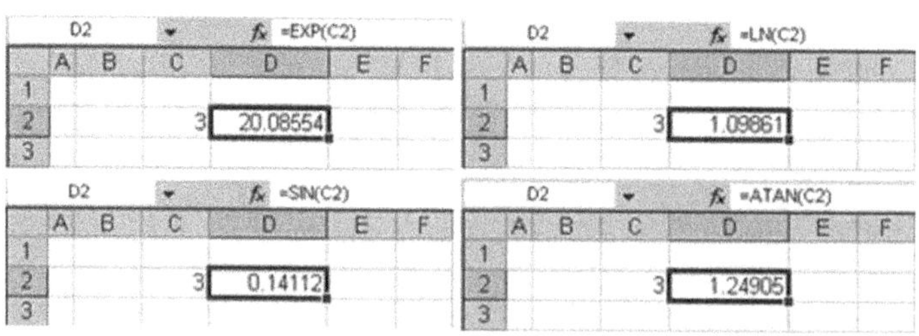

Figure 7.8 Simple functions.

The function consists of a name and a list of arguments embraced in parentheses and separated by commas (or by another list separator). For example, the function in Fig. 7.9 (see also worksheet Function) calculates the value of the cumulative (`cumulative` = TRUE), standard (`mean` = 0, `standard_dev` = 1), and normal distribution (see Section 6.2.3) for the argument value given in cell A1.

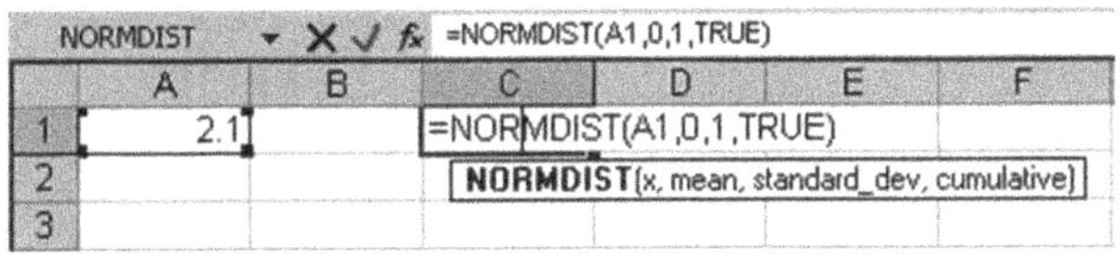

Figure 7.9 Entering the function by means of **Formula Bar**.

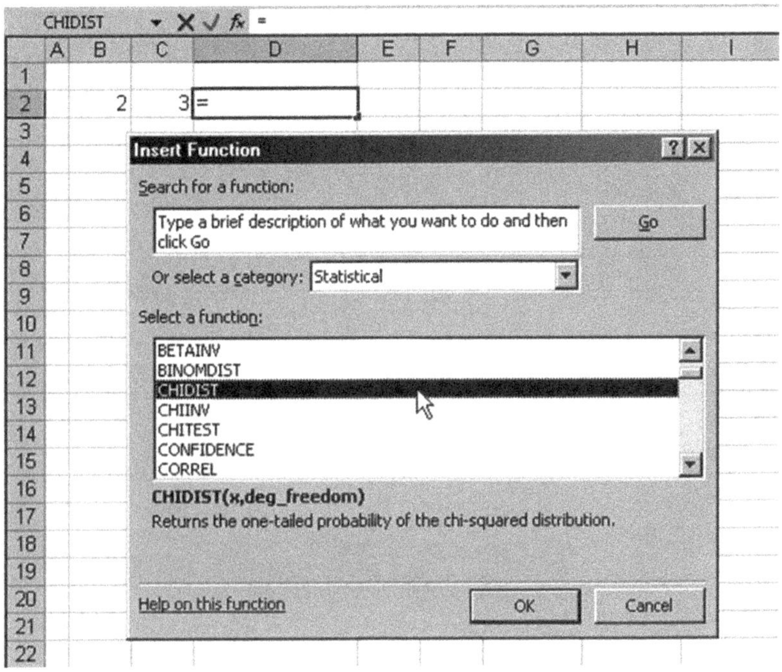

Figure 7.10 Entering function by means of **Insert Function**.

There are different ways to enter a formula. The most straightforward way is to type the formula in the **Formula Bar** window (see Fig. 7.2). Beforehand, it is necessary to make the formula window visible with the help of the **View-Formula Bar** menu item. This method is suitable if you remember the function syntax. A formula can be entered very quickly as the arguments are added by clicking the cells with the corresponding values.

The second method is helpful when we do not remember a function's name and/or the list of arguments. In this case, one can use the button **Insert Function** (see Fig. 7.2). This button opens the dialog window (Fig. 7.10) with the list of accessible functions.

As soon as the specific function is selected, the second dialog box (Fig. 7.11) appears. This box displays the list of the function's arguments.

7.1.7 Important Functions

Excel provides numerous standard worksheet functions, which cannot be outlined here. Only functions that are repeatedly used in chemometric applications will be considered. They are also presented in worksheet SUM.

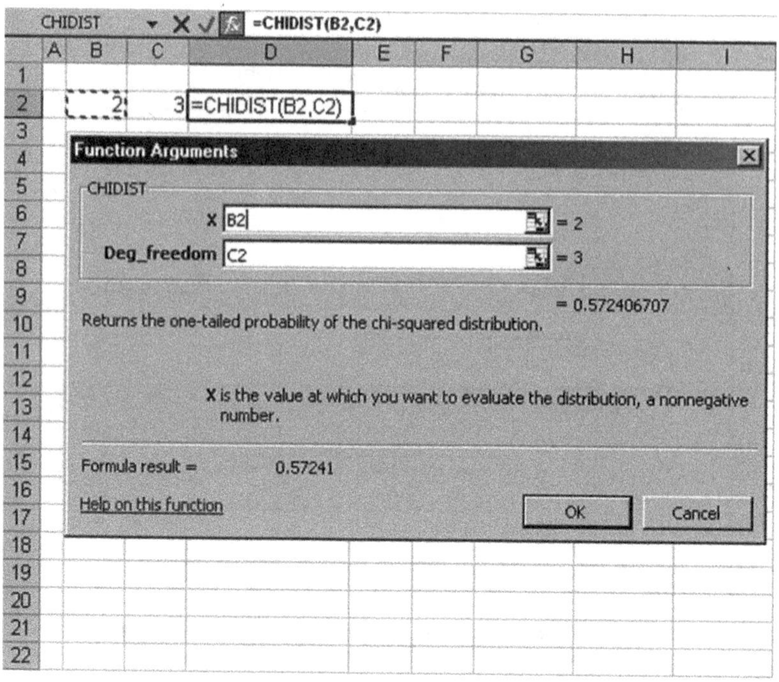

Figure 7.11 Selecting the arguments values.

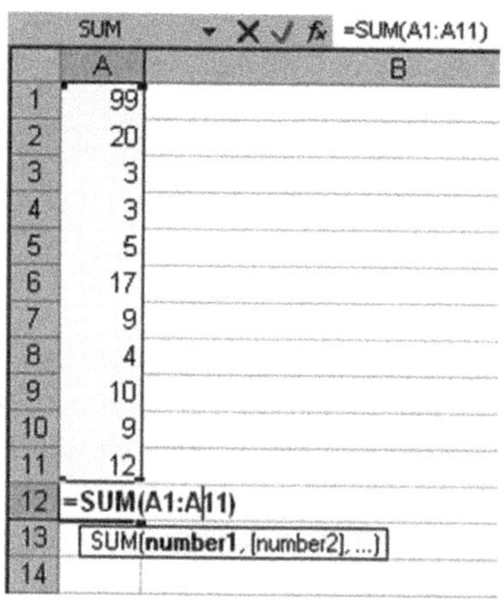

Figure 7.12 Function SUM.

SUM

Sums up all values in the argument list, or in a range, and returns the sum.

Syntax
`SUM(number1 [,number2] [, ...])`

Example (Fig. 7.12)

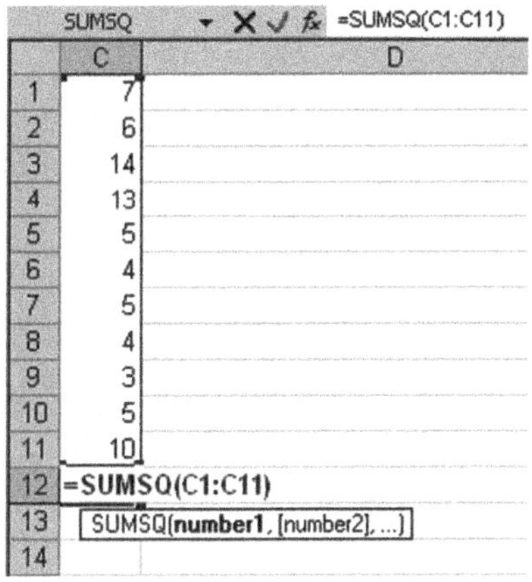

Figure 7.13 Function **SUMSQ**.

SUMSQ

Returns the sum of squares of the arguments or the cells in a range.

Syntax
SUMSQ(**number1** [,number2] [,...])

Example (Fig. 7.13)

SUMPRODUCT

Performs a pair-wise multiplication of the corresponding components in the given ranges and returns the sum of these products.

Syntax
SUMPRODUCT (array1, array2, ...)

Example (Fig. 7.14)

AVERAGE

Returns the sample mean of the arguments or the average of the cell values in a range (see Section 6.3.5).

Syntax
AVERAGE(**number1** [,**number2**] [,...])

Example (Fig. 7.15)

VAR

Calculates the sample variance $\sum (x - \overline{x})^2/(n - 1)$ (see Section 6.3.5).

Syntax

`VAR(number1,number2, ...)`

Example (Fig. 7.16)

STDEV

Calculates the sample standard deviation $\sqrt{\sum(x-\bar{x})^2/(n-1)}$ (see Section 6.3.5).

Syntax

`STDEV(number1,number2, ...)`

Example (Fig. 7.17)

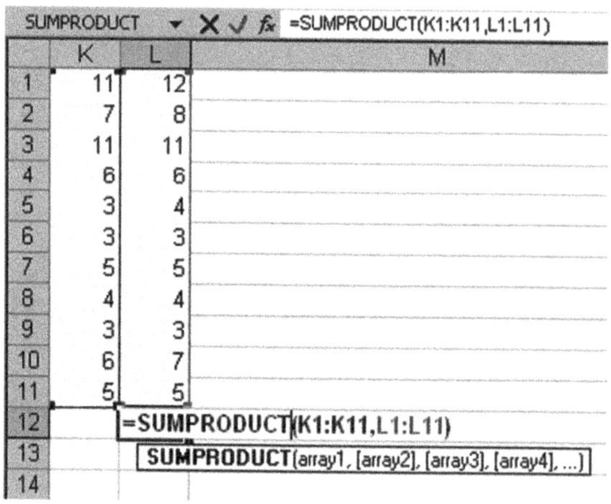

Figure 7.14 Function **SUMPRODUCT**.

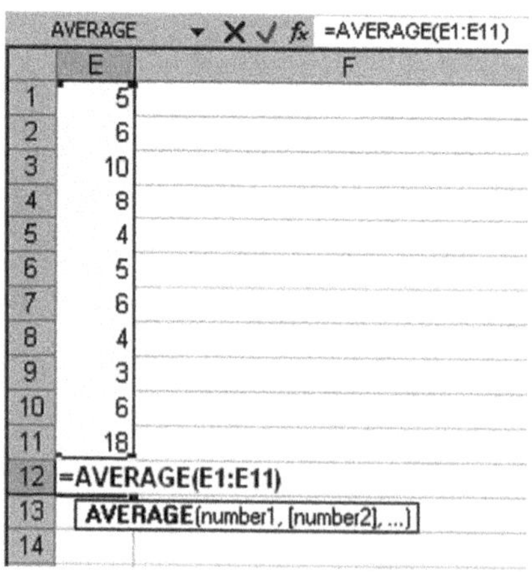

Figure 7.15 Function **AVERAGE**.

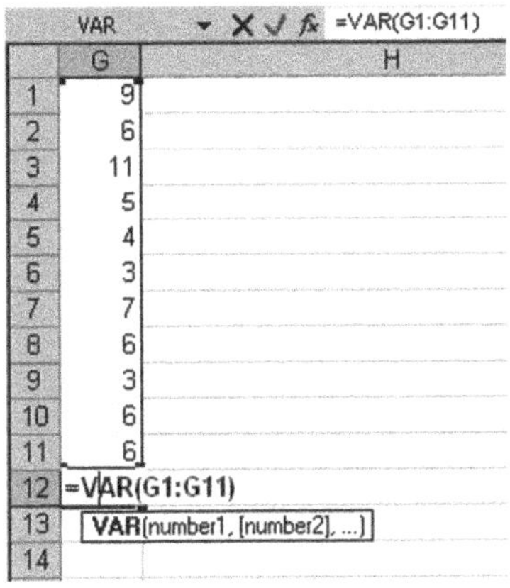

Figure 7.16 Function **VAR**.

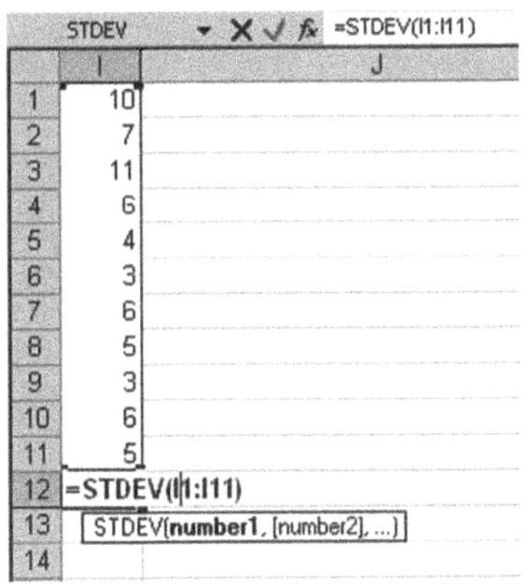

Figure 7.17 Function **STDEV**.

`CORREL`

Returns the correlation coefficient $\rho_{xy} = \text{cov}(X, Y)/\sigma_X\sigma_Y$ calculated for the cell ranges **array1** and **array2** (see Section 6.3.6).

Syntax
`CORREL (array1, array2)`

Example (Fig. 7.18)

Functions can be combined in one formula, as shown in the example in Fig. 7.19.

7.1.8 Errors in Formulas

Errors can occur both while typing in a formula and further in the course of a worksheet modification. When it happens, the cell does not contain the expected result but displays

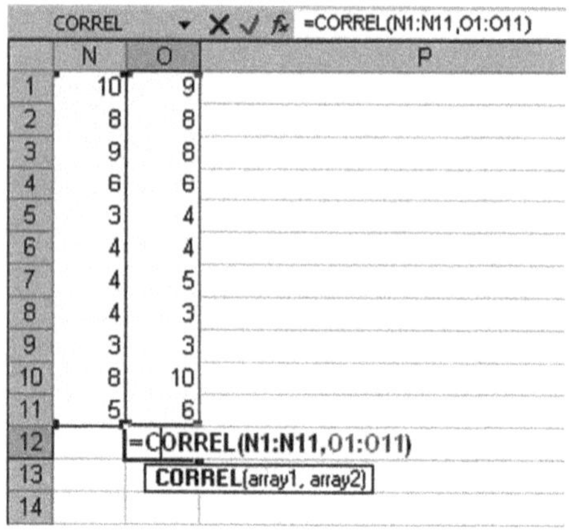

Figure 7.18 Function **CORREL**.

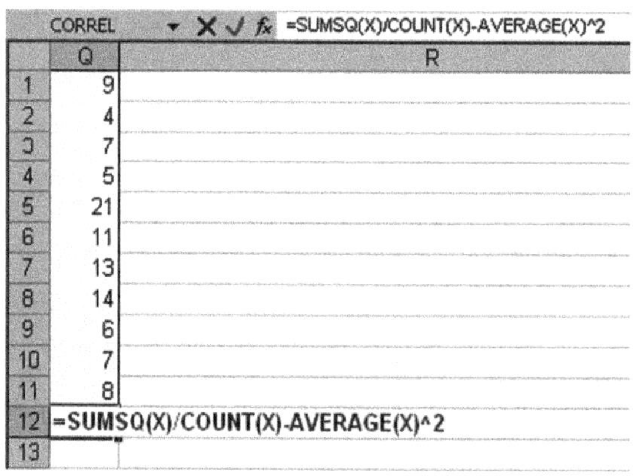

Figure 7.19 Composite formula.

special symbols indicating a certain type of error. The description of various error types is presented in worksheet Errors and shown in Fig. 7.20.

Menu item **Formula Auditing** in the **Tool** menu helps to reveal the source of an error.

7.1.9 Formula Dragging

Often, while working with matrices, there is a necessity to enter a range of formulae. For example, while performing the Standard Normal Variate (SNV) preprocessing of spectral data, it is necessary to calculate the sample mean and standard deviation values for each row. Even for a relatively small example, such as one shown in Fig. 7.22, it would be tiresome to enter the same formula repeatedly, changing only one argument. In a real dataset, the number of rows may reach several thousands. Fortunately, Excel has a special dragging mechanism for such operations.

Let us explain the *dragging technique* with the example provided in worksheet Expand. Let us start with typing a formula in cell J3.

	A	B	C
1	**Errors in formulas**		
2	**Formula**	**Result**	**Reason**
3	=1234567890	#########	column is not wide enough
4	=SUM(B4 B5)	#NULL!	areas do not intersect
5	=1/0	#DIV/0!	division by zero (0).
6	=SIN("text")	#VALUE!	wrong type of argument
7	=EXP(Name)	#REF!	refernce not valid
8	=ixp(2)	#NAME?	wrong function name
9	=EXP(EXP(100))	#NUM!	number is too large
10	=MATCH(1,2)	#N/A	missing data
11			

Figure 7.20 Errors in formulas.

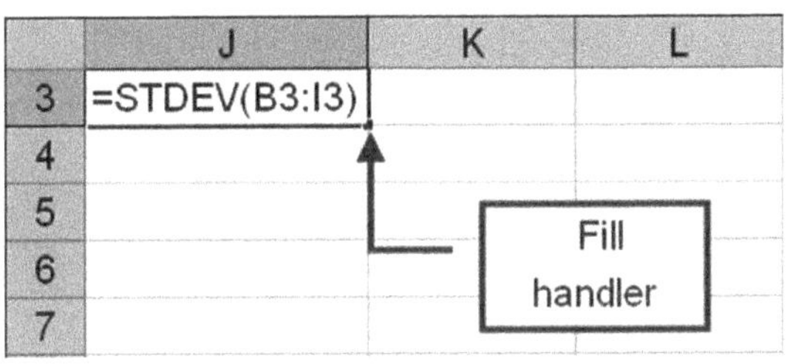

Figure 7.21 Fill handle.

Formulas in the adjacent cells can be entered with the help of a fill handle, a small black point at the bottom right-hand corner of the selection, as shown in Fig. 7.21. This handle transforms into a black cross when moused over. Afterward the active cell may be dragged to the adjustment region. Dragging direction may be vertical (down or up), as in Fig. 7.22 and horizontal (right or left).

A formula may be duplicated in another way. In step one, we copy the cell containing the formula. Then we select a range of cells where this formula should be entered. Afterward we use **Paste Special** operation with **Formulas** option selected as shown in Fig. 7.23.

Owing to the usage of relative referencing, such as B3:I3, we end up with proper references of the function's arguments regardless of the duplication method.

7.1.10 Create a Chart

Excel allows creating charts of different types. Two types are of most importance. They are the *scatter* and the *line* plots. Figure 7.24 presents an example of a scatter plot.

Such charts are used to present scores plots, relations between predicted versus measured values, etc. There is a substantial difference between scatter and line charts. A scatter chart

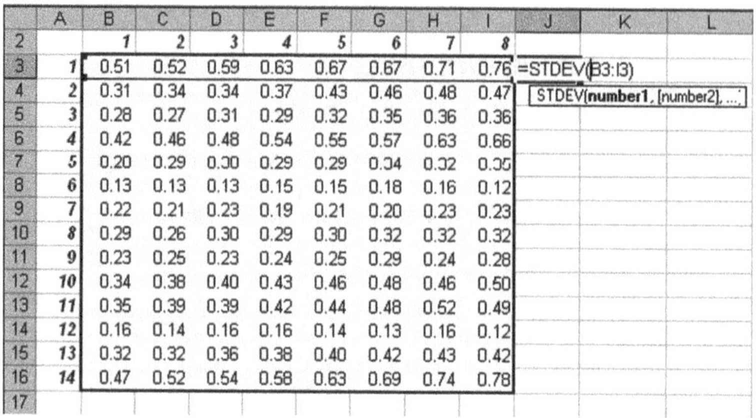

A B C D E F G H I J K L
1 2 3 4 5 6 7 8
1 0.51 0.52 0.59 0.63 0.67 0.67 0.71 0.76 =STDEV(B3:I3)
2 0.31 0.34 0.34 0.37 0.43 0.46 0.48 0.47
STDEV(number1, [number2], ...)
3 0.28 0.27 0.31 0.29 0.32 0.35 0.36 0.36
4 0.42 0.46 0.48 0.54 0.55 0.57 0.63 0.66
5 0.20 0.29 0.30 0.29 0.29 0.34 0.32 0.35
6 0.13 0.13 0.13 0.15 0.15 0.18 0.16 0.12
7 0.22 0.21 0.23 0.19 0.21 0.20 0.23 0.23
8 0.29 0.26 0.30 0.29 0.30 0.32 0.32 0.32
9 0.23 0.25 0.23 0.24 0.25 0.29 0.24 0.28
10 0.34 0.38 0.40 0.43 0.46 0.48 0.46 0.50
11 0.35 0.39 0.39 0.42 0.44 0.48 0.52 0.49
12 0.16 0.14 0.16 0.16 0.14 0.13 0.16 0.12
13 0.32 0.32 0.36 0.38 0.40 0.42 0.43 0.42
14 0.47 0.52 0.54 0.58 0.63 0.69 0.74 0.78

Figure 7.22 Formula dragging.

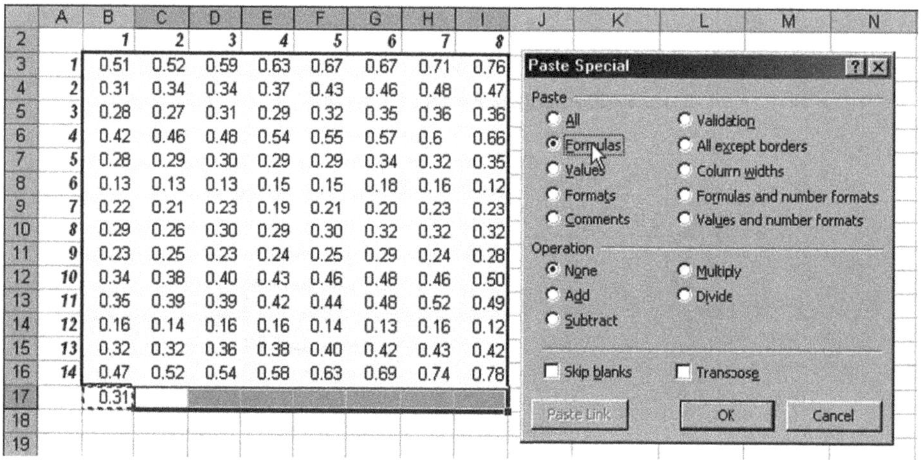

Figure 7.23 Duplicating a set of similar formulas.

has two peer value axes. In the line chart, the x-axis is used only as a category axis, which displays just the order of categories rather than the scaled numeric values. Therefore, the line charts are suitable for presenting dependences on the number of principal components, for example, diagrams of root mean square error of calibration (RMSEC) and root mean square error of prediction (RMSEP) versus model complexity.

Chart construction is performed differently in 2003 and 2007 versions of Excel. We will not be discussing this here and leave it for self-study.

7.2 MATRIX OPERATIONS

7.2.1 Array Formulas

Many matrix operations are performed with the help of so-called *array functions*. The result of calculation of such a function is not a single value (number) but a set of numbers, that

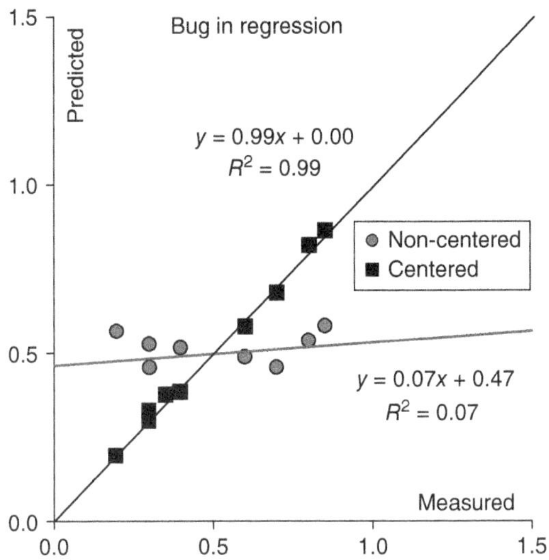

Figure 7.24 Scatter plot.

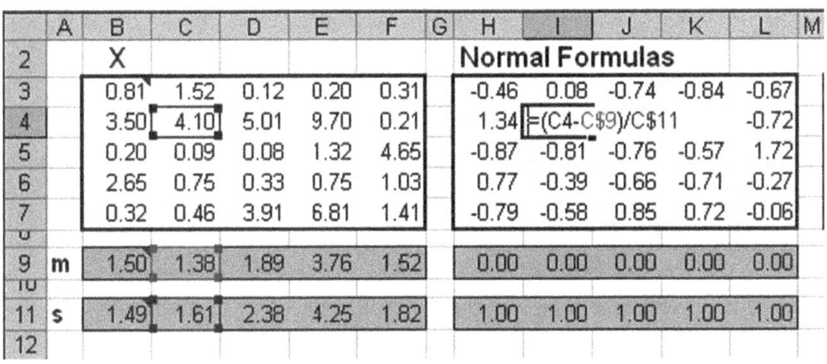

	A	B	C	D	E	F	G	H	I	J	K	L	M
2		X						Normal Formulas					
3		0.81	1.52	0.12	0.20	0.31		-0.46	0.08	-0.74	-0.84	-0.67	
4		3.50	4.10	5.01	9.70	0.21		1.34	=(C4-C$9)/C$11			-0.72	
5		0.20	0.09	0.08	1.32	4.65		-0.87	-0.81	-0.76	-0.57	1.72	
6		2.65	0.75	0.33	0.75	1.03		0.77	-0.39	-0.66	-0.71	-0.27	
7		0.32	0.46	3.91	6.81	1.41		-0.79	-0.58	0.85	0.72	-0.06	
9	m	1.50	1.38	1.89	3.76	1.52		0.00	0.00	0.00	0.00	0.00	
11	s	1.49	1.61	2.38	4.25	1.82		1.00	1.00	1.00	1.00	1.00	
12													

Figure 7.25 Normal formulas.

is, an array, even when it is a single value. Array formulas are created in the same way as any other Excel formula with one difference. To enter an array formula, it is necessary to press **CTRL+SHIFT+ENTER** for its completion.

Let us explain the application of array formulas with a simple example presented in worksheet Standard. Suppose we should perform autoscaling (column-wise centering and scaling) of the data in matrix **X** (see Section 8.1.4). For this purpose, the mean values m_j and the standard deviation values s_j should be calculated for each column j of matrix **X**. Afterward, we should subtract values m_j from each column and divide the result by s_j.

$$\tilde{x}_{ij} = (x_{ij} - m_j)/s_j$$

This procedure can be performed by the normal (not array) Excel functions as it is shown in Fig. 7.25.

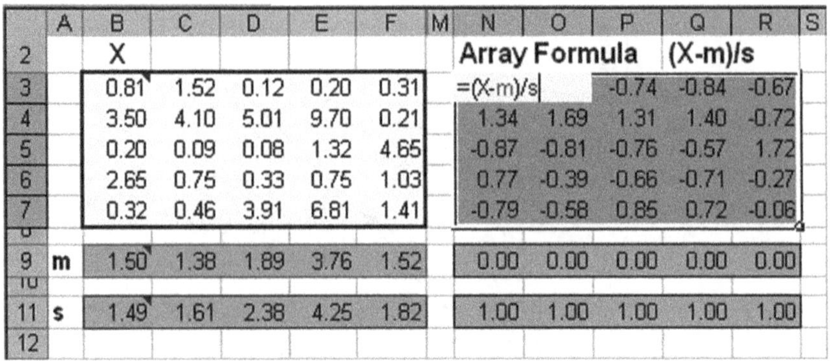

Figure 7.26 Array formula.

In this case, it is important to use symbol $ in front of the rows with number 9 (**m**) and 11 (**s**) to freeze the position of the corresponding values in the formulas.

For a large matrix **X**, it is convenient to apply an array formula (Fig. 7.26). Let us name the corresponding ranges in the worksheet as X, m, and s. Firstly, we select an empty range N3:R7 that has the same size as the anticipated result. Then, we enter the formula =(X-m)/s in the **Formula Bar**. Finally, we complete the operation by pressing **CTRL+SHIFT+ENTER**. If all these actions were performed properly, we obtain { =(X-m)/s} in the **Formula Bar**. The braces, {}, indicate an array formula.

7.2.2 Creating and Editing an Array Formula

To enter an array formula properly, it is necessary to select a region whose size corresponds to the size of the expected result. When the selected area is larger, the redundant cells are filled with an error symbol #N/A after the calculation. If the selected region is smaller, a part of the result values will be missing. After the selection of the resulting region, the formula is placed in the **Formula Bar** and the whole operation is completed by pressing **CTRL+SHIFT+ENTER**.

Alternatively, one can enter a formula in one cell and then select a region keeping this cell at the top left of the selected range, switch to the **Formula Bar**, and press **CTRL+SHIFT+ENTER**.

To change an array formula, it is necessary to select the whole range containing the formula result and switch to the Formula Bar. The braces that frame the array formula will disappear. After that you can edit the formula and press **CTRL+SHIFT+ENTER** to calculate the result.

To extend the output range of an array formula, it is sufficient to select a new range, go to the **Formula Bar**, and press **CTRL+SHIFT+ENTER**. It is a little bit more complicated to reduce the resulting region, that is, to delete redundant cells that contain the #N/A symbols. Firstly, it is recommended to select one cell from the resulting region, switch to the **Formula Bar**, and copy the formula by pressing **Ctrl+Ins**. Then, you should clear the old resulting range and select a smaller one. Afterward you should switch to the **Formula Bar**, paste the formula by pressing **Shift+Ins**, and finish with **CTRL+SHIFT+ENTER**.

Changes in the cells inside the resulting region are forbidden. When attempting to change the content of such a cell, the following warning box (Fig. 7.27) is displayed.

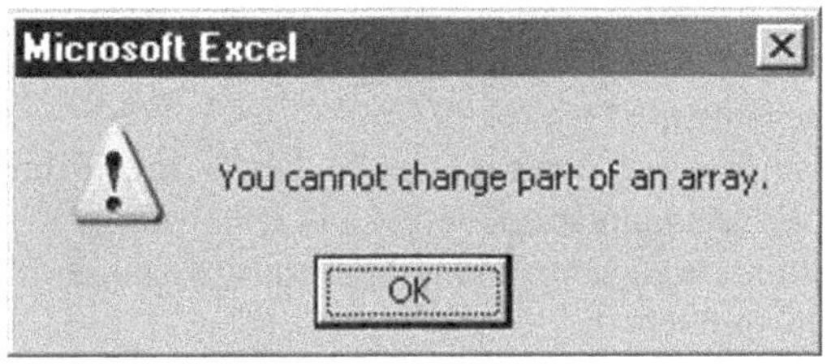

Figure 7.27 Warning concerning forbidden operations with an array formula.

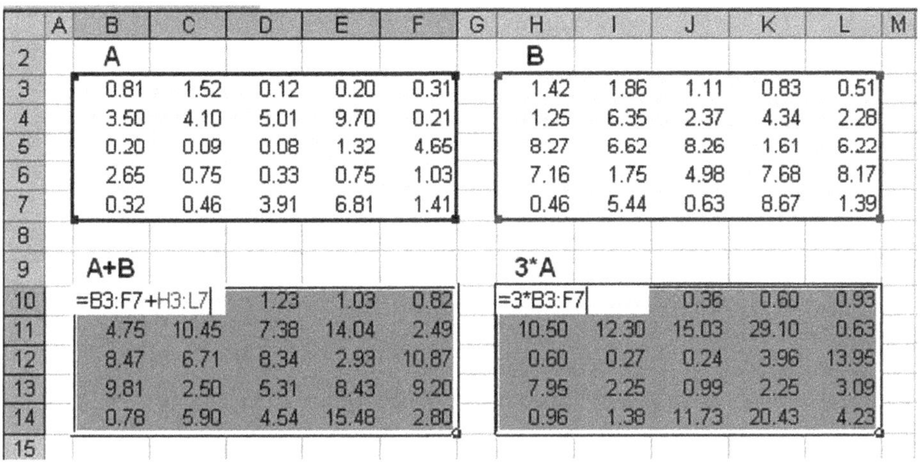

	A	B	C	D	E	F	G	H	I	J	K	L	M
2		A						B					
3		0.81	1.52	0.12	0.20	0.31		1.42	1.86	1.11	0.83	0.51	
4		3.50	4.10	5.01	9.70	0.21		1.25	6.35	2.37	4.34	2.28	
5		0.20	0.09	0.08	1.32	4.65		8.27	6.62	8.26	1.61	6.22	
6		2.65	0.75	0.33	0.75	1.03		7.16	1.75	4.98	7.68	8.17	
7		0.32	0.46	3.91	6.81	1.41		0.46	5.44	0.63	8.67	1.39	
8													
9		A+B						3*A					
10		=B3:F7+H3:L7		1.23	1.03	0.82		=3*B3:F7		0.36	0.60	0.93	
11		4.75	10.45	7.38	14.04	2.49		10.50	12.30	15.03	29.10	0.63	
12		8.47	6.71	8.34	2.93	10.87		0.60	0.27	0.24	3.96	13.95	
13		9.81	2.50	5.31	8.43	9.20		7.95	2.25	0.99	2.25	3.09	
14		0.78	5.90	4.54	15.48	2.80		0.96	1.38	11.73	20.43	4.23	
15													

Figure 7.28 Matrices addition and multiplication by a number.

7.2.3 Simplest Matrix Operations

It is easy to sum up the matrices or multiply a matrix by a number with the help of array functions in Excel illustrated in worksheet Binary and in Fig. 7.28.

For matrix multiplication, the array function **MMULT** is used. This is explained in Section 7.2.6.

7.2.4 Access to the Part of a Matrix

Two standard worksheet functions are used for the access and manipulation with a part of a matrix. They are illustrated in worksheet Lookup.

OFFSET

Returns a reference to a range specified by the number of rows and columns away from a cell or a range of cells.

Syntax

`OFFSET(reference, rows, cols [, height] [, width])`

reference is a reference to the first cell in the region from which the offset starts;

rows is the number of rows the result refers to (down, if this number is positive and up if the number is negative);

cols works in the same way but it refers to columns (the offset is to the right from the base cell for a positive argument and left if it is negative);

`height` is an optional argument and must be a positive number. This is the number of rows in the resulting reference;

`width` is an optional argument and must be a positive number. This is the number of columns in the resulting reference

Remarks

- If the arguments height and/or width are omitted, they are assumed to be equal to the height or width of the `reference`;
- The argument **reference** is a reference to a real (located in a worksheet) range and not to a virtual array.

Example (Fig. 7.29)
OFFSET is an array function and must be completed by **CTRL+SHIFT+ENTER**.

INDEX

Returns the values in an array selected by the indices of rows and columns.

Syntax
INDEX (**reference** [, row_num] [, column_num])

reference is a numerical array (a matrix);

`row_num` is an optional argument, the row number from which the values are selected;

`col_num` is an optional argument, the column number from which the values are selected;

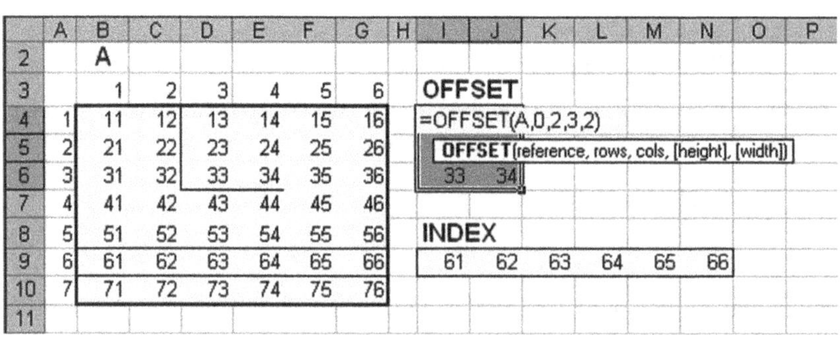

Figure 7.29 Function **OFFSET**.

Remarks

- if argument `row_num` is omitted, then the whole column is selected;
- if argument `col_num` is omitted, then the whole row is selected;
- if both arguments are used, the function returns a value in the cell located on the intercept of the corresponding row and column;
- the argument **`reference`** can be both a reference to a region located in a worksheet and to a virtual array.

Example (Fig. 7.30)
`INDEX` is an array function, which must be completed by **CTRL+SHIFT+ENTER**.

7.2.5 Unary Operations

The following unary operations can be performed on matrices. See worksheet Unary for illustration.

MINVERSE

Returns an inverse matrix.

Syntax
`MINVERSE (array)`
  `array` is a numerical array (a matrix).

Remarks

- **`array`** must be a square matrix;
- For a singular matrix, the symbol #VALUE is returned.

Example (Fig. 7.31)
`MINVERSE` is an array function and must be completed by **CTRL+SHIFT+ENTER**.

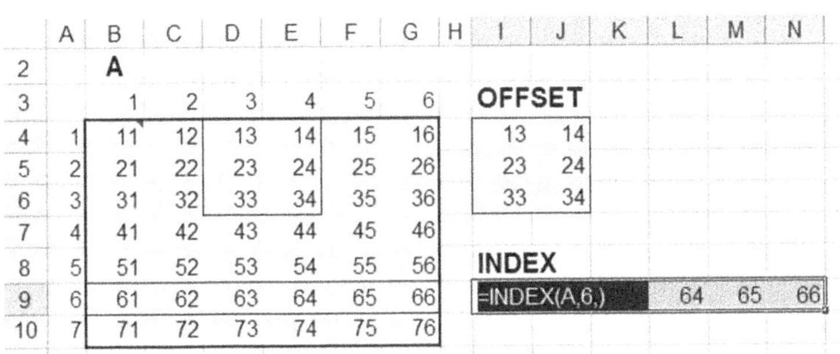

Figure 7.30 Function **`INDEX`**.

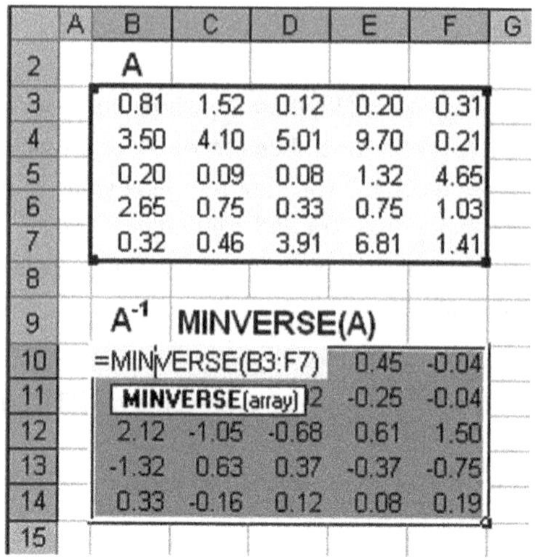

A B C D E F G
A
0.81 1.52 0.12 0.20 0.31
3.50 4.10 5.01 9.70 0.21
0.20 0.09 0.08 1.32 4.65
2.65 0.75 0.33 0.75 1.03
0.32 0.46 3.91 6.81 1.41
A-1 MINVERSE(A)
=MINVERSE(B3:F7) 0.45 -0.04
MINVERSE(array) -0.25 -0.04
2.12 -1.05 -0.68 0.61 1.50
-1.32 0.63 0.37 -0.37 -0.75
0.33 -0.16 0.12 0.08 0.19

Figure 7.31 Function **MINVERSE**.

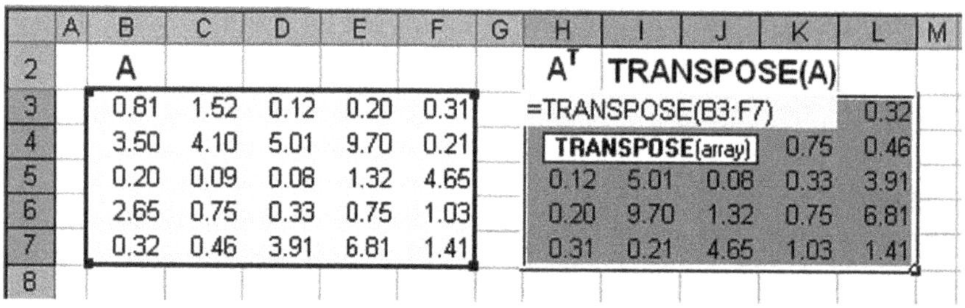

A B C D E F G H I J K L M
A
AT TRANSPOSE(A)
0.81 1.52 0.12 0.20 0.31
=TRANSPOSE(B3:F7) 0.32
3.50 4.10 5.01 9.70 0.21
TRANSPOSE(array) 0.75 0.46
0.20 0.09 0.08 1.32 4.65
0.12 5.01 0.08 0.33 3.91
2.65 0.75 0.33 0.75 1.03
0.20 9.70 1.32 0.75 6.81
0.32 0.46 3.91 6.81 1.41
0.31 0.21 4.65 1.03 1.41

Figure 7.32 Function **TRANSPOSE**.

TRANSPOSE

Returns the transposed matrix.

Syntax
TRANSPOSE (array)
array is a numerical array (a matrix).

Example (Fig. 7.32)
TRANSPOSE is an array function and must be completed by **CTRL+SHIFT+ENTER**.

MDETERM

Returns the matrix determinant.

Syntax

`MDETERM (array)`
`array` is a numerical array (a matrix).

Remarks

- **`array`** must be a square matrix.

`MDETERM` is not an array function and must be entered by pressing **ENTER**.

7.2.6 Binary Operations

The following *binary* operation can be performed on matrices. See worksheet MMULT for illustrations.

MMULT

Returns the product of two matrices.

Syntax
`MMULT (array1, array2)`
`array1`, **`array2`** are the numerical arrays (matrices).

Remarks

- The number of columns in **`array1`** must be equal to the number of rows in **`array2`**, otherwise symbol #VALUE! is returned;
- The number of elements in the resulting matrix must be less than or equal to 5461 (Excel 2003).

Example (Fig. 7.33)
`MMULT` is an array function and must be completed by **CTRL+SHIFT+ENTER**.

7.2.7 Regression

Several standard worksheet functions involve linear regression.

TREND

Builds linear regression

$$y = b + m_1 x_1 + \cdots + m_J x_J + e$$

and approximates the known response values of the vector **known_y's** for a given predictor matrix known_x's and returns response values for a given matrix new_x's.

Syntax

TREND(**known_y's** [,known_x's] [,new_x's] [,const])

known_y's is the vector of the known response values **y** (calibration set);

known_x's is an optional argument, for example, matrix **X** of the known values of predictors (a calibration set);

new_x's is an optional argument, for example, matrix $\mathbf{X}_{new}$ of the new predictor values (a test set or a new matrix) for which the results are calculated and returned;

const is an optional argument. It is a Boolean value, which indicates whether the parameter b equals zero. If const equals TRUE or omitted, then b is calculated in an ordinary way. Otherwise $b = 0$.

Remarks

- Vector **known_y's** must be located in one column. In this case, each column of array known_x's is treated as a separate variable;
- If argument known_x's is omitted, it is supposed to be the vector of numbers {1;2;3; ... } with the same size as **known_y's**;
- Matrix of new values new_x's must have the same number of columns (variables) as matrix known_x's;
- If argument new_x's is omitted, it is assumed to be the same as known_x's. The result is a vector with the same number of rows as in array new_x's.

Example

This function is illustrated in worksheet Regression and in Fig. 7.34.

TREND is an array function and must be completed by **CTRL+SHIFT+ENTER**.

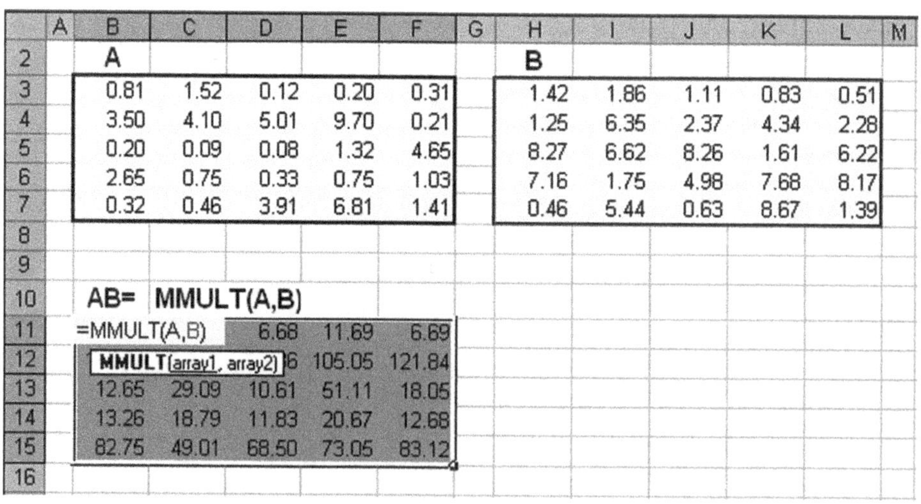

A
0.81 1.52 0.12 0.20 0.31
3.50 4.10 5.01 9.70 0.21
0.20 0.09 0.08 1.32 4.65
2.65 0.75 0.33 0.75 1.03
0.32 0.46 3.91 6.81 1.41
B
1.42 1.86 1.11 0.83 0.51
1.25 6.35 2.37 4.34 2.28
8.27 6.62 8.26 1.61 6.22
7.16 1.75 4.98 7.68 8.17
0.46 5.44 0.63 8.67 1.39
AB= MMULT(A,B)
=MMULT(A,B) 6.68 11.69 6.69
MMULT(array1, array2) 6 105.05 121.84
12.65 29.09 10.61 51.11 18.05
13.26 18.79 11.83 20.67 12.68
82.75 49.01 68.50 73.05 83.12

Figure 7.33 Function **MMULT**.

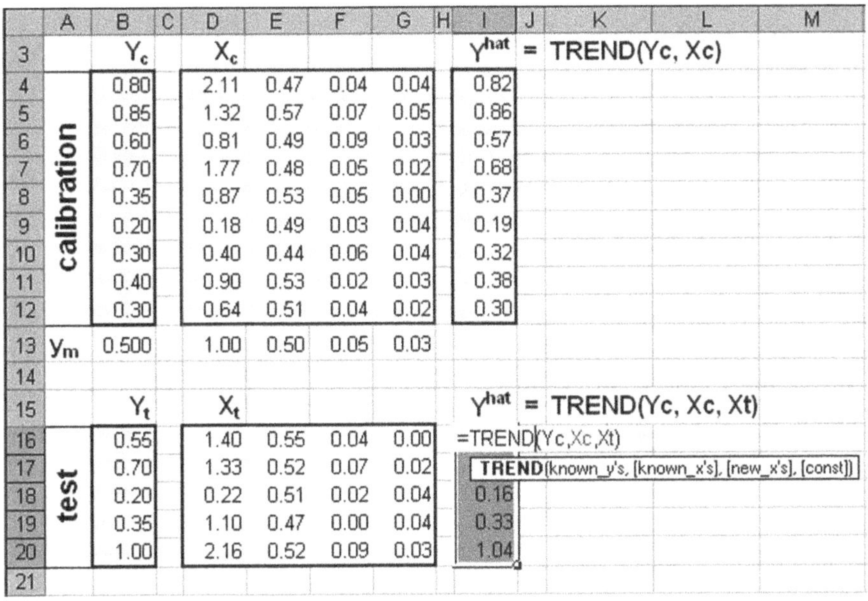

A B C D E F G H I J K L M
Yc Xc Yhat = TREND(Yc, Xc)
calibration
3
4 0.80 2.11 0.47 0.04 0.04 0.82
5 0.85 1.32 0.57 0.07 0.05 0.86
6 0.60 0.81 0.49 0.09 0.03 0.57
7 0.70 1.77 0.48 0.05 0.02 0.68
8 0.35 0.87 0.53 0.05 0.00 0.37
9 0.20 0.18 0.49 0.03 0.04 0.19
10 0.30 0.40 0.44 0.06 0.04 0.32
11 0.40 0.90 0.53 0.02 0.03 0.38
12 0.30 0.64 0.51 0.04 0.02 0.30
13 Ym 0.500 1.00 0.50 0.05 0.03
14
15 Yt Xt Yhat = TREND(Yc, Xc, Xt)
test
16 0.55 1.40 0.55 0.04 0.00 =TREND(Yc,Xc,Xt)
17 0.70 1.33 0.52 0.07 0.02 TREND(known_y's, [known_x's], [new_x's], [const])
18 0.20 0.22 0.51 0.02 0.04 0.16
19 0.35 1.10 0.47 0.00 0.04 0.33
20 1.00 2.16 0.52 0.09 0.03 1.04
21

Figure 7.34 Function **TREND**.

LINEST

Calculates the statistics for the linear regression

$$y = b + m_1x_1 + \cdots + m_Jx_J + e$$

Syntax

LINEST(known_y's [,known_x's] [,new_x's] [,const] [,stats])

known_y's is the vector of the known response values **y** (calibration set);

known_x's is an optional argument. Matrix **X** of the known values of predictors (a calibration set);

new_x's is an optional argument. Matrix $\mathbf{X}_{\text{new}}$ of the new predictor values (a test set or a new matrix) for which the results are calculated and returned;

const is an optional argument. It is a Boolean value, which indicates whether parameter b equals zero. If const equals TRUE or omitted, then b is calculated in an ordinary way. Otherwise $b = 0$.

stats is an optional argument. It is a Boolean value, which indicates the necessity to display additional statistics information. If stats is equal to FALSE or omitted, the estimates of the regression parameters $m_J, \ldots, m_2, m_1$ and b are returned. Otherwise, the table shown in Fig. 7.35 is returned.

$\mathbf{m}_J, \ldots, \mathbf{m}_2, \mathbf{m}_1$ and **b** are the estimates of the regression parameters;

$\mathbf{s}_J, \ldots, \mathbf{s}_2, \mathbf{s}_1$ and $\mathbf{s}_b$ are the standard errors for the regression parameters estimates;

R^2 is the coefficient of determination;
s_y is the standard error for the *y* estimate;
F is the *F*-statistics;
DoF is the number of the degrees of freedom;
SS_{reg} is the regression sum of squares;
SS_{res} is the residual sum of squares.

Remarks

- **LINEST** is a badly designed function and it is very inconvenient in application;
- Remarks included in the **TREND** description are applicable for **LINEST** function too.

Example

This function is illustrated in worksheet Regression and in Fig. 7.36.
LINEST is an array function and must be completed by **CTRL+SHIFT+ENTER**.

m_J	m_{J-1}	
s_J	s_{J-1}	
R^2	s_y	#N/A
F	DoF	#N/A
SS_{reg}	SS_{res}	#N/A

m_2	m_1	b
s_2	s_1	s_b
#N/A	#N/A	#N/A
#N/A	#N/A	#N/A
#N/A	#N/A	#N/A

Figure 7.35 Table provided by **LINEST** function.

A	B	C	D	E	F	G	L	M	N	O	P	Q	R
	Y_c		X_c					m_4	m_3	m_2	m_1	b	
calibration	0.80		2.11	0.47	0.04	0.04		=LINEST(Yc,Xc,,TRUE)			0.33	-0.74	
	0.85		1.32	0.57	0.07	0.05		LINEST(known_y's, [known_x's], [const], [stats])					
	0.60		0.81	0.49	0.09	0.03		0.99	0.03	#N/A	#N/A	#N/A	
	0.70		1.77	0.48	0.05	0.02		149.21	4.00	#N/A	#N/A	#N/A	
	0.35		0.87	0.53	0.05	0.00		0.46	0.00	#N/A	#N/A	#N/A	
	0.20		0.18	0.49	0.03	0.04							
	0.30		0.40	0.44	0.06	0.04							
	0.40		0.90	0.53	0.02	0.03							
	0.30		0.64	0.51	0.04	0.02							
y_m	0.500		1.00	0.50	0.05	0.03							

Figure 7.36 Function **LINEST**.

7.2.8 Critical Bug in Excel 2003

Functions **TREND** and **LINEST** in Excel 2003 provide wrong results in some circumstances. Wrong results are returned when simultaneously

- the mean value for each predictor variable in **X** is equal to zero;
- the mean value of the response vector **y** does not equal zero.

Worksheet Bug and Fig. 7.37 demonstrate this. Mean values for all columns of the matrix X_c are equal to zero but the mean value for the vector Y_c does not equal zero.

Example

The situation may be corrected by using a special trick. You can apply **TREND** function to the centered response values and correct the result afterward. For this purpose, the following formula =TREND(Yc-ym, Xc)+ym, is used (see the same figure).

It is strange that the bug was not mentioned by users; however, in Excel 2007 this error is corrected.

7.2.9 Virtual Array

Often in the course of data processing, there is a problem of storing the intermediate computations, which are not important by themselves but have to be calculated in order to reach the result. For example, the residuals in the principal component analysis (PCA) decomposition are rarely analyzed per se but used for calculation of the explained variance, orthogonal distances, etc. At the same time, such intermediate arrays may be very large and must be calculated for various numbers of principal components. They cause flooding of the worksheets by the unnecessary intermediate information. One can avoid this situation by applying virtual arrays. Let us explain this on a simple example.

Suppose that we have a matrix **A** and the goal is to calculate the determinant of the matrix $\mathbf{A}^t\mathbf{A}$. Worksheet Virtual and Fig. 7.38 demonstrate two methods of calculations. The first

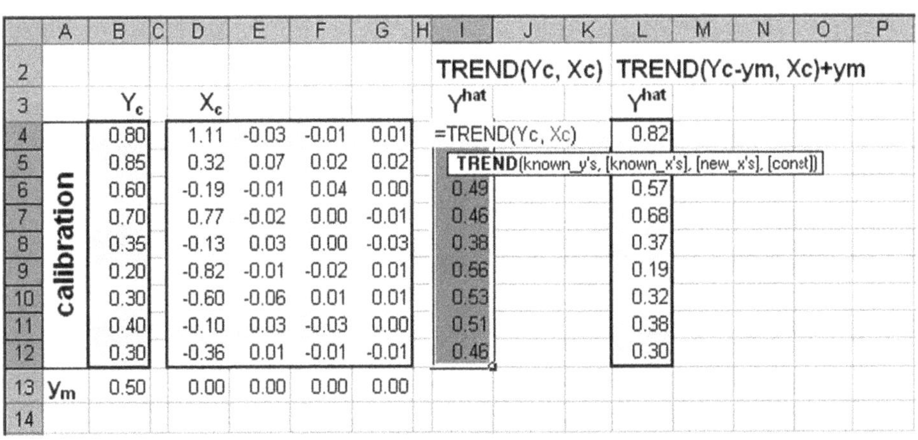

TREND(Yc, Xc)
TREND(Yc-ym, Xc)+ym
Yc
Xc
Yhat
Yhat
calibration
=TREND(Yc, Xc)
TREND(known_y's, [known_x's], [new_x's], [const])
0.80 1.11 -0.03 -0.01 0.01 0.82
0.85 0.32 0.07 0.02 0.02
0.60 -0.19 -0.01 0.04 0.00 0.49 0.57
0.70 0.77 -0.02 0.00 -0.01 0.46 0.68
0.35 -0.13 0.03 0.00 -0.03 0.38 0.37
0.20 -0.82 -0.01 -0.02 0.01 0.56 0.19
0.30 -0.60 -0.06 0.01 0.01 0.53 0.32
0.40 -0.10 0.03 -0.03 0.00 0.51 0.38
0.30 -0.36 0.01 -0.01 -0.01 0.46 0.30
ym 0.50 0.00 0.00 0.00 0.00

Figure 7.37 Bug in the regression functions in Excel 2003.

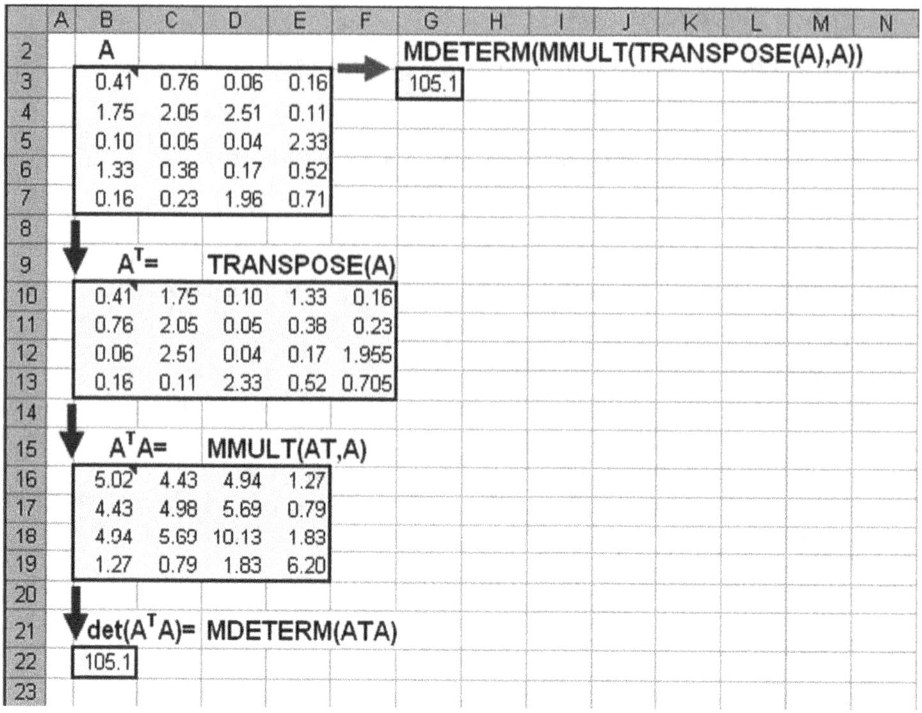

Figure 7.38 Virtual array application.

method uses a sequence of intermediate calculations indicated by the arrows pointing down. The second method uses only one formula and is indicated by the arrow pointing right. Both methods provide the same result, but the first one occupies a lot of space in the worksheet, whereas the latter one uses several intermediate virtual arrays. All these virtual arrays are the same, as calculated by the first method, but they are not returned on the worksheet explicitly.

The first virtual array is the transposed matrix $\mathbf{A}^t$ calculated by function **TRANSPOSE**(A). The second virtual array is the product of the first virtual array and the matrix **A** calculated by function **MMULT**(**TRANSPOSE**(A), A). Function **MDETERM** applied to the second virtual array returns the result.

Virtual arrays are useful when calculating various auxiliary characteristics as a part of multivariate data analysis, such as residuals, eigenvalues, leverages, etc. This is explained in detail in Chapter 13.

7.3 EXTENSION OF EXCEL POSSIBILITIES

7.3.1 VBA Programming

Sometimes, the standard Excel capabilities are not sufficient and we have to add some user-defined features. A specially designed programming tool, Microsoft Visual Basic for Applications (VBA) is used for this purpose. VBA helps to create *macros* (a special set of instructions for performing a sequence of operations) and user-defined *functions* (a special

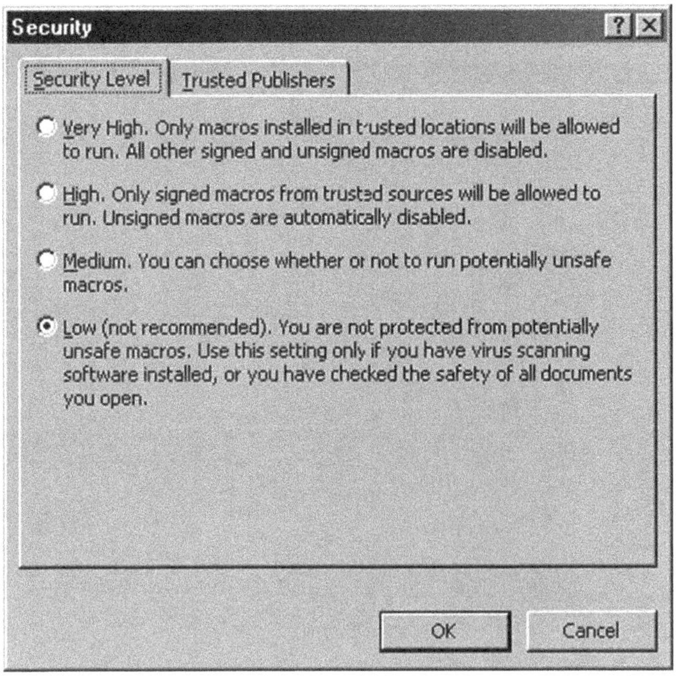

Figure 7.39 Selection the security level in Excel 2003.

set of instructions for performing calculations in a worksheet). Macros are used for automation of standard procedures. Once a macro is created, it can be repeatedly used for a routine operation. To run a macro from the menu, select **Tools-Macro-Macros** (item). Sometimes, it is more convenient to assign a macro to a special new button, placed in the **Tool**, or in a worksheet. The user-defined functions are run in the same way as the standard Excel worksheet functions via the **Formula Bar**.

To make macros and user-defined functions accessible, one should set up a certain level of security via menu item **Tools-Macro-Security** (Excel 2003) shown in Fig. 7.39.

In Excel 2007, the security level selection is performed via **Office Button-Excel Options-Trust Center** as presented in Fig. 7.40.

On each opening of an Excel file, the system will ask permission for running macros, if the **Medium** level (Excel 2003) or **Disable all macros with notification** (Excel 2007) has been selected. We recommend using the security level as shown in Figs 7.39 and 7.40, with a caution to use reliable antivirus software to scan alien Excel files. The capabilities of the VBA application are substantially limited on the initial installation of Excel 2007. To restore these properties, one should follow the sequence **Office Button-Excel Options-Popular** and switch on the option **Show Developer Tab in the Ribbon**.

7.3.2 Example

Let us see the application of VBA in an example. For modeling the nonisothermic kinetics (differential scanning calorimetry (DSC), thermogravimetric analysis (TGA), etc.), one

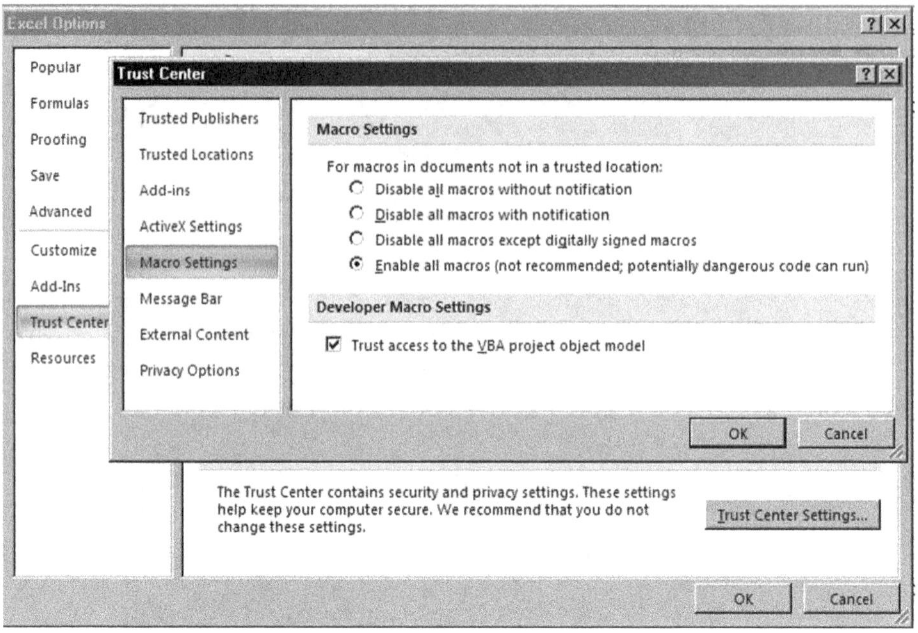

Figure 7.40 Selection the security level in Excel 2007.

should calculate the integral exponential function $E_1(x)$. By definition

$$E_1(x) = \int_x^{\infty} (e^{-t}/t)dt$$

An infinite series can be used for $E_1(x)$ calculation

$$E_1(x) = -\gamma - \ln(x) - \sum_{n=1}^{\infty} R_n,$$

$$\gamma = 0.57721567, \quad R_n = (-x)^n/(nn!)$$

Each series term can be placed in its own cell in a worksheet. Afterward one can sum up all these cells. This approach is presented in worksheet VBA and in Fig. 7.41.

We cannot consider this approach as a wise one. First of all, such calculations occupy a rather large area in the worksheet. The main drawback, however, is that we do not know in advance how many terms should be included in the summation. Sometimes 10 terms are sufficient; in other cases even 50 terms are not enough.

	A	B	C	D	E	F
2	x=	4		Gamma=	0.5772156	
3			n	R_n	$E_1(x)$	
4			0	-1.96351	***0.0037794***	
5			1	=-POWER(-x,C5)/C5/FACT(C5)		
6			2	-4		
7			3	3.5555556		
8			4	-2.666667		
9			5	1.7066667		
10			6	-0.948148		
11			7	0.4643991		
12			8	-0.203175		
13			9	0.0802665		
14			10	-0.028896		
15			11	0.0095524		
16			12	-0.002919		
17			13	0.000829		
18			14	-0.00022		
19			15	5.474E-05		
20			16	-1.28E-05		
21			17	2.841E-06		
22			18	-5.96E-07		
23			19	1.189E-07		
24			20	-2.26E-08		
25						

Figure 7.41 Calculation of $E_1(x)$ function in a worksheet.

	A	B	G	H	I	J	K	L	M	N
2	x=	4	n=	1	n=	1	n+1=	2		
3			R_1=	4	R_n=	4	R_{n+1}=	=-J3*x*(L2-1)*POWER(L2,-2)		
4			S_1=	2.03649	S_n=	2.03649	S_{n+1}=	***-1.96351***	=E1(x)	
5										
6					Iterations=	1		Repeat		
7										

Figure 7.42 Calculation of $E_1(x)$ function by recurrence method.

7.3.3 Macro Example

The second way of calculation is to apply a recurrent formula that links two neighboring terms of the series

$$R_n(x) = (x(n-1)/n^2)R_{n-1}(x).$$

To apply this formula, it is necessary to manage the recurrence calculations in a worksheet. The example is shown in worksheet VBA and in Fig. 7.42.

Figure 7.43 Macro recording.

The first iteration step calculates the values in range L2:L4 using values from range J2:J4. To start the next iteration step, one should copy the values from range L2:L4 and paste them into range J2:J4 as values. A copy-paste operation should insert only values without formulas and a **Copy-Paste Special** command should be used for formulas. Range H2:H4 contains the initial values used to start the iteration process. By numerous repetition of the **Copy-Paste Special** operation, one can yield the target value in cell L4, although the process is boring.

It is better to make a macro in order to automate the procedure. The simplest way to design a macro is to record the operations performed in a worksheet. Go to the menu item **Tools-Macro-Record New Macro**. In the dialog window (Fig. 7.43), you should give a name to the macro and indicate a book where this macro will be stored.

After the **OK** button is pressed, all operations performed in a worksheet are recorded in the macro procedure. When the recording process is over, it must be terminated by the **Tools-Macro-Stop Recording** command. The result can be found in the Visual Basic editor.

Figure 7.44 shows the macro after a minor correction. We added a cycle for repeating the **Copy-Paste** operation for nIter times. The value of nIter is determined in cell J6 in the worksheet. Cell J6 has a local name n. To finalize the automation, we use the button Repeat with macro `Iteration` assigned to it. Such an approach was applied to develop the algorithms iterative target transformation factor analysis (ITTFA) and alternating least squares (ALS) (Sections 12.4.1 and 12.4.2) used in the Multivariate Curve Resolution problem.

7.3.4 User-Defined Function Example

Last but not least, the most elegant solution for the calculation of an integral exponential function is the development of a user-defined function with the help of VBA.

Figure 7.45 illustrates a VBA code and a way to call the function. We will not consider the VBA programming here as this is a very large and complex issue. For self-study, one can refer to the abundant literature on this matter.

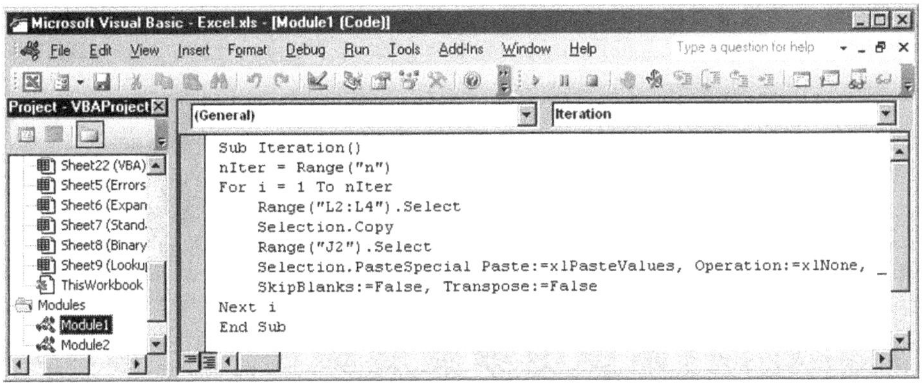

Figure 7.44 Visual Basic editor.

```
Function IntExp(dX)
      GammaConst = 0.5772156
      Eps = 0.00000000001
      IntExp = -GammaConst -Log(dX)
      nFact = 1
      For n = 1 To 1000
          nFact = nFact * n
          dR = (-dX) ^ n / n / nFact
          IntExp = IntExp - dR
          If (Abs(dR) < Eps) Then Exit Function
      Next n
End Function
```

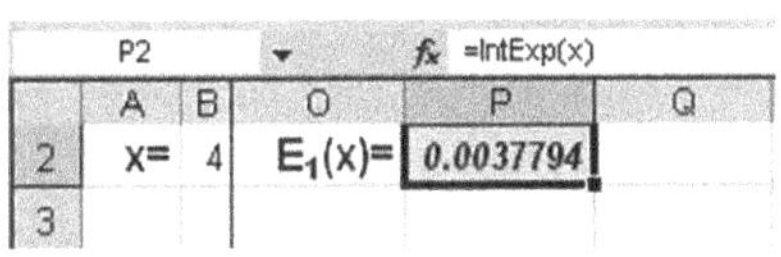

Figure 7.45 Function IntExp.

The VBA macros are rather slow and not entirely fit for large calculations. For example, we do not recommend using the VBA programming for the PCA decomposition. For large arrays, such a procedure will take a lot of time. One should rather consider both Excel and VBA as a front end tool for input and output of data. The date, in turn, is passed to the *dynamic link library* (DLL), and coded in a fast programming language, such as C++ (back end). This specific approach was used in Fitter and Chemometrics Add-In. More about user-defined functions can be found in Chapter 13.

7.3.5 Add-Ins

The VBA programs written by a user are stored in the same Excel workbook in which they are created. To make the macros available for other workbooks, one can either copy the macros into these workbooks or create an Add-In application. An *Add-In* is a special Excel file containing several VBA modules linked to the required DLL libraries. An Add-In may be added to Excel to extend its capabilities. The standard Excel configuration includes several Add-Ins. The most interesting among them are Solver and Analysis Toolpak Add-Ins.

Solver is designed to optimize a value in the target cell. The result is calculated with respect to other cells functionally linked to the target one. Analysis Toolpak contains a set of statistical functions for data analysis.

There are various Excel Add-Ins available on the Internet. Some of them may be downloaded free of charge, others are commercial. The J-Walk Chart Tools Add-In is a free utility for chart management. At allows adding a legend, names, etc. to the chart elements; that is, all features that are absent in the standard Excel package. XLStat is a large and expensive package for statistical analysis, including partial least square (PLS) regression. The Multivariate Analysis Add-In is a shareware package for multivariate data analysis from Bristol University. Fitter is an Add-In for nonlinear regression analysis.

Chemometrics is a collection of the worksheet functions for multivariate data analysis, which is explained in Chapter 8. Chemometrics Add-In is the main tool used in this book.

7.3.6 Add-In Installation

Before applying an Add-In program, one should perform an *installation* consisting of two steps. Firstly, all files included in the Add-In *package* must be copied onto the computer.

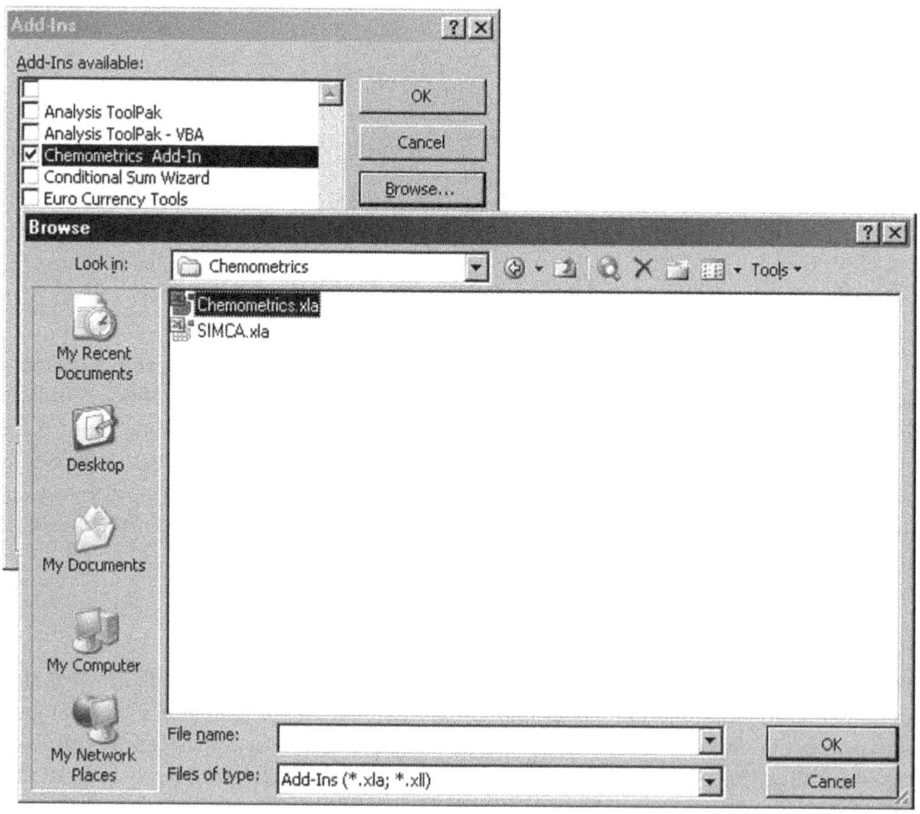

Figure 7.46 Add-In installation.

Some Add-Ins provide a special Setup program for an automatic installation. In other cases, the files should be located by a user. We will explain how this should be performed.

A package always includes a file with extension XLA and several additional files with extensions *DLL*, *HLP*, etc. All these additional files should be placed in folder **C:\Windows** or **C:\Windows\System** or **C:\Windows\System32**. The main XLA file may be located in any folder. However, the two following folders are preferable.

Microsoft recommends placing an XLA file in the folder **C:\Documents and Settings\User\Application Data\Microsoft\AddIns** (here **User** is the login name under which the Windows is currently running). Placing the XLA file in this directory provides an easy access to the file in the second step of installation. At the same time when the same workbooks are run on several computers under different **User** names, the links to the XLA files are lost. Therefore, it is necessary to update the links continuously. We recommend to copy the file Chemometrics.xla in the folder that has the same name on different computers, for example, **C:\Program Files\Chemometrics**. An automated installation of Chemometrics Add-In is presented in Chapter 8.

The second step of installation is performed from the opened Excel workbook. For Excel 2003, one should go to **Tools-Add-Ins** and in Excel 2007 to **Office Button-Excel Options-Add-Ins-Go**. In the open dialog window (see Fig. 7.46), you must press **Browse** and find the pertinent XLA file.

As soon as the Add-In is installed, it may be activated and deactivated by ticking the corresponding mark in front of the **Add-In** name. To uninstall an Add-In program, one should remove the mark next to its name, close Excel, and delete all files related to this Add-In from the computer.

CONCLUSION

We have considered just the main features of the matrix calculation in Excel. Many details are left out. Chapter 8 will fill this gap.

8

PROJECTION METHODS IN EXCEL*

This chapter describes the Chemometrics Add-In, which is an Add-In designed especially for Microsoft Excel. The Chemometrics Add-In enables data analysis on the basis of the following projection methods: Principal Component Analysis (PCA) and Projection on Latent Structures (PLS1 and PLS2).

Preceding chapters	2, 7
Dependent chapters	9–13, 15
Matrix skills	Basics
Statistical skills	Low
Excel skills	Basic
Chemometric skills	Basic
Chemometrics Add-In	Used
Accompanying workbook	Projection.xls

8.1 PROJECTION METHODS

8.1.1 Concept and Notation

Projection methods are widely used for multivariate data analysis, especially in chemometrics. They are applied both to one-block data **X** (classification, e.g., PCA) and to double-block data such as **X** and **Y** (calibration, e.g., principal component regression (PCR) and partial least squares (PLS)).

*With contributions from Oxana Rodionova.

Chemometrics in Excel, First Edition. Alexey L. Pomerantsev.

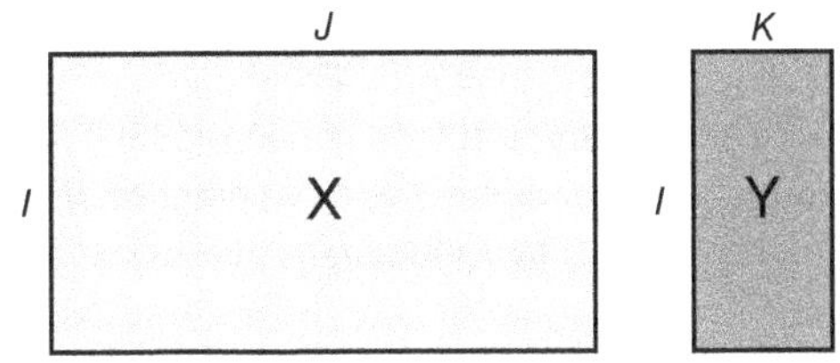

Figure 8.1 Multivariate data.

Let us consider a $(I \times J)$ data matrix **X**, where I is the number of objects (rows) and J is the number of independent variables (columns). Ordinarily, the number of variables is rather high ($J \gg 1$). We can simultaneously analyze the $(I \times K)$ matrix **Y**, where I is the same number of objects and K is the number of responses. See Fig. 8.1.

The essence of the projection techniques is a considerable reduction of data dimensionality for both blocks **X** and **Y**. There are many reviews on projection methods, and the reader is invited to refer to them for more detail.

8.1.2 PCA

PCA is the oldest projection method. This method uses new formal (or latent) variables $\mathbf{t}_a (a = 1, \ldots, A)$, which are linear combinations of the original variables $\mathbf{x}_j (j = 1, \ldots, J)$, that is,

$$\mathbf{t}_a = \mathbf{p}_{a1}\mathbf{x}_1 + \cdots + \mathbf{p}_{aJ}\mathbf{x}_J,$$

or in the matrix notation

$$\mathbf{X} = \mathbf{T}\mathbf{P}^t + \mathbf{E} = \sum_{a=1}^{A} \mathbf{t}_a \mathbf{p}_a^t + \mathbf{E} \tag{8.1}$$

In this equation, **T** is called the *scores matrix* or *scores*. Its dimension is $(I \times A)$. Matrix **P** is called the *loadings matrix*, or *loadings*, and it has a dimension $(A \times J)$. **E** is the *residuals* $(I \times J)$ matrix. See Fig. 8.2.

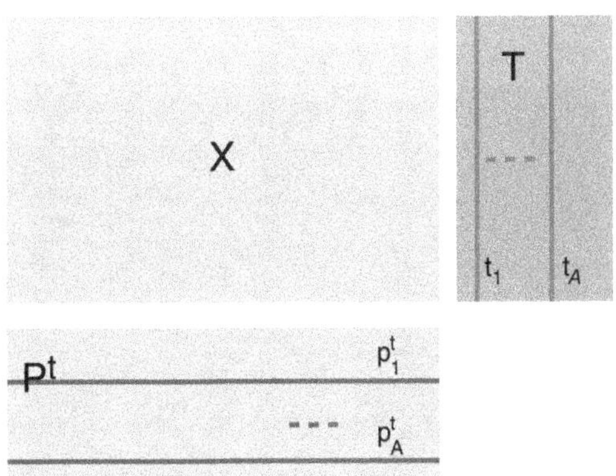

Figure 8.2 Graphic representation of PCA.

New variables $\mathbf{t}_a$ are often called *principal components* (*PCs*); therefore, the method is called *PCA*. The number of columns $\mathbf{t}_a$ in matrix $\mathbf{T}$ and columns $\mathbf{p}_a$ in matrix $\mathbf{P}$ is equal to A, which is the *number of PCs*. It defines the projection model complexity. The value of A is certainly less than the number of variables J and the number of objects I.

Scores and loadings matrices have the following properties

$$\mathbf{T}^t\mathbf{T} = \Lambda = \mathrm{diag}\{\lambda_1, \ldots, \lambda_A\}, \quad \mathbf{P}^t\mathbf{P} = \mathbf{I}$$

The recurrent algorithm, non-linear iterative partial least squares (NIPALS) (see Section 15.5.3) is often used for calculation of PCA scores and loadings. After constructing the PC space, new objects $\mathbf{X}_{\text{new}}$ can be projected onto this space. In other words, the scores matrix $\mathbf{T}_{\text{new}}$ can be calculated. In PCA, this is easily done by equation

$$\mathbf{T}_{\text{new}} = \mathbf{X}_{\text{new}}\mathbf{P}.$$

Naturally, $\mathbf{X}_{\text{new}}$ matrix should be preprocessed in the same way as the training matrix $\mathbf{X}$, which was used for PCA decomposition

8.1.3 PLS

PLS can be considered as a generalization of PCA. In PLS, the decomposition of matrices $\mathbf{X}$ and $\mathbf{Y}$ is conducted simultaneously

$$\mathbf{X} = \mathbf{T}\mathbf{P}^t + \mathbf{E}, \quad \mathbf{Y} = \mathbf{U}\mathbf{Q}^t + \mathbf{F}, \quad \mathbf{T} = \mathbf{X}\mathbf{W}\left(\mathbf{P}^t\mathbf{W}\right)^{-1} \tag{8.2}$$

A projection is built in order to maximize correlation between corresponding vectors of $\mathbf{X}$-scores $\mathbf{t}_a$ and $\mathbf{Y}$-scores $\mathbf{u}_a$. See Fig. 8.3 for illustration.

If block $\mathbf{Y}$ consists of several responses (i.e., $K > 1$), two types of projections can be built; they are PLS1 and PLS2. In the first case, the projection space is built separately for each response variable $\mathbf{y}_k$. Scores $\mathbf{T}$ ($\mathbf{U}$) and loadings $\mathbf{P}$ ($\mathbf{W}$, $\mathbf{Q}$) depend on the response $\mathbf{y}_k$

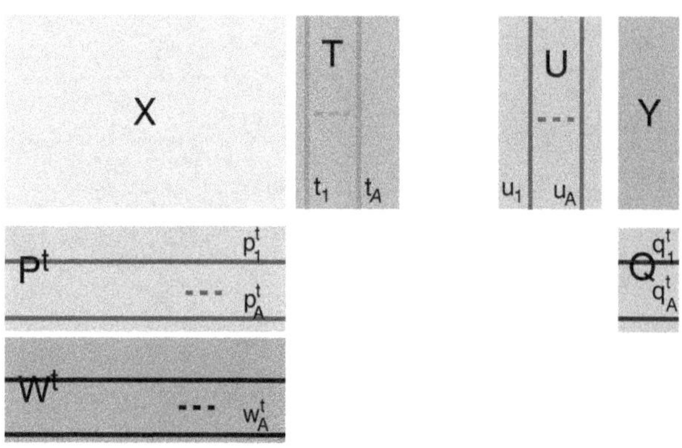

Figure 8.3 PLS2 graphic representation.

in use. Such an approach is called *PLS1*. In the *PLS2* method, one common space is built for all responses. The following expressions show the properties of PLS projection matrices

$$\mathbf{T}^t\mathbf{T} = \Lambda = \mathrm{diag}\{\lambda_1, \ldots, \lambda_A\}, \quad \mathbf{W}^t\mathbf{W} = \mathbf{I}, \quad \mathbf{T} = \mathbf{XR}, \quad \mathbf{R} = \mathbf{W}(\mathbf{P}^t\mathbf{W})^{-1}$$

$$\mathbf{R}^t\mathbf{P} = \mathbf{I}, \quad \mathbf{Y}^{hat} = \mathbf{XB} = \mathbf{TQ}^t, \quad \mathbf{B} = \mathbf{RQ}^t, \quad \mathbf{Q}^t = \mathbf{\Lambda}^{-1}\mathbf{T}^t\mathbf{Y}, \quad \mathbf{W}^t = \mathbf{\Lambda}^{-1}\mathbf{U}^t\mathbf{X}$$

A recurrent algorithm is used for the calculation of PLS scores and loadings. This algorithm calculates one PLS component at a time. It is presented in Section 15.5.4. A similar algorithm for PLS2 is described in Section 15.5.5.

8.1.4 Data Preprocessing

It is worth mentioning that in the course of **X** and **Y** decomposition, the PCA and PLS methods do not take into consideration the free term. This could be seen from Eqs 8.1 and 8.2. It is initially supposed that all columns in matrices **X** и **Y** have zero mean values, that is,

$$\sum_{i=1}^{I} x_{ij} = 0 \quad \text{and} \quad \sum_{i=1}^{I} y_{ik} = 0, \quad \text{for each } j = 1, \ldots, J \text{ and } k = 1, \ldots, K$$

This condition can be easily satisfied by data *centering*.

Centering implied a subtraction of matrix **M** from the original matrix **X**, that is,

$$\widetilde{\mathbf{X}} = \mathbf{X} - \mathbf{M}$$

Centering is a column-wise operation. For each vector $\mathbf{x}_j$, the mean value is calculated by

$$m_j = (x_{1j} + \cdots + x_{Ij})/I.$$

In that case, $\mathbf{M} = (m_1\mathbf{1}, \ldots, m_J\mathbf{1})$, where $\mathbf{1}$ is the $(I \times 1)$ vector of ones. Centering is a compulsory procedure that precedes the application of projection methods. The second simplest preprocessing technique is scaling.

Scaling is not as indispensable as centering. Compared to centering, scaling does not change the data structure but simply modifies the weights of different parts of data. The most widely used scaling is a column-wise one. This can be expressed as a right multiplication of matrix **X** by matrix **W**, that is,

$$\widetilde{\mathbf{X}} = \mathbf{XW}$$

Matrix **W** is the $(J \times J)$ diagonal matrix. As a rule, the diagonal elements w_{jj} are equal to inverse values of the standard deviations, that is,

$$d_j = \sqrt{\sum_{i=1}^{I} (x_{ij} - m_j)^2 / I}$$

calculated for each column $\mathbf{x}_j$. Row-wise scaling (also called *normalization*) is a left multiplication of matrix **X** by a diagonal matrix **W**, that is,

$$\widetilde{\mathbf{X}} = \mathbf{WX}$$

In this formula, $\mathbf{W}$ is the $(I \times I)$ diagonal matrix, and its elements $\mathbf{w}_{ii}$ are ordinarily the inverse values of the standard deviations calculated for each row $\mathbf{x}_i^t$

Combination of centering and column-wise scaling

$$\tilde{x}_{ij} = \left(x_{ij} - m_j\right)/d_j$$

is called *autoscaling*.

Data scaling is often used to make the contribution of various variables to a model more equal (i.e., in hybrid methods like LC–MS), to account for heterogeneous errors, or when different data blocks should be combined in one model. Scaling can also be seen as a method for stabilization of numerical calculations. At the same time, scaling should be used with caution as such preprocessing can essentially change the results of quality analysis. When the raw data structure is a priori assumed homogeneous and homoscedastic, data preprocessing is not only unnecessary but also harmful. This is discussed in Chapter 12 in the example of HPLC–DAD data.

Every type of preprocessing (centering, scaling, etc.) is first applied to the calibration dataset. This set is used to calculate the values of m_j and d_j, which, in turn, are used for preprocessing of both training and test sets later on.

In Chemometrics Add-In, preprocessing is done automatically. If there is a need in any particular preprocessing, it can be done using standard worksheet functions, or special user-defined functions, which are explained in Sections 13.1.2 and 13.2.3.

8.1.5 Didactic Example

File Projection.xls is used to illustrate the Chemometrics Add-In facilities. The performance of all the above-mentioned methods is illustrated with a simulated dataset (**X**, **Y**), which should be centered but not scaled.

File Projection.xls includes the following worksheets:

Intro: short introduction

Data: data used in the example. Block **X** consists of 14 objects (9 in the calibration set and 5 in the test set) and 50 variables. Block **Y** includes two responses, which correspond to the 14 objects. The worksheet also presents a legend that explains the names of the arrays

PCA: application of the worksheet functions **ScoresPCA** and **LoadingsPCA**

PLS1: application of the worksheet functions **ScoresPLS**, **UScoresPLS**, **LoadingsPLS**, **WLoadingsPLS**, and **QLoadingsPLS**

PLS2: application of the worksheet functions **ScoresPLS2**, **UScoresPLS2**, **LoadingsPLS2**, **WLoadingsPLS2**, and **QLoadingsPLS2**

Plus: application of the additional worksheet functions **MIdent**, **MIdentD2**, **MTrace**, and **MCutRows**

Unscrambler: comparison of the results obtained by Chemometrics Add-In and by the Unscrambler program

SIMCA-P: comparison of the results obtained by Chemometrics Add-In and by the SIMCA program

8.2 APPLICATION OF CHEMOMETRICS ADD-IN

8.2.1 Installation

Before starting to work with Chemometrics Add-In, the software should be properly installed. The instructions for Chemometrics Add-In installation are presented in Chapter 3.

8.2.2 General

Chemometrics Add-In uses all input data from an active Excel workbook. This information should be pasted directly in the worksheet area (data **X** and **Y**). A user may organize the working space the way he/she likes, that is, to place all data onto one worksheet, or split them between several worksheets, or even use worksheets from different workbooks. The results are returned as the worksheet arrays. The software does not have any limitations on the size of input arrays, that is, the number of objects (I), the number of variables (J), and the number of responses (K). The dimension is only limited by the supported computer memory and limitations of Excel itself.

Chemometrics Add-In includes user-defined functions, which can be used as standard worksheet functions. For this purpose, a user can employ **Insert Function** dialog box and select the **User Defined** category. All functions, described below can be found in the **Select a function** window.

The returned values are arrays; therefore, the functions should be entered in array formula style, that is, applying **CTRL+SHIFT+ENTER** at the end of the formula input (see Section 7.2.1).

Function arguments can be numbers, names, arrays, or references containing numbers. To insert an array formula, it is necessary to select a range, the size of which corresponds to the expected output array. If a selected range is larger than the output array, the superfluous cells are filled with #N/A symbols. On the contrary, when the selected range is smaller than the output array, a part of the output information is lost.

Number of Principal/PLS Components

Each function has an optional argument `PC` that defines the number of PCs (A). If `PC` is omitted, the output corresponds to the selected area. If PC value is more than $min\ (I, J)$, the decomposition is done for the maximum possible number of PCs and superfluous cells are filled with #N/A symbols.

Centering and/or Scaling

Each function has optional arguments **`CentWeightX`** and **`CentWeightY`**, which define whether centering and/or scaling for **X** and **Y** arrays is performed. These arguments can be as follows:

- **0** – no centering and no scaling (default value)
- **1** – only centering, that is, subtraction of column-wise mean values
- **2** – only scaling by column-wise standard deviations
- **3** – centering and scaling, that is, autoscaling

If any argument **`CentWeightX`** or **`CentWeightY`** is omitted it is assumed to be (equal to) `0`.

8.3 PCA

8.3.1 `ScoresPCA`

`ScoresPCA` performs decomposition of the matrix **X** using the PCA method (Eq. 8.1) and then returns the array of score values $\mathbf{T}_{new}$ calculated for matrix $\mathbf{X}_{new}$.

Syntax

`ScoresPCA` `(X [, PC] [,CentWeightX] [, Xnew])`

`X` is the array of **X**-values (calibration set);

`PC` is an optional argument (integer), which defines the number of PCs (*A*), used in the PCA decomposition;

`CentWeightX` is an optional argument (integer) that indicates whether centering and/or scaling is done;

`Xnew` is an optional argument that presents an array of new values $\mathbf{X}_{new}$ (test set) for which the score values $\mathbf{T}_{new}$ are calculated.

Remarks

- The arrays `Xnew` and `X` must have the same number of columns;
- If argument `Xnew` is omitted, it is assumed to be the same as `X`, and thus the calibration score values **T** are returned;
- The result is an array (matrix) where the number of rows equals the number of rows in array `Xnew`, and the number of columns equals the number of PCs (*A*);
- **`TREND`** is a similar standard worksheet function (Section 7.2.7).

Example

An example is given in worksheet PCA and shown in Fig. 8.4.

`ScoresPCA` is an array function which must be completed by **CTRL+SHIFT+ENTER**.

8.3.2 `LoadingsPCA`

Performs decomposition of matrix **X** using the PCA method (Eq. 8.1) and returns an array of loading values **P**.

Syntax

`LoadingsPCA` `(X [, PC] [, CentWeightX])`

`X` is the array of **X**-values (calibration set);

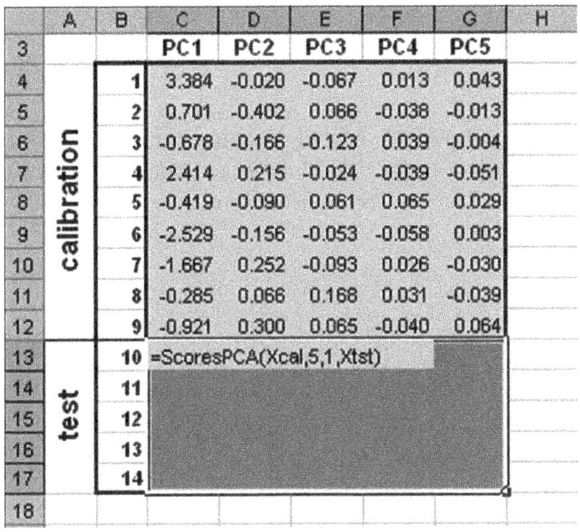

			PC1	PC2	PC3	PC4	PC5
4	calibration	1	3.384	-0.020	-0.067	0.013	0.043
5		2	0.701	-0.402	0.066	-0.038	-0.013
6		3	-0.678	-0.166	-0.123	0.039	-0.004
7		4	2.414	0.215	-0.024	-0.039	-0.051
8		5	-0.419	-0.090	0.061	0.065	0.029
9		6	-2.529	-0.156	-0.053	-0.058	0.003
10		7	-1.667	0.252	-0.093	0.026	-0.030
11		8	-0.285	0.066	0.168	0.031	-0.039
12		9	-0.921	0.300	0.065	-0.040	0.064
13	test	10	=ScoresPCA(Xcal,5,1,Xtst)				
14		11					
15		12					
16		13					
17		14					

Figure 8.4 Example of **`ScoresPCA`** function.

`PC` is an optional argument (integer), which defines the number of PCs (*A*), used in PCA decomposition;

`CentWeightX` is an optional argument (integer) that indicates whether centering and/or scaling is done.

Remarks

- The result is an array (matrix) where the number of rows is equal to the number of columns (*J*) in array **X**, and the number of columns is equal to the number of PCs (*A*);
- **`MINVERSE`** is a similar standard worksheet function (Section 7.2.5).

Example

An example is given in worksheet PCA and shown in Fig. 8.5.

`LoadingsPCA` is an array function, which must be completed by **CTRL+SHIFT+ENTER**.

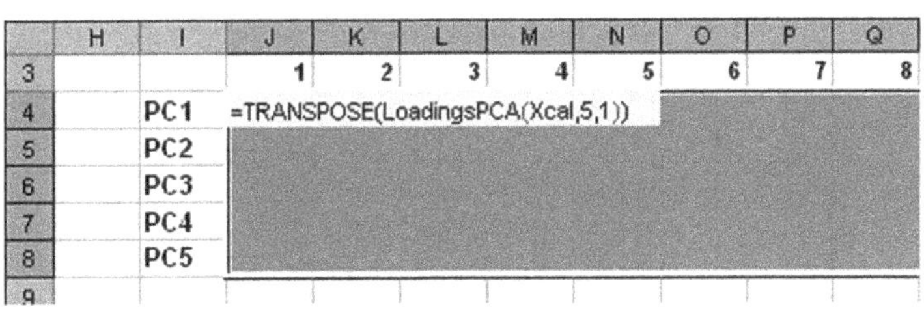

		1	2	3	4	5	6	7	8
4	PC1	=TRANSPOSE(LoadingsPCA(Xcal,5,1))							
5	PC2								
6	PC3								
7	PC4								
8	PC5								

Figure 8.5 Example of **`LoadingsPCA`** function.

8.4 PLS

8.4.1 ScoresPLS

Performs decomposition of matrices **X** and **Y** using the method of PLS (Eq. 8.2) and then returns an array that presents the PLS score values $\mathbf{T}_{new}$ calculated for matrix $\mathbf{X}_{new}$.

Syntax

ScoresPLS (**X**, **Y** [, PC] [, CentWeightX] [, CentWeightY] [, Xnew])

- **X** is the array of **X**-values (calibration set);
- **Y** is the array of **Y**-values (calibration set);
- PC is an optional argument (integer), which defines the number of PLS components (A), used in PLS decomposition;
- CentWeightX is an optional argument (integer) that indicates whether centering and/or scaling for matrix **X** is done;
- CentWeightY is an optional argument (integer) that indicates whether centering and/or scaling for matrix **Y** is done;
- Xnew is an optional argument that presents an array of new values $\mathbf{X}_{new}$ (test set) for which the PLS score values $\mathbf{T}_{new}$ are calculated.

Example

An example is given in worksheet PLS1 and shown in Fig. 8.6.

	A	B	C	D	E	F	G	H
3			PC1	PC2	PC3	PC4	PC5	
4	calibration	1	3.364	0.026	0.065	-0.016	-0.023	
5		2	0.742	0.384	-0.085	-0.042	-0.019	
6		3	-0.652	0.189	0.126	0.057	0.023	
7		4	2.373	-0.213	0.030	0.000	0.045	
8		5	-0.406	0.079	-0.058	0.048	-0.021	
9		6	-2.492	0.167	0.043	-0.043	0.037	
10		7	-1.683	-0.230	0.101	-0.008	-0.058	
11		8	-0.294	-0.095	-0.161	0.041	0.003	
12		9	-0.951	-0.308	-0.061	-0.037	0.013	
13	test	10	=ScoresPLS(Xcal,YcalA,5,1,1,Xtst)					
14		11						
15		12						
16		13						
17		14						
18								

Figure 8.6 Example of **ScoresPLS** function.

Remarks

- Array **Y** must have only one column ($K = 1$);
- Arrays **Y** and **X** must have the same number of rows (I);
- Arrays Xnew and **X** must have the same number of columns (J);
- If argument Xnew is omitted, it is assumed to be the same as **X** and thus the PLS calibration score values **T** are returned;
- The result is an array (matrix) where the number of rows equals the number of rows in array Xnew, and the number of columns equals the number of PCs (A);
- **TREND** is a similar standard worksheet function (Section 7.2.7).

ScoresPLS is an array function, which must be completed by **CTRL+SHIFT+ ENTER**.

8.4.2 UScoresPLS

Performs decomposition of matrices **X** and **Y** using the method of PLS (Eq. 8.2) and then returns an array that presents the PLS score values $\mathbf{U}_{\mathrm{new}}$ calculated for matrices $\mathbf{X}_{\mathrm{new}}$ and $\mathbf{Y}_{\mathrm{new}}$.

Syntax

UScoresPLS (**X**, **Y**, [, PC] [, CentWeightX] [, CentWeightY] [, Xnew] [, Ynew])

X is the array of **X**-values (calibration set);

Y is the array of **Y**-values (calibration set);

PC is an optional argument (integer), which defines the number of PLS components (A), used in PLS decomposition;

CentWeightX is an optional argument (integer) that indicates whether centering and/or scaling for matrix **X** is done;

CentWeightY is an optional argument (integer) that indicates whether centering and/or scaling for matrix **Y** is done;

Xnew is an optional argument that presents an array of new values $\mathbf{X}_{\mathrm{new}}$ (test set) for which PLS score values $\mathbf{U}_{\mathrm{new}}$ are calculated;

Ynew is an optional argument that presents an array of new values $\mathbf{Y}_{\mathrm{new}}$ (test set) for which PLS score values $\mathbf{U}_{\mathrm{new}}$ are calculated.

Remarks

- Arrays **Y** and Ynew must have only one column ($K = 1$);
- Arrays **Y** and **X** must have the same number of rows (I);
- Arrays Xnew and **X** must have the same number of columns (J);

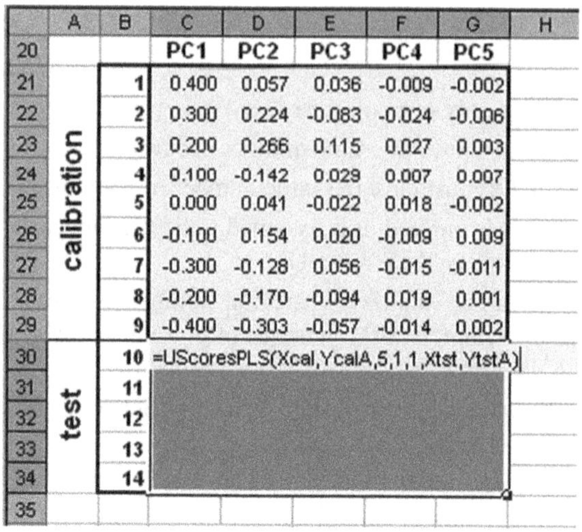

	A	B	C	D	E	F	G	H
20			PC1	PC2	PC3	PC4	PC5	
21	calibration	1	0.400	0.057	0.036	-0.009	-0.002	
22		2	0.300	0.224	-0.083	-0.024	-0.006	
23		3	0.200	0.266	0.115	0.027	0.003	
24		4	0.100	-0.142	0.029	0.007	0.007	
25		5	0.000	0.041	-0.022	0.018	-0.002	
26		6	-0.100	0.154	0.020	-0.009	0.009	
27		7	-0.300	-0.128	0.056	-0.015	-0.011	
28		8	-0.200	-0.170	-0.094	0.019	0.001	
29		9	-0.400	-0.303	-0.057	-0.014	0.002	
30	test	10	=UScoresPLS(Xcal,YcalA,5,1,1,Xtst,YtstA)					
31		11						
32		12						
33		13						
34		14						
35								

Figure 8.7 Example of **`UScoresPLS`** function.

- If argument `Xnew` is omitted, it is assumed to be the same as **`X`** and thus the PLS calibration score values **U** are returned;
- If argument `Ynew` is omitted, it is assumed to be the same as **`Y`**, and thus the PLS calibration score values **U** are returned;
- The result is an array (matrix) where the number of rows equals the number of rows in array `Xnew`, and the number of columns equals the number of PCs (*A*);
- **`TREND`** is a similar standard worksheet function (Section 7.2.7).

Example

An example is given in worksheet PLS1 and shown in Fig. 8.7.

`UScoresPLS` is an array function, which must be completed by **CTRL+SHIFT+ENTER**.

8.4.3 `LoadingsPLS`

Performs decomposition of matrices **X** and **Y** using the method of PLS (Eq. 8.2) and then returns an array that presents the loadings values **P**.

Syntax

`LoadingsPLS` `(`**`X`**`,` **`Y`** `[, PC] [, CentWeightX] [, CentWeightY])`

`X` is the array of **X**-values (calibration set);

`Y` is the array of **Y**-values (calibration set);

`PC` is an optional argument (integer), which defines the number of PLS components (A), used in the PLS decomposition;

`CentWeightX` is an optional argument (integer) that indicates whether centering and/or scaling for matrix **X** is done;

`CentWeightY` is an optional argument (integer) that indicates whether centering and/or scaling for matrix **Y** is done.

Remarks

- Array **`Y`** must have only one column ($K = 1$);
- Arrays **`Y`** and **`X`** must have the same number of rows (I);
- The result is an array (matrix) where the number of rows equals the number of columns (J) in array **`X`**, and the number of columns equals the number of PCs (A);
- **`MMULT`** is a similar standard worksheet function (Section 7.2.6).

Example

An example is given in worksheet PLS1 and shown in Fig. 8.8.

`LoadingsPLS` is an array function, which must be completed by **CTRL+SHIFT+ ENTER**.

8.4.4 `WLoadingsPLS`

Performs decomposition of matrices **X** and **Y** using the method of PLS (Eq. 8.2) and then returns an array that presents the loading weights values **W**.

Syntax

`WLoadingsPLS (X, Y` `[, PC] [, CentWeightX] [, CentWeightY])`

`X` is the array of **X**-values (calibration set);

`Y` is the array of **Y**-values (calibration set);

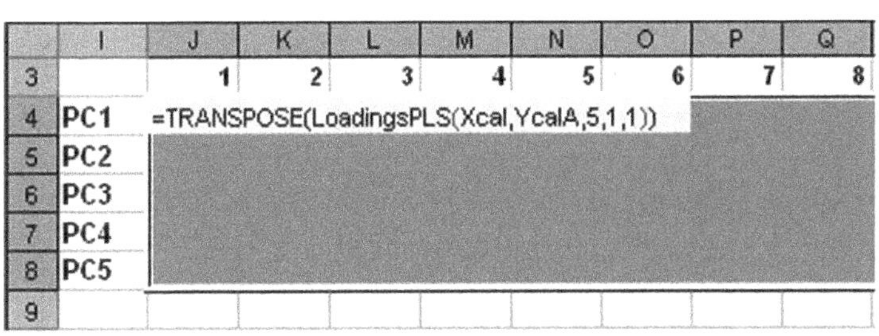

Figure 8.8 Example of **`LoadingsPLS`** function.

`PC` is an optional argument (integer), which defines the number of PLS components (A), used in the PLS decomposition;

`CentWeightX` is an optional argument (integer) that indicates whether centering and/or scaling for matrix **X** is done;

`CentWeightY` is an optional argument (integer) that indicates whether centering and/or scaling for matrix **Y** is done.

Remarks

- Array **`Y`** must have only one column ($K = 1$);
- Arrays **`Y`** and **`X`** must have the same number of rows (I);
- The result is an array (matrix) where the number of rows equals the number of columns (J) in array **`X`** and the number of columns equals the number of PCs (A);
- **`MMULT`** is a similar standard worksheet function (Section 7.2.6).

Example

An example is given in worksheet PLS1 and shown in Fig. 8.9.

`WLoadingsPLS` is an array function, which must be completed by **CTRL+SHIFT+ENTER**.

8.4.5 `QLoadingsPLS`

Performs decomposition of matrices **X** and **Y** using the method of PLS (Eq. 8.2) and then returns an array that presents the loadings values **Q**.

Syntax

`QLoadingsPLS` `(`**`X`**`,` **`Y`** `[, PC] [, CentWeightX] [, CentWeightY])`

`X` is the array of **X**-values (calibration set);

`Y` is the array of **Y**-values (calibration set);

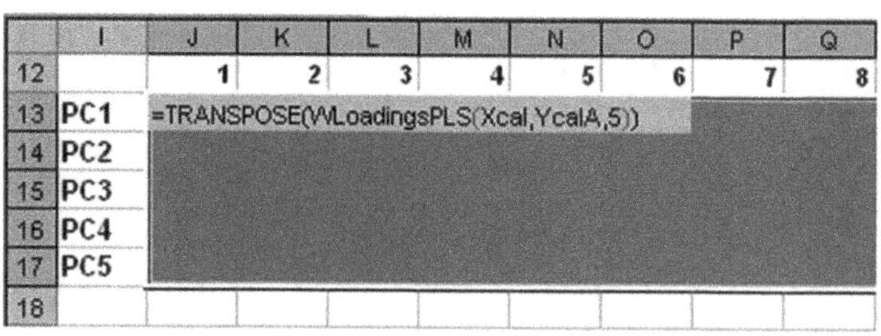

Figure 8.9 Example of **`WLoadingsPLS`** function.

`PC` is an optional argument (integer), which defines the number of PLS components (A), used in the PLS decomposition;

`CentWeightX` is an optional argument (integer) that indicates whether centering and/or scaling for matrix **X** is done;

`CentWeightY` is an optional argument (integer) that indicates whether centering and/or scaling for matrix **Y** is done.

Remarks

- Array **`Y`** must have only one column ($K = 1$);
- Arrays **`Y`** and **`X`** must have the same number of rows (I);
- The result is an array (vector) where the number of columns equals the number of PCs (A) and the number of rows equals the number of responses K, that is, 1;
- **`MMULT`** is a similar standard worksheet function (Section 7.2.6).

Example

An example is given in worksheet PLS1 and shown in Fig. 8.10.

`QLoadingsPLS` is an array function, which must be completed by **CTRL+SHIFT+ENTER**.

8.5 PLS2

8.5.1 `ScoresPLS2`

Performs decomposition of matrices **X** and **Y** using the method of PLS2 (Eq. 8.2) and then returns an array that presents the PLS2 score values $\mathbf{T}_{new}$ calculated for matrix $\mathbf{X}_{new}$.

Syntax

`ScoresPLS2``(`**`X`**`, `**`Y`**` [, PC] [, CentWeightX] [, CentWeightY] [, Xnew] )`

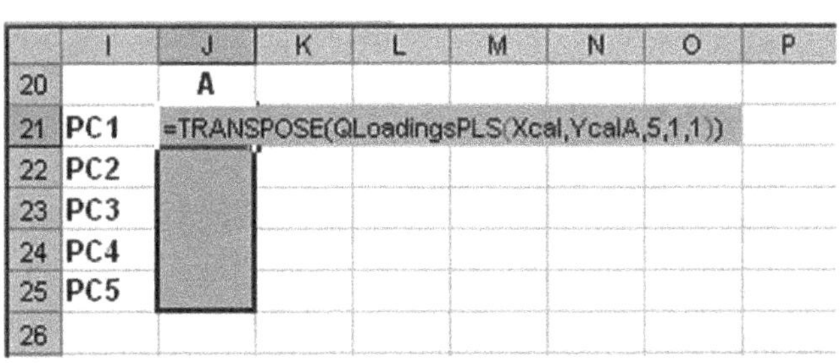

Figure 8.10 Example of **`QLoadingsPLS`** function.

X is the array of **X**-values (calibration set);

Y is the array of **Y**-values (calibration set);

`PC` is an optional argument (integer), which defines the number of PLS components (A), used in the PLS2 decomposition;

`CentWeightX` is an optional argument (integer) that indicates whether centering and/or scaling for matrix **X** is done;

`CentWeightY` is an optional argument (integer) that indicates whether centering and/or scaling for matrix **Y** is done;

`Xnew` is an optional argument that presents an array of new values $\mathbf{X}_{new}$ (test set) for which PLS2 score values $\mathbf{T}_{new}$ are calculated;

`Ynew` is an optional argument that presents an array of new values $\mathbf{Y}_{new}$ (test set) for which PLS2 score values $\mathbf{T}_{new}$ are calculated.

Remarks

- Arrays **Y** and **X** must have the same number of rows (I);
- Arrays `Xnew` and **X** must have the same number of columns (J);
- If argument `Xnew` is omitted, it is assumed to be the same as **X**, and thus the calibration PLS2 score values **T** are returned;
- The result is an array (matrix) where the number of rows equals the number of rows in array `Xnew` and the number of columns equals the number of PCs (A);
- **TREND** is a similar standard worksheet function (Section 7.2.7).

Example

An example is given in worksheet PLS2 and shown in Fig. 8.11.

ScoresPLS2 is an array function, which must be completed by **CTRL+SHIFT+ENTER**.

8.5.2 UScoresPLS2

Performs decomposition of matrices **X** and **Y** using the method of PLS2 (Eq. 8.2) and then returns an array that presents the PLS score values $\mathbf{U}_{new}$ calculated for matrices $\mathbf{X}_{new}$ and $\mathbf{Y}_{new}$.

Syntax

```
UScoresPLS2 (X, Y, [, PC] [, CentWeightX] [, CentWeightY]
[, Xnew] [, Ynew])
```

X is the array of **X**-values (calibration set);

Y is the array of **Y**-values (calibration set);

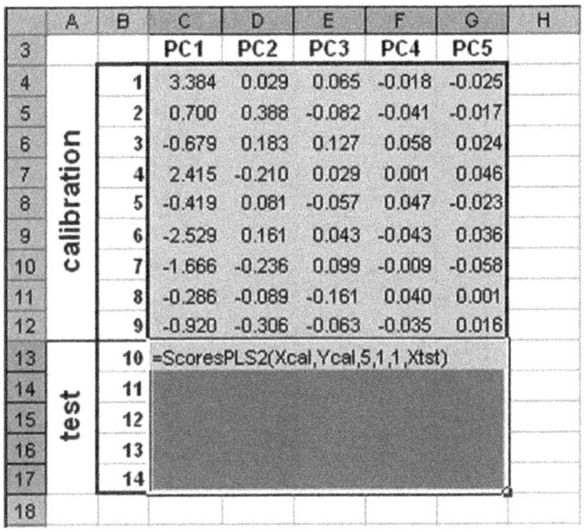

	A	B	C	D	E	F	G	H
3			PC1	PC2	PC3	PC4	PC5	
4	calibration	1	3.384	0.029	0.065	-0.018	-0.025	
5		2	0.700	0.388	-0.082	-0.041	-0.017	
6		3	-0.679	0.183	0.127	0.058	0.024	
7		4	2.415	-0.210	0.029	0.001	0.046	
8		5	-0.419	0.081	-0.057	0.047	-0.023	
9		6	-2.529	0.161	0.043	-0.043	0.036	
10		7	-1.666	-0.236	0.099	-0.009	-0.058	
11		8	-0.286	-0.089	-0.161	0.040	0.001	
12		9	-0.920	-0.306	-0.063	-0.035	0.016	
13	test	10	=ScoresPLS2(Xcal,Ycal,5,1,1,Xtst)					
14		11						
15		12						
16		13						
17		14						
18								

Figure 8.11 Example of **`ScoresPLS2`** function.

`PC` is an optional argument (integer), which defines the number of PLS components (A), used in the PLS2 decomposition;

`CentWeightX` is an optional argument (integer) that indicates whether centering and/or scaling for matrix **X** is done;

`CentWeightY` is an optional argument (integer) that indicates whether centering and/or scaling for matrix **Y** is done;

`Xnew` is an optional argument that presents an array of new values $\mathbf{X}_{new}$ (test set) for which PLS2 score values $\mathbf{U}_{new}$ are calculated;

`Ynew` is an optional argument that presents an array of new values $\mathbf{Y}_{new}$ (test set) for which PLS2 score values $\mathbf{U}_{new}$ are calculated.

Example

An example is given in worksheet PLS2 and shown in Fig. 8.12.

Remarks

- Arrays **`Y`** and `Ynew` must have the same number of columns (K);
- Arrays **`Y`** and **`X`** must have the same number of rows (I);
- Arrays `Xnew` and **`X`** must have the same number of columns (J);
- If argument `Xnew` is omitted, it is assumed to be the same as **`X`**, and thus the calibration PLS2 score values **U** are returned;
- If argument `Ynew` is omitted, it is assumed to be the same as **`Y`**, and thus the PLS2 calibration score values **U** are returned;
- The result is an array (matrix) where the number of rows equals the number of rows in array `Xnew` and the number of columns equals the number of PCs (A);
- **`TREND`** is a similar standard worksheet function (Section 7.2.7).

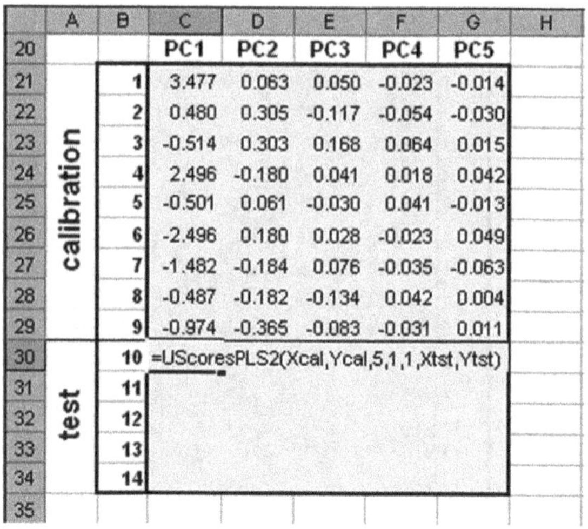

	A	B	C	D	E	F	G	H
20			PC1	PC2	PC3	PC4	PC5	
21	calibration	1	3.477	0.063	0.050	-0.023	-0.014	
22		2	0.480	0.305	-0.117	-0.054	-0.030	
23		3	-0.514	0.303	0.168	0.064	0.015	
24		4	2.496	-0.180	0.041	0.018	0.042	
25		5	-0.501	0.061	-0.030	0.041	-0.013	
26		6	-2.496	0.180	0.028	-0.023	0.049	
27		7	-1.482	-0.184	0.076	-0.035	-0.063	
28		8	-0.487	-0.182	-0.134	0.042	0.004	
29		9	-0.974	-0.365	-0.083	-0.031	0.011	
30	test	10	=UScoresPLS2(Xcal,Ycal,5,1,1,Xtst,Ytst)					
31		11						
32		12						
33		13						
34		14						
35								

Figure 8.12 Example of **`UScoresPLS2`** function.

`UScoresPLS2` is an array function, which must be completed by **CTRL+SHIFT+ENTER**.

8.5.3 `LoadingsPLS2`

Performs decomposition of matrices **X** and **Y** using the method of PLS2 (Eq. 8.2) and then returns an array that presents the loadings values **P**.

Syntax

`LoadingsPLS2` `(X, Y [, PC] [, CentWeightX] [, CentWeightY])`

`X` is the array of **X**-values (calibration set);

`Y` is the array of **Y**-values (calibration set);

`PC` is an optional argument (integer), which defines the number of PLS components (A), used in the PLS2 decomposition;

`CentWeightX` is an optional argument (integer) that indicates whether centering and/or scaling for matrix **X** is done;

`CentWeightY` is an optional argument (integer) that indicates whether centering and/or scaling for matrix **Y** is done.

Remarks

- Arrays **`Y`** and **`X`** must have the same number of rows (I);

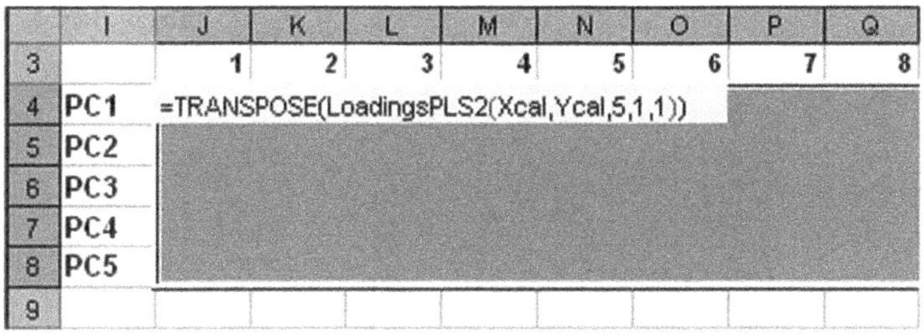

Figure 8.13 Example of `LoadingsPLS2` function.

- The result is an array (matrix) where the number of rows equals the number of columns (J) in array `X` and the number of columns equals the number of PCs (A);
- `MMULT` is a similar standard worksheet function (Section 7.2.6).

Example

An example is given in worksheet PLS2 and shown in Fig. 8.13.

`LoadingsPLS2` is an array function, which must be completed by **CTRL+SHIFT+ENTER**.

8.5.4 `WLoadingsPLS2`

Performs decomposition of matrices **X** and **Y** by using the method PLS2 (Eq. 8.2) and returns the loadings weights values **W**.

Syntax

`WLoadingsPLS2 (X, Y [, PC] [, CentWeightX] [, CentWeightY])`

`X` is the array of **X**-values (calibration set);

`Y` is the array of **Y**-values (calibration set);

`PC` is an optional argument (integer), which defines the number of PLS components (A), used in the PLS2 decomposition;

`CentWeightX` is an optional argument (integer) that indicates whether centering and/or scaling for matrix **X** is done;

`CentWeightY` is an optional argument (integer) that indicates whether centering and/or scaling for matrix **Y** is done.

Remarks

- Arrays `Y` and `X` must have the same number of rows (I);
- The result is an array (matrix) where the number of rows equals the number of columns (J) in array `X` and the number of columns equals the number of PCs (A);

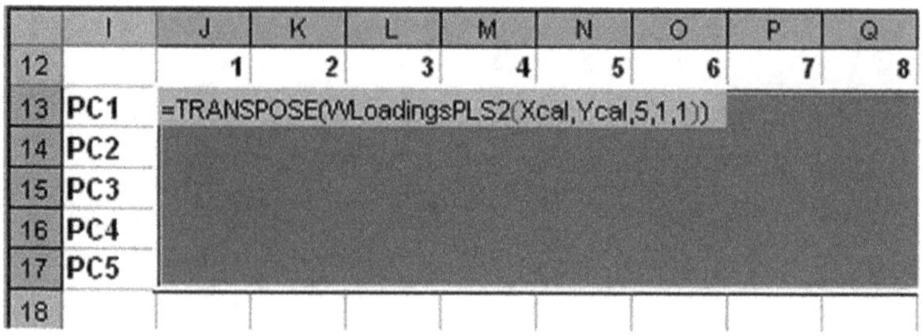

Figure 8.14 Example of **`WLoadingsPLS2`** function.

- **`MMULT`** is a similar standard worksheet function (Section 7.2.6).

Example

An example is given in worksheet PLS2 and shown in Fig. 8.14.

`WLoadingsPLS2` is an array function, which must be completed by **CTRL+SHIFT+ENTER**.

8.5.5 `QLoadingsPLS2`

Performs decomposition of matrices **X** and **Y** using the method of PLS2 (Eq. 8.2) and then returns the loadings values **Q**.

Syntax

`QLoadingsPLS2` `(X, Y [, PC] [, CentWeightX] [, CentWeightY])`

`X` is the array of **X**-values (calibration set);

`Y` is the array of **Y**-values (calibration set);

`PC` is an optional argument (integer), which defines the number of PLS components (A), used in the PLS2 decomposition;

`CentWeightX` is an optional argument (integer) that indicates whether centering and/or scaling of matrix **X** is done;

`CentWeightY` is an optional argument (integer) that indicates whether centering and/or scaling of matrix **Y** is done.

Remarks

- Arrays `Y` and `X` must have the same number of rows (I);
- The result is an array (matrix) where the number of columns equals the number of PCs (A) and the number of rows equals the number of columns in array `Y`, that is, the number of responses K;
- **`MMULT`** is a similar standard worksheet function (Section 7.2.6).

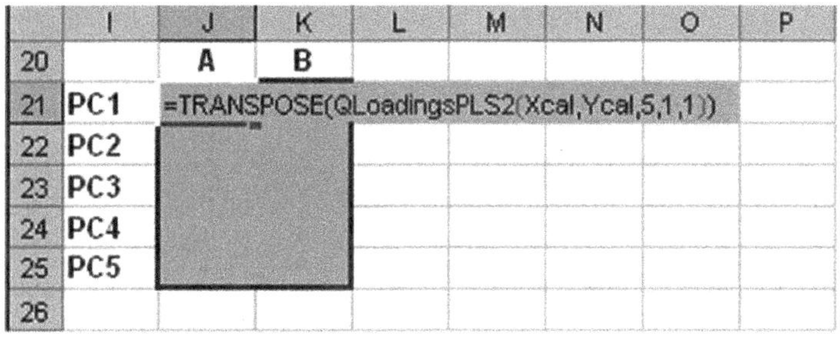

Figure 8.15 Example of function `QLoadingsPLS2`.

Example

An example is given in worksheet PLS2 and shown in Fig. 8.15.

`QLoadingsPLS2` is an array function, which must be completed by **CTRL+SHIFT+ENTER**.

8.6 ADDITIONAL FUNCTIONS

8.6.1 `MIdent`

Returns a square identity matrix

Syntax

`MIdent(Size)`

`Size` is an integer, which defines matrix dimension.

Remarks

- The result is a square array (matrix) with the number of rows and columns equal to `Size`;
- No similar standard worksheet function can be found.

Example

An example is given in worksheet Plus and shown in Fig. 8.16.

`MIdent` is an array function, which must be completed by **CTRL+SHIFT+ENTER**.

8.6.2 `MIdentD2`

Returns a two-diagonal rectangular matrix of zeros and ones, which is a result of cutting a row-block from an identity matrix.

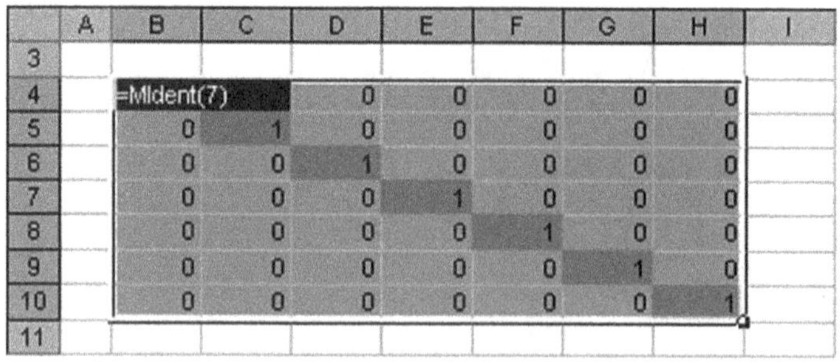

Figure 8.16 Example of **`MIdent`** function.

Syntax

`MIdentD2``(`**`Size`**`, `**`CutFrom`**`, `**`CutOff`**`)`

`Size` is an integer (**`Size`** > 1), which defines the dimension of the initial identity matrix;

`CutFrom` is an integer (**`CutFrom`** > 0), which defines the row number where the cut starts;

`CutOff` is an integer (**`CutOff`** $\geq$ 0), which defines the number of rows to be deleted from the initial identity matrix.

Remarks

- The result is a rectangular array (matrix) where the number of rows equals **`Size-CutOff`** and the number of columns equals **`Size`**;
- When (**`CutFrom`** + **`CutOff`** − 1) > **`Size`**, the function returns #VALUE!;
- No similar standard worksheet function can be found.

Example

An example is given in worksheet Plus and shown in Fig. 8.17.

`MIdentD2` is an array function, which must be completed by **CTRL+SHIFT+ ENTER**.

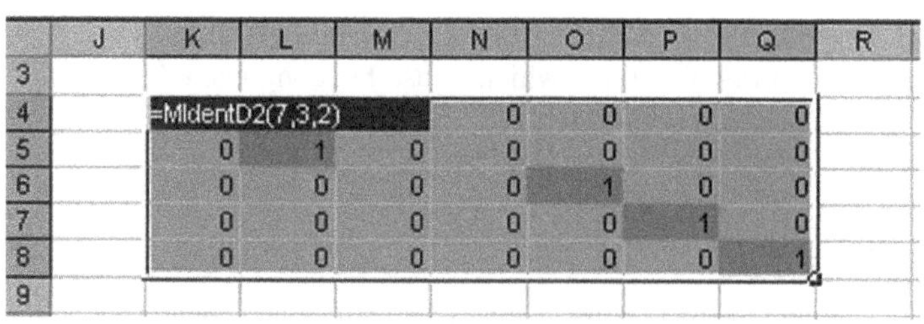

Figure 8.17 Example of **`MIdentD2`** function.

8.6.3 `MCutRows`

Returns a matrix with removed rows.

Syntax
`MCutRows(X, CutFrom, CutOff)`

`X` is an array (matrix) to be cut;

`CutFrom` is an integer (**`CutFrom`** > 0), which defines the row number where cut starts;

`CutOff` is integer (**`CutOff`** ≥ 0), which defines the number of rows that should be deleted from array **`X`**.

Remarks

- The function calculates the initial dimension of array **`X`**: `nRows` and `nColumns`. The result is a rectangular array (matrix) where the number of rows equals `nRows`−**`CutOff`** and the number of columns equals `nColumns`;
- When (**`CutFrom`** + **`CutOff`** − 1) > `nRows`, the function returns #VALUE!;
- **`OFFSET`** is a similar standard worksheet function (Section 7.2.4).

Example

An example is given in worksheet Plus and shown in Fig. 8.18.

`MCutRows` is an array function, which must be completed by **CTRL+SHIFT+ENTER**.

8.6.4 `MTrace`

Returns the trace of a matrix (array).

Syntax
`MTrace(X)`

	A	B	C	D	E	F	G	H	I	J	K
21		1	2	3	4	5	6	7	8	9	10
22	1	=MCutRows(Xcal,7,1)			0.63	0.67	0.67	0.71	0.76	0.82	0.84
23	2	0.31	0.34	0.34	0.37	0.43	0.46	0.48	0.47	0.53	0.53
24	3	0.28	0.27	0.31	0.29	0.32	0.35	0.36	0.36	0.40	0.41
25	4	0.42	0.46	0.48	0.54	0.55	0.57	0.63	0.66	0.67	0.71
26	5	0.28	0.29	0.30	0.29	0.29	0.34	0.32	0.35	0.37	0.39
27	6	0.13	0.13	0.13	0.15	0.15	0.18	0.16	0.12	0.15	0.17
28	8	0.29	0.26	0.30	0.29	0.30	0.32	0.32	0.32	0.35	0.35
29	9	0.23	0.25	0.23	0.24	0.25	0.29	0.24	0.28	0.27	0.28
30											

Figure 8.18 Example of **`MCutRows`** function.

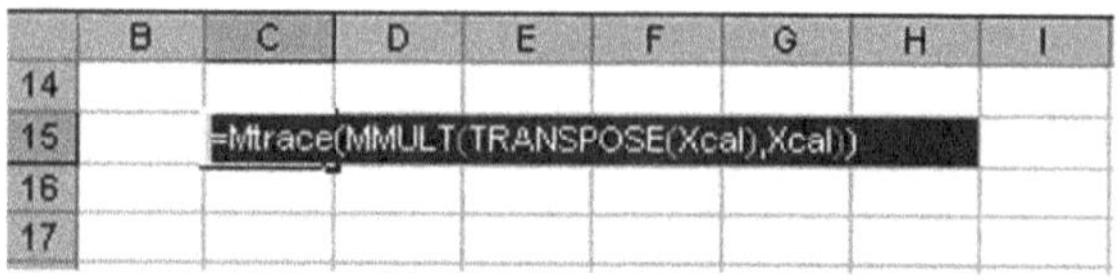

Figure 8.19 Example of **`MTrace`** function.

`X` is a square array (matrix).

Remarks

- **`X`** must be a square matrix;
- **`MDETERM`** is a similar standard worksheet function (Section 7.2.5).

Example

An example is given in worksheet Plus and shown in Fig. 8.19.
`MTrace` is not as an array formula and can be completed by **ENTER**.

CONCLUSION

Chemometrics Add-In provides many possibilities to calculate various scores and loadings for projection methods using original data. Furthermore, employing these results, a typical graphic representation used in multivariate data analysis can be obtained, among other score plots and loading plots. Such an approach also provides possibilities for the construction of calibration models, calculation of root mean square error of calibration (RMSEC), and root mean square error of prediction (RMSEP), etc.

PART III

CHEMOMETRICS

9

PRINCIPAL COMPONENT ANALYSIS (PCA)

This chapter describes the principal component analysis (PCA), which is a basic approach used in chemometrics to solve various problems.

Preceding chapters	5–8
Dependent chapters	10–13, 15
Matrix skills	Basic
Statistical skills	Basic
Excel skills	Basic
Chemometric skills	Low
Chemometrics Add-In	Yes
Accompanying workbook	People.xls

9.1 THE BASICS

9.1.1 Data

PCA is applied to data collected in a matrix, which is a rectangular numerical table of dimension of I rows by J columns. An example is given in Fig. 9.1. Matrices are considered in more detail in Chapter 5.

Traditionally, the rows of such a matrix stand for samples. They are indexed by i, which runs from 1 to I. The columns represent variables and they are indexed by $j = 1, \dots, J$.

PCA extracts relevant information from the data. What can be considered as information depends on the nature of a problem being solved. Sometimes the data contain the information we need, and, in some cases, the information can be absent. The data dimension, that is, the number of samples and variables, is of great importance for the successful extraction

Chemometrics in Excel, First Edition. Alexey L. Pomerantsev.

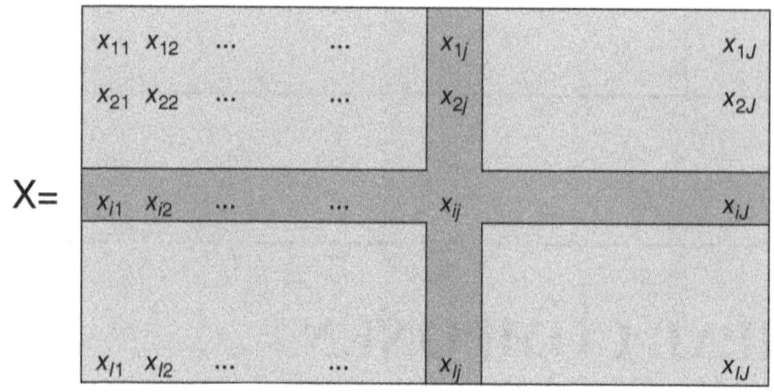

Figure 9.1 Data matrix.

of information. As a rule, there are no redundant data; it is better to have too many values than too few. In practice, this means that after a sample spectrum is obtained, one should use all data, or, at least, a significant part thereof. It would be a mistake to discard all readings except a few characteristic wavelengths.

Data always (or almost always) contain an unwanted component called *noise*. The nature of noise varies, but in many cases, noise is the part of the data that does not contain relevant information. What part of data should be considered as noise and what part as information always depend on the goals of the investigation and methods used to achieve them. The noise and the redundancy in the data always manifest themselves via correlations between the variables. The noise (data errors) leads to nonsystematic, random relations between the variables. The notion of the effective (chemical) rank and the hidden, aka latent, variables, which number is equal to this rank, is the most important concept of PCA.

9.1.2 Intuitive Approach

We try to explain the essence of the principal component (PC) method using an intuitive geometric interpretation. Let us start with the simplest case when the data consist of only two variables x_1 and x_2 (i.e., $J = 2$). Such data can be easily represented in a two-dimensional plane as shown in Fig. 9.2.

Each row of the data table (i.e., each sample) is plotted using the corresponding coordinates as an empty dot on the chart as shown in Fig. 9.2. Let us draw a line through these points in such a way that maximum data variation is observed along the line. This line is called the *first principal component* (PC1). In Fig. 9.2, this line is marked by the label "PC1." Let us project the original data points onto this axis. The resulting values are presented by solid dots. We can assume that all original data points should, in fact, be located on this new axis and that they deviated from their true position as a result of intervention of an unknown force. By projecting them onto the PC1 axis, we simply returned them back to their proper place. In this case, all deviations from the new axis can be considered as noise, that is, irrelevant information. However, we have to be sure about the interpretation of this noise and verify that the residuals are indeed noise. Such validation can be done in the same way as in case with the original data, that is, by finding an axis of the maximum residuals variation. This axis is called the *second principal component* (PC2). We should continue in this manner until the residuals become true noise, that is, a random chaotic set of values.

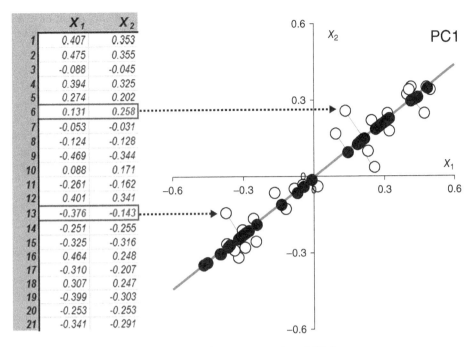

	X_1	X_2
1	0.407	0.353
2	0.475	0.355
3	-0.088	-0.045
4	0.394	0.325
5	0.274	0.202
6	0.131	0.258
7	-0.053	-0.031
8	-0.124	-0.128
9	-0.469	-0.344
10	0.088	0.171
11	-0.261	-0.162
12	0.401	0.341
13	-0.376	-0.143
14	-0.251	-0.255
15	-0.325	-0.316
16	0.464	0.248
17	-0.310	-0.207
18	0.307	0.247
19	-0.399	-0.303
20	-0.253	-0.253
21	-0.341	-0.291

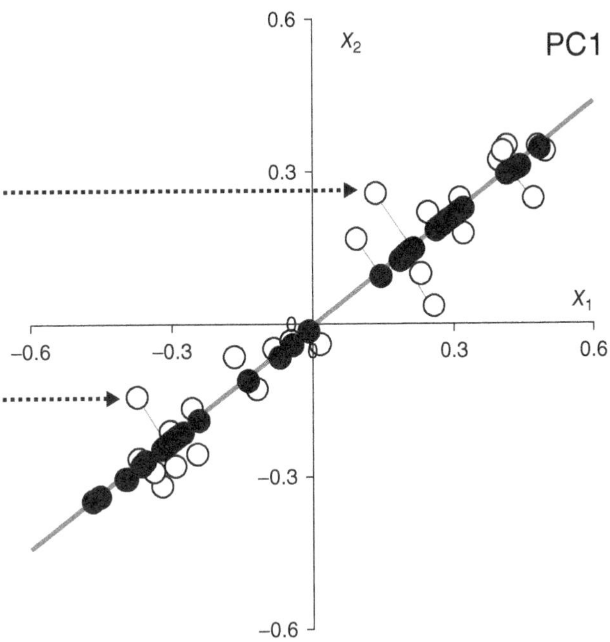

Figure 9.2 Graphical representation of 2D data.

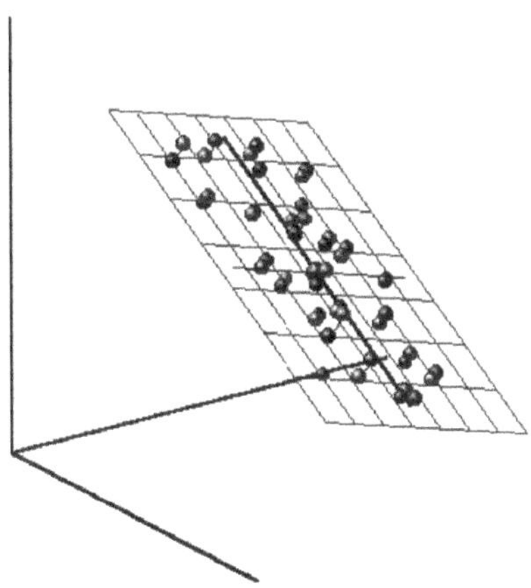

Figure 9.3 Graphical representation of the principal components.

In a multivariate case (see Fig. 9.3), the process of the PCs selection is as follows:

1. The center of the data cloud is found and the coordinates' origin is transferred toward this point. This is the zero principal component (PC0).
2. The direction of the maximum data variation is sought. This is the PC1.

3. If the data are not fully described (the noise part is large), another direction (PC2) orthogonal to the first PC axis is determined in order to describe the maximum variation in the residuals.
4. Etc.

As a result, the data are represented in a new space the dimension of which is much lower than the original number of variables. Often, the dataset is materially simplified, for example, a thousand of variables is reduced to merely two or three PCs. Moreover, nothing essential is discarded; all variables are still accounted for. At the same time, an insignificant part of the data is separated and treated as noise. The PCs are the desired latent variables that control the data.

9.1.3 Dimensionality Reduction

The essence of PCA is a significant decrease of the data dimensionality. The original matrix **X** is replaced by two matrices **T** and **P**, whose common dimension, A, is smaller than the number of variables (columns) J in the matrix **X**. An illustration of this is presented in Fig. 9.4.

The second dimension of **T**, that is, the number of samples (rows) I, is preserved. If the decomposition is performed correctly, that is, the dimensionality A is properly determined, the matrix **T** will include the same information that was contained in the initial matrix **X**. Yet, the matrix **T** is smaller and therefore less complex than **X**.

9.2 PRINCIPAL COMPONENT ANALYSIS

9.2.1 Formal Specifications

Let **X** be the $(I \times J)$ matrix, where I is the number of rows (samples) and J is the number of columns (variables), which, as a rule, are numerous ($J >> 1$). The method of PCs uses new formal variables $\mathbf{t}_a (a = 1, \ldots, A)$, which are linear combinations of the original variables $\mathbf{x}_j$, $(j = 1, \ldots, J)$, that is,

$$\mathbf{t}_a = \mathbf{p}_{a1}\mathbf{x}_1 + \cdots + \mathbf{p}_{aJ}\mathbf{x}_J. \tag{9.1}$$

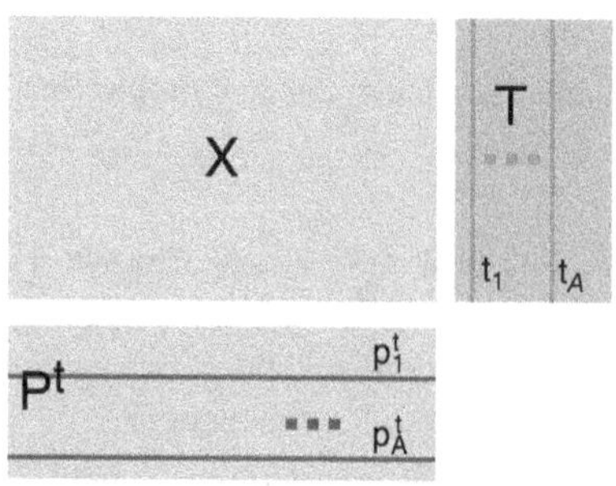

Figure 9.4 Decomposition of matrix **X**.

Using these new variables, matrix **X** is decomposed into a product of two matrices **T** and **P**,

$$\mathbf{X} = \mathbf{T}\mathbf{P}^{\mathrm{t}} + \mathbf{E} = \sum_{a=1}^{A} \mathbf{t}_a \mathbf{p}_a^{\mathrm{t}} + \mathbf{E}. \tag{9.2}$$

Matrix **T** is called the *scores* matrix. It is of the dimension $(I \times A)$. Matrix **P** is called the *loadings* matrix. It is of the dimension $(J \times A)$. **E** is the *residuals* matrix of the dimension $(I \times J)$.

New variables $\mathbf{t}_a$ are called *PCs*; therefore, the method is known as the *PCA*. The number of columns $\mathbf{t}_a$ in matrix **T** and the number of columns $\mathbf{p}_a$ in matrix **P** are equal to A, which is called the *number of PCs*. This value is certainly smaller than the number of variables J or the number of samples I.

An important property of PCA is the *orthogonality* (independence) of the PCs. This means that the scores matrix **T** is not rebuilt when the number of components increases. Matrix **T** is only augmented by one more column that corresponds to the new PC direction. The same happens with the loadings matrix **P**. See Fig. 9.5 for illustration.

9.2.2 Algorithm

Oftentimes, the PCA scores and loadings are obtained using a recursive algorithm nonlinear iterative partial least squares (NIPALS), which is presented in Section 15.5.3. In this book, PCA is performed in Excel using a special Chemometrics Add-In, which extends the list of the standard Excel functions and gives a possibility to conduct the PCA decomposition in the Excel worksheets. A detailed description of the add-in can be found in Chapter 8.

Once the PCs' space is constructed, new samples $\mathbf{X}_{\text{new}}$ can be projected onto the space, or, in other words, their scores matrix $\mathbf{T}_{\text{new}}$ can be found using the formula

$$\mathbf{T}_{\text{new}} = \mathbf{X}_{\text{new}}\mathbf{P}. \tag{9.3}$$

9.2.3 PCA and SVD

PCA is closely related to another decomposition method, being the singular value decomposition (SVD), which is presented in Section 5.2.11. In the latter case, the original matrix **X** is decomposed into a product of three matrices

$$\mathbf{X} = \mathbf{U}\mathbf{S}\mathbf{V}^{\mathrm{t}}$$

Here, **U** is the matrix composed of orthonormal eigenvectors $\mathbf{u}_r$ of matrix $\mathbf{X}\mathbf{X}^{\mathrm{t}}$

$$\mathbf{X}\mathbf{X}^{\mathrm{t}}\mathbf{u}_r = \lambda_r \mathbf{u}_r;$$

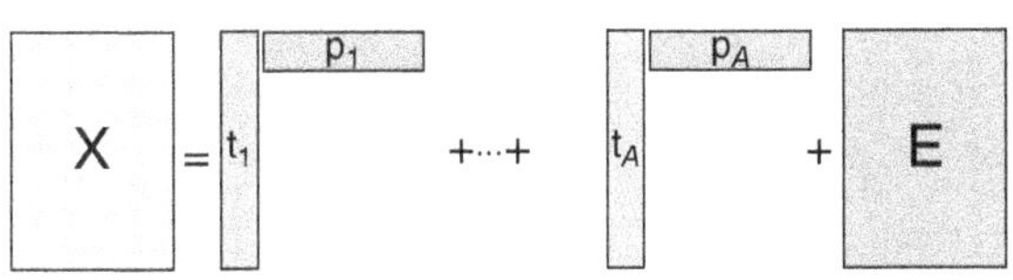

Figure 9.5 Principal components decomposition.

V is the matrix composed of the orthonormal eigenvectors $\mathbf{v}_r$ of matrix $\mathbf{X}^t\mathbf{X}$

$$\mathbf{X}^t\mathbf{X}\mathbf{v}_r = \lambda_r \mathbf{v}_r;$$

S is the positive definite diagonal matrix, whose elements are the singular values $\sigma_1 \geq \ldots \geq \sigma_R \geq 0$, which are equal to the square root of the eigenvalues λ_r

$$\sigma_r = \sqrt{\lambda_r}.$$

PCA and SVD methods are connected by the following simple equations

$$\mathbf{T} = \mathbf{US}, \quad \mathbf{P} = \mathbf{V}.$$

9.2.4 Scores

The scores matrix **T** contains the projections of the samples (the J-dimensional row vectors $\mathbf{x}_1, \ldots, \mathbf{x}_I$) onto the A-dimensional PCs' subspace. Rows $\mathbf{t}_1, \ldots, \mathbf{t}_I$ of matrix **T** are the samples' coordinates in the new coordinate system. Columns $\mathbf{t}_1, \ldots, \mathbf{t}_A$ of matrix **T** are orthogonal. Each column $\mathbf{t}_a$ holds the projections of all samples onto the a-th axis of the new coordinates.

The scores plot is the main tool of the PC analysis. The scores represent information that is useful for understanding the data structure. In this plot, each sample is shown in the $(\mathbf{t}_a, \mathbf{t}_b)$ coordinates plane. Very often the $(\mathbf{t}_1, \mathbf{t}_2)$ plane is used, whose axes are denoted by PC1 and PC2, respectively. An example is presented in Fig. 9.6. If two score points are located close to each other, the related samples are similar, that is, they are positively correlated. The samples positioned at right angles to the coordinates' origin are uncorrelated. Finally, the diametrically opposite samples have a negative correlation.

The example below will give more color on the scores plot analysis.

The following relation holds for the matrix of scores

$$\mathbf{T}^t\mathbf{T} = \Lambda = \text{diag}\{\lambda_1, \ldots, \lambda_A\},$$

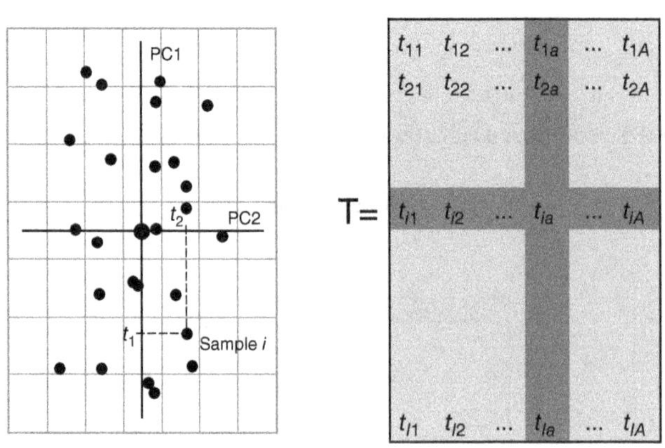

Figure 9.6 Score plot.

where $\lambda_1 \geq \ldots \geq \lambda_A \geq 0$ are the eigenvalues (see Section 9.2.7), which characterize each component's importance.

$$\lambda_a = \sum_{i=1}^{I} t_{ia}^2, \quad a = 1, \ldots, A \tag{9.4}$$

The zero eigenvalue λ_0 is defined as the sum of all eigenvalues, that is,

$$\lambda_0 = \sum_{a=1}^{A} \lambda_a = \mathrm{sp}(\mathbf{X}^t\mathbf{X}) = \sum_{i=1}^{I} \sum_{j=1}^{J} x_{ij}^2 \tag{9.5}$$

A special Chemometrics Add-In worksheet function **`ScoresPCA`** is employed for the PCA scores calculations. This function is explained in Section 8.3.1.

9.2.5 Loadings

The loadings matrix **P** is a transition matrix from the original J-dimensional space of variables $\mathbf{x}_1, \ldots, \mathbf{x}_J$ to the A-dimensional space of PCs. Each row of matrix **P** comprises the coefficients relating to variables **t** and **x**, as it is presented in Eq. 9.1. For example, the a-th row is a projection of all variables $\mathbf{x}_1, \ldots, \mathbf{x}_J$ onto the *a-th* PC axis. Each column of matrix **P** is projection of the related variable $\mathbf{x}_j$ onto the new coordinate space.

The loadings plot is used to study the role of the variables. In this plot, each variable $\mathbf{x}_j$ is represented by a point in the coordinate plane $(\mathbf{p}_i, \mathbf{p}_j)$, for example, in the $(\mathbf{p}_1, \mathbf{p}_2)$ plane, with the axes labeled "PC1" and "PC2" as shown in the example in Fig. 9.7. The analysis of the loadings plot is similar to the scores plot. For example, it is possible to see which variables are correlated and which are independent. A joint study of the paired scores and loadings plots (to be called the *bi-plot*) can also provide useful information about the data.

In PCA, the loadings are normalized orthogonal vectors, that is,

$$\mathbf{P}^t\mathbf{P} = \mathbf{I}$$

To calculate the PCA loadings in Chemometrics Add-In, a special worksheet function **`LoadingsPCA`** is used. It is presented in Section 8.3.2.

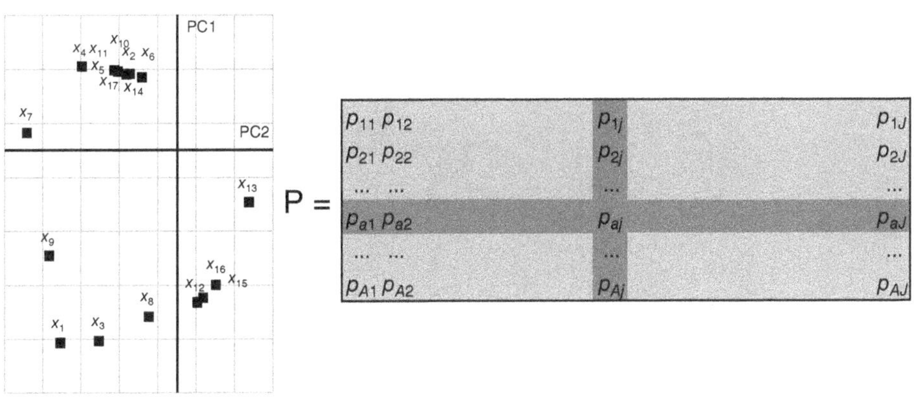

Figure 9.7 Loadings plot.

9.2.6 Data of Special Kind

The result of PCA modeling does not depend on the order in which samples or variables are presented. In other words, the rows or columns of the original matrix **X** can be rearranged without causing a change in the resulting matrices of scores and loadings. However, in some cases, it is recommended to preserve and track this order as it helps to understand the data structure better.

Consider a simple example such as modeling of the data collected by means of the high-performance liquid chromatography with diode array detection (HPLC-DAD). The data are represented by a matrix with 30 samples (I) and 28 variables (J). The samples correspond to the retention times varying between 0 and 30. The variables are the wavelengths in the range of 220–350 nm used for detection. The HPLC-DAD data are shown in Fig. 9.8.

These data are well modeled by PCA with two PCs. It is obvious that in this example, the sequence of samples and variables is important as it reflects the natural pace of time and the continuous spectral range. It is useful to plot the scores and loadings in connection with the relevant factor, that is, the scores against time and the loadings against the wavelengths (see Fig. 9.9).

This example is also discussed in Chapter 12.

9.2.7 Errors

The PCA decomposition of matrix **X** is a sequential iterative process that can be terminated at any step $a = A$. The resulting matrix

$$\hat{\mathbf{X}} = \mathbf{T}\mathbf{P}^t$$

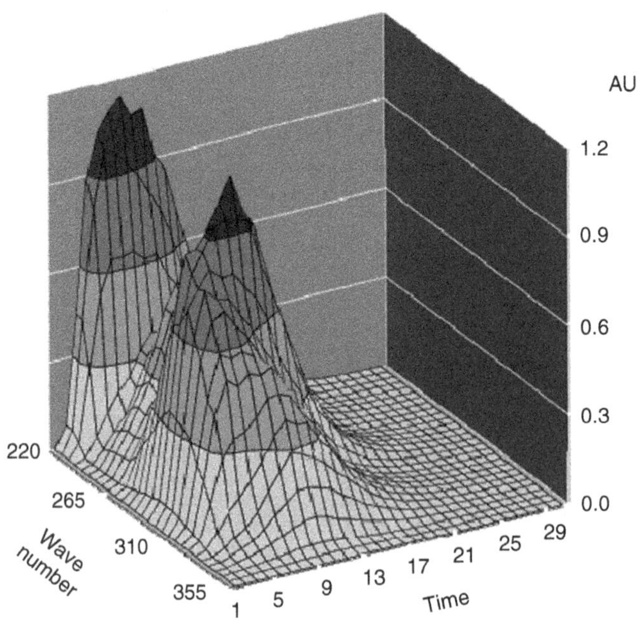

Figure 9.8 HPLC-DAD (high-performance liquid chromatography with diode array detection) data.

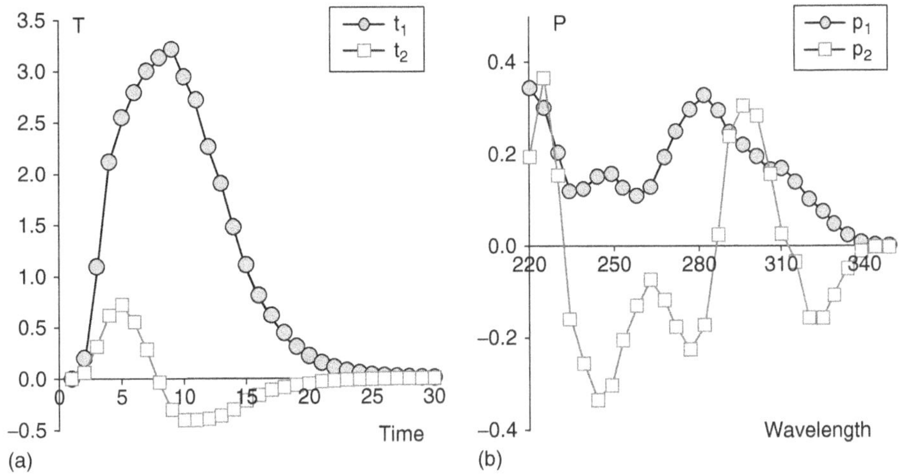

Figure 9.9 Scores (**T**) and loadings (**P**) for HPLC-DAD data.

generally differs from matrix **X**. The difference

$$\mathbf{E} = \mathbf{X} - \hat{\mathbf{X}}$$

is called the *residual matrix*.

Consider a geometric interpretation of the residuals. Each original sample $\mathbf{x}_i$ (a row of matrix **X**) can be represented in the J-dimensional space as a row vector with coordinates $\mathbf{x}_i = (x_{i1}, x_{i2}, \ldots, x_{iJ})$ as shown in Fig. 9.10.

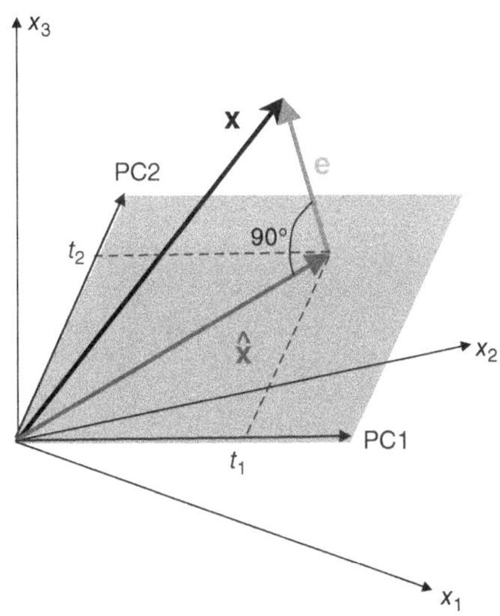

Figure 9.10 PCA geometry.

PCA projects $\mathbf{x}_i$ onto a vector $\mathbf{t}_i = (t_{i1}, t_{i2}, \dots, t_{iA})$ belonging to the A-dimensional space of PCs. In the original space, the coordinates of vector $\mathbf{t}_i$ are

$$\hat{\mathbf{x}}_i = (\hat{x}_{i1}, \hat{x}_{i2}, \dots, \hat{x}_{iJ}).$$

The residual vector is the difference between the original vector and its projection

$$\mathbf{x}_i - \hat{\mathbf{x}}_i = \mathbf{e}_i = (e_{i1}, e_{i2}, \dots, e_{iJ})$$

Vector $\mathbf{e}_i$ constitutes the i-th row of the residual matrix $\mathbf{E}$.

The residuals analysis (Fig. 9.11) helps to understand the data structure and the quality of the PCA model, that is, whether PCA represents the data well enough. The PCA residuals calculation can be performed using methods presented in Sections 13.1.5 and 13.2.4.

The value

$$v_i = \sum_{j=1}^{J} e_{ij}^2 \tag{9.6}$$

defines the squared distance between the initial vector $\mathbf{x}_i$ and its projection onto the PCA space. The smaller it is, the better the approximation of the *i-th* sample. Two methods can be applied to calculate the distance. The first method makes use of the standard worksheet functions (see Section 13.1.7). The second one employs a special user-defined function (see Section 13.2.6).

The sum of the squared residuals divided by the number of variables

$$d_i = (1/J)\sum_{j=1}^{J} e_{ij}^2$$

provides an estimate of the i-th sample variance.

A sample mean distance v_0 is calculated using the formula

$$v_0 = (1/I)\sum_{i=1}^{I} v_i$$

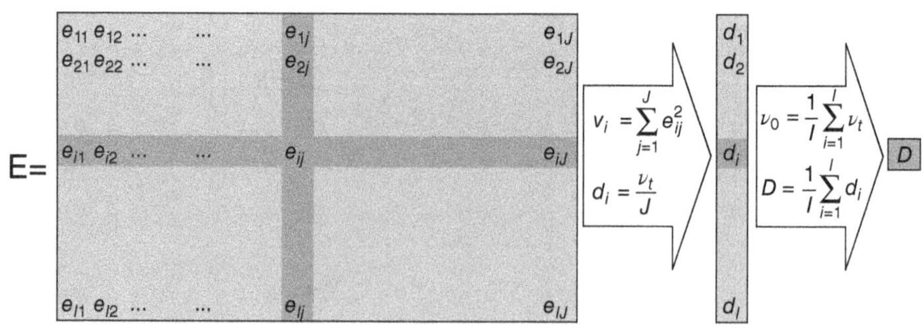

Figure 9.11 Residuals calculation.

A sample mean variance D is calculated as follows

$$D = (1/I)\sum_{i=1}^{I} d_i$$

9.2.8 Validation

If the use of the PCA model is not limited to data exploration only, but also includes prediction or classification, the model should be validated. In the *test set validation* method, the initial dataset consists of two independently derived subsets, both of which are sufficiently representative. The first subset is called the *training set* and used for the modeling. The second subset is called the *test set*. It is only employed for the validation. The PCA model developed using the training set is then applied to the test set, and the results are compared. In this manner, the decisions about the model complexity and accuracy are made. See Fig. 9.12.

In some cases, the sample number is too small to perform such a validation. Therefore, an alternative approach, such as *cross-validation* method can be used (see Section 10.1.3). Moreover, the leverage correction method can also be employed for validation. We leave this method out for self-study.

9.2.9 Decomposition "Quality"

The result of the PCA modeling is the estimates $\hat{\mathbf{X}}_c$, calculated by applying the model to the training set $\mathbf{X}_c$. In addition, we obtain a set of $\hat{\mathbf{X}}_t$ values, which are the estimates calculated by the same model using the formula in Eq. 9.3 and applied to the test set $\mathbf{X}_t$. The latter is considered as a set of new samples. Deviations are represented by the residual matrix, which is calculated as follows.

At the training stage:

$$\mathbf{E}_c = \mathbf{X}_c - \hat{\mathbf{X}}_c,$$

At the validation stage:

$$\mathbf{E}_t = \mathbf{X}_t - \hat{\mathbf{X}}_t.$$

The following values represent the mean "quality" of the PCA modeling.

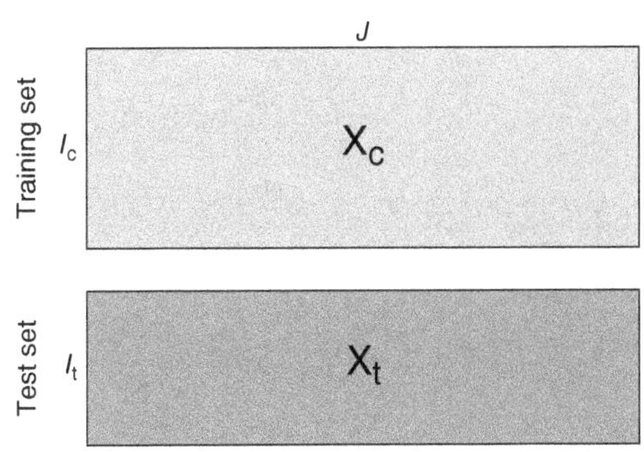

Figure 9.12 Training and test sets.

Total residual variance in training (TRVC) and in validation (TRVP)

$$\text{TRVC} = (1/I_c/J)\sum_{i=1}^{I_c}\sum_{j=1}^{J} e_{ij}^2 = D_c, \quad \text{TRVP} = (1/I_t/J)\sum_{i=1}^{I_t}\sum_{j=1}^{J} e_{ij}^2 = D_t$$

The total variance is expressed in the same units (or rather, in their squares) as the initial **X** values.

Explained residual variance in training (ERVC) and in validation (ERVP)

$$\text{ERVC} = 1 - \sum_{i=1}^{I_c}\sum_{j=1}^{J} e_{ij}^2 \Big/ \sum_{i=1}^{I_c}\sum_{j=1}^{J} x_{ij}^2, \quad \text{ERVC} = 1 - \sum_{i=1}^{I_t}\sum_{j=1}^{J} e_{ij}^2 \Big/ \sum_{i=1}^{I_t}\sum_{j=1}^{J} x_{ij}^2$$

The explained variance is a relative value. It is calculated using the natural normalization, which is the sum of the squares of all initial values x_{ij}. The value is expressed in percentage units or as a fraction unit. In all formulas above, e_{ij} values are the elements of matrix $\mathbf{E}_c$ or matrix $\mathbf{E}_t$. For all characteristics, the notation of which ends on C (e.g., TRVC), one should use matrix $\mathbf{E}_c$ (training). Similarly, for the characteristics with names ending on P (e.g., TRVP), one should employ matrix $\mathbf{E}_t$ (validation).

9.2.10 Number of Principal Components

As mentioned above, the method of PCs is an iterative procedure that appends the new components sequentially, one by one. It is important to decide when this process should be terminated, that is, one should determine the appropriate PCs number, A.

If the PCs number is too small, the data description will be incomplete. On the other hand, the excessive number of PCs results in overfitting, that is, in modeling of noise rather than substantive information. Typically, the number of PCs is selected using a plot in which the explained residual variance (ERV) is depicted as a function of the PCs number. An example is shown in Fig. 9.13a.

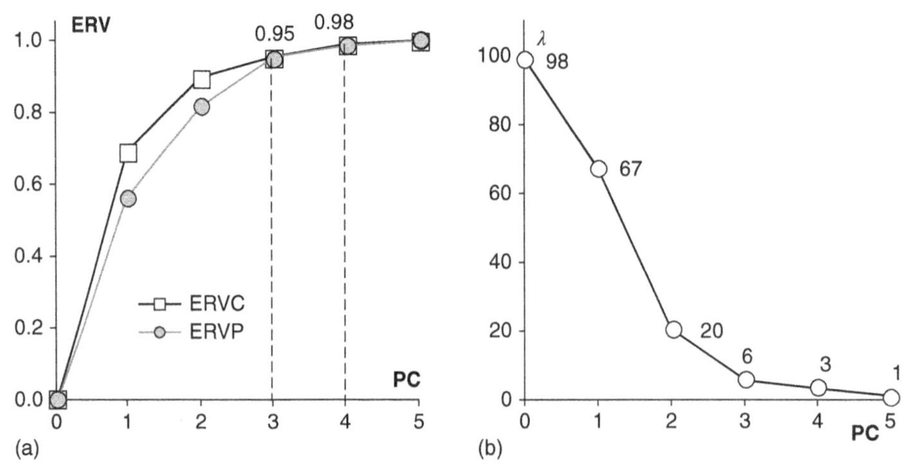

Figure 9.13 Selection of the PC number.

This plot clearly demonstrates that the correct PCs number could be 3 or 4. Three components explain 95% of initial variation and four PCs explain 98% of it. The final decision on the number A can be taken only after the analysis of the data. Another useful tool is a plot depicting the eigenvalues (Eq. 9.4) in dependence with the PCs number. An example is shown in Fig. 9.13b. This figure shows a sharp change in the curvature of the line (break) at PC = 3. Therefore, the appropriate number of PCs is three or four.

9.2.11 The Ambiguity of PCA

The PCA decomposition

$$\mathbf{X} = \mathbf{T}\mathbf{P}^{\mathrm{t}} + \mathbf{E}$$

is not unique. Matrices $\mathbf{T}$ and $\mathbf{P}$ can be replaced by other matrices $\widetilde{\mathbf{T}}$ and $\widetilde{\mathbf{P}}$, which give a similar decomposition

$$\mathbf{X} = \widetilde{\mathbf{T}}\widetilde{\mathbf{P}}^{\mathrm{t}} + \mathbf{E}$$

with the same residual matrix $\mathbf{E}$. The simplest example is a simultaneous change of signs in the corresponding components $\mathbf{t}_a$ and $\mathbf{p}_a$, with which the product

$$\widetilde{\mathbf{t}}_a\widetilde{\mathbf{p}}_a^{\mathrm{t}} = \mathbf{t}_a\mathbf{p}_a^{\mathrm{t}}$$

remains unchanged. It is interesting that NIPALS algorithm (Section 15.5.3) displays the same feature, that is, it can provide the result up to the sign. Therefore, its implementation in different programs can lead to the inversion of the PCs directions.

A more complex case is a simultaneous rotation of matrices $\mathbf{T}$ and $\mathbf{P}$. Let $\mathbf{R}$ be an orthogonal rotation matrix of dimension $A \times A$, that is, a matrix such that $\mathbf{R}^{\mathrm{t}} = \mathbf{R}^{-1}$(Section 5.1.4). Then

$$\mathbf{T}\mathbf{P}^{\mathrm{t}} = \mathbf{T}\mathbf{R}\mathbf{R}^{-1}\mathbf{P}^{\mathrm{t}} = (\mathbf{T}\mathbf{R})(\mathbf{P}\mathbf{R})^{\mathrm{t}} = \widetilde{\mathbf{T}}\widetilde{\mathbf{P}}^{\mathrm{t}}$$

Note that the new scores and loadings matrices preserve all properties of the old matrices, that is,

$$\widetilde{\mathbf{T}}^{\mathrm{t}}\widetilde{\mathbf{T}} = \mathbf{T}^{\mathrm{t}}\mathbf{T} = \mathbf{\Lambda}$$

$$\widetilde{\mathbf{P}}^{\mathrm{t}}\widetilde{\mathbf{P}} = \mathbf{P}^{\mathrm{t}}\mathbf{P} = \mathbf{I}.$$

This property of PCA is called the *rotational ambiguity*. It is widely used for solving the curve resolution problems, in particular in the method of *Procrustes rotation*, which is discussed in Section 12.3.1. If one steps away from the condition of the PC orthogonality, the matrix decomposition becomes even more general. Now, let $\mathbf{R}$ be a nonsingular $(A \times A)$ matrix. Then,

$$\widetilde{\mathbf{T}} = \mathbf{T}\mathbf{R}, \quad \widetilde{\mathbf{P}} = \mathbf{P}(\mathbf{R}^{\mathrm{t}})^{-1}$$

These scores and loadings matrices do not satisfy the orthogonality and normality conditions any more. But they may consist of the nonnegative elements only and also comply with other restrictions imposed in the multivariate curve resolution task. A more detailed analysis of this problem is given in Chapter 12.

9.2.12 Data Preprocessing

In many cases, before applying PCA, the original data should be preprocessed, for example, centered and/or scaled. These transformations are performed on the columns, that is, the variables. The data preprocessing procedures are described in Section 8.1.4.

9.2.13 Leverage and Deviation

For a given PCs number, A, a value

$$h_i = \mathbf{t}_i^t(\mathbf{T}_A^t\mathbf{T}_A)^{-1}\mathbf{t}_i = \sum_{a=1}^{A} t_{ia}^2/\lambda_a$$

is called *leverage*. It is equal to the squared *Mahalanobis distance* (Section 6.2.7) to the *i-th* sample in the scores space; therefore, the leverage characterizes how far a sample is located in the scores space. The leverages conform with a relation

$$h_0 = (1/I)\sum_{i=1}^{I} h_i \equiv A/I$$

that holds identically by the PCA definition. The leverages are calculated using the standard worksheet functions (Section 13.1.8) or by a special user-defined function (Section 13.2.7).

Another important PCA feature is *deviation* v_i, which is calculated as the sum of the squared residuals as in Eq. 9.6. This is the squared Euclidean distance from the sample i to the subspace of the PCA scores. The deviations are calculated using the standard worksheet functions (Section 13.1.7) or by a special user-defined function (Section 13.2.6).

The two values, the score distance h_i, and the orthogonal distance v_i, determine the position of an object (sample) with regard to the developed PCA model. Too large a score and/or orthogonal distance indicates a sample, which can be an extreme object or an outlier. The analysis of the h_i and v_i values gives a basis for soft independent modeling of class analogy (SIMCA), which is a supervised classification method explained in Section 11.3.4.

9.3 PEOPLE AND COUNTRIES

9.3.1 Example

PCA is illustrated by an example that is presented in workbook People.xls. The workbook contains the following sheets:

Intro: short introduction
Layout: scheme that explains the range names used in the example
Data: the example data
PCA: PCA decomposition developed by Chemometrics Add-In
Scores 1–2: the first PC1–PC2 scores analysis
Scores 3–4: the elder PC3–PC4 scores analysis
Loadings: the loadings analysis
Residuals: the residuals analysis

9.3.2 Data

The example uses the European demographic survey data published in the CAMO book by Kim Esbensen.[1] For didactic reasons, only a small subset is used. It includes the data on 32 persons, 16 of which are from Northern Europe (Scandinavia) and 16 are from the South of Europe (Mediterranean). There are both 16 men and 16 women so they are equally represented in the dataset. The data are presented in worksheet Data and shown in Fig. 9.14.

The persons are characterized by 12 variables listed in Table 9.1.

Note that *Sex, Hair*, and *Region* are the discrete variables with two possible values: −1 or +1, whereas the remaining nine variables can take continuous numeric values.

9.3.3 Data Exploration

First of all, it is interesting to look at the plots linking all of these variables. Does *Height* depend on *Weight*? Do women differ from men in the *Wine* consumption? Does *Income* relate to *Age*? Does *Weight* depend on the consumption of *Beer*?

Some of these relationships are shown in Fig. 9.15. For clarity, all plots use the same marks: women (F) are shown by round dots, males (M) by squares, the North (N) is represented in gray shading, and the South (S) is in black.

Raw Data

		Height	Weight	Hair	Shoes	Age	Income	Beer	Wine	Sex	Strength	Region	IQ
1	MN	198	92	-1	48	48	45	420	115	-1	98	-1	100
2	MN	184	84	-1	44	33	33	350	102	-1	92	-1	130
3	MN	183	83	-1	44	37	34	320	98	-1	91	-1	127
4	MN	182	80	-1	42	35	30	398	65	-1	85	-1	140
5	MN	180	80	-1	43	36	30	388	63	-1	84	-1	129
6	MN	183	81	-1	42	37	35	345	45	-1	90	-1	105
7	MN	180	82	-1	44	43	37	355	82	-1	88	-1	109
8	MN	180	81	-1	44	46	42	362	90	-1	86	-1	113
9	MS	185	82	-1	45	26	16	295	180	-1	92	1	109
10	MS	187	84	-1	46	27	16.5	299	178	-1	95	1	119
11	MS	177	65	-1	41	26	18	209	160	-1	86	1	120
12	MS	180	72	-1	43	33	19	236	175	-1	85	1	115
13	MS	181	75	-1	43	42	31	198	161	-1	83	1	105
14	MS	176	68	-1	42	50	36	195	177	-1	82	1	96
15	MS	175	67	1	42	55	38	185	187	-1	80	1	105
16	MS	178	75	-1	42	30	24	203	208	-1	81	1	118
17	FN	166	47	-1	36	32	28	270	78	1	75	-1	112
18	FN	170	60	1	38	23	20	312	99	1	81	-1	110
19	FN	172	64	1	39	24	22	308	91	1	82	-1	102
20	FN	169	51	1	36	24	23	250	89	1	78	-1	98
21	FN	168	52	1	37	27	23.5	260	86	1	78	-1	100
22	FN	157	47	1	36	32	32	235	92	1	70	-1	127
23	FN	164	50	1	38	41	34	255	134	1	76	-1	101
24	FN	162	49	1	37	40	34	265	124	1	75	-1	108
25	FS	168	50	1	37	49	34	170	162	1	76	1	135
26	FS	166	49	1	36	21	14	150	245	1	75	1	123
27	FS	158	46	1	34	30	18	120	120	1	70	1	119
28	FS	163	50	1	36	18	11	143	136	1	75	1	102
29	FS	162	50	1	36	20	11.5	133	146	1	74	1	132
30	FS	165	51	1	36	36	26	121	129	1	76	1	126
31	FS	161	48	1	35	41	31.5	116	196	1	75	1	120
32	FS	160	48	1	35	40	31	118	198	1	74	1	129
mean		173.1	64.5	0.0	39.9	34.4	27.4	249.5	131.6	0.0	81.5	0.0	115.1
STD		10.1	15.2	1.0	3.9	9.5	8.9	90.6	49.5	1.0	7.3	1.0	12.2

Figure 9.14 Raw data in the "People" example.

[1] K.H. Esbensen. *Multivariate Data Analysis – In Practice: An Introduction to Multivariate Data Analysis and Experimental Design*, 5th Edition, CAMO AS Publ., Oslo 2002.

TABLE 9.1 Variables used in the Demographic Survey. Reproduced from Esbensen 2002, with permission of CAMO Software

Height	Centimeter
Weight	Kilogram
Hair	Short (−1) or long (+1)
Shoes	European standard size
Age	Years
Income	Thousands of Euros per year
Beer	Consumption in liters per year
Wine	Consumption in liters per year
Sex	Male (−1) or female (+1)
Strength	An index based on the physical abilities test
Region	North (−1) or South (+1)
IQ	As measured by the standard test

The relationship between *Weight* and *Height* is shown in Fig. 9.15a. A direct (positive) correlation is evident. Accounting for the point shapes, we can also conclude that the majority of men (M) are heavier and taller than women (F).

Figure 9.15b shows another pair of variables: *Weight* and *Beer*. Here, besides a trivial fact that big people drink more, and women drink less, two distinct groups of the Southerners and the Northerners can be observed. The former drink less beer while weighing the same.

The same groups are evident in Fig. 9.15c, which shows a relationship between *Wine* and *Beer* consumption. The plot demonstrates a negative correlation between these variables, that is, the consumption of wine decreases with the increase in beer consumption. *Wine* is more popular in the *South*, but *Beer* is in the *North*. It is interesting that in both groups women are located on the left side, but not on the bottom side, in relation to men. This means that while consuming less beer, the fair sex is not inferior in wine consumption. The last plot in Fig. 9.15d shows the connection between *Age* and *Income*.

It is easy to see that even in this relatively small dataset, both positively and negatively correlated variables are present. Is it possible to plot all variable pairs in this set? Hardly. The problem is that for 12 variables there are $12\,(12 - 1)/2 = 66$ such pairs.

9.3.4 Data Pretreatment

Before the data are subjected to PCA, they should be prepared. Simple statistical calculations show that autoscaling is necessary (see Fig. 9.16).

The mean values (dots in Fig. 9.16) are far from zero for some variables. Moreover, the standard deviations (depicted by bars in Fig. 9.16) are very different. Preprocessing is performed in sheet *Data* using an array formula explained in Section 13.1.3. After autoscaling, the mean values of all new variables become equal to zero and all deviations are equal to one.

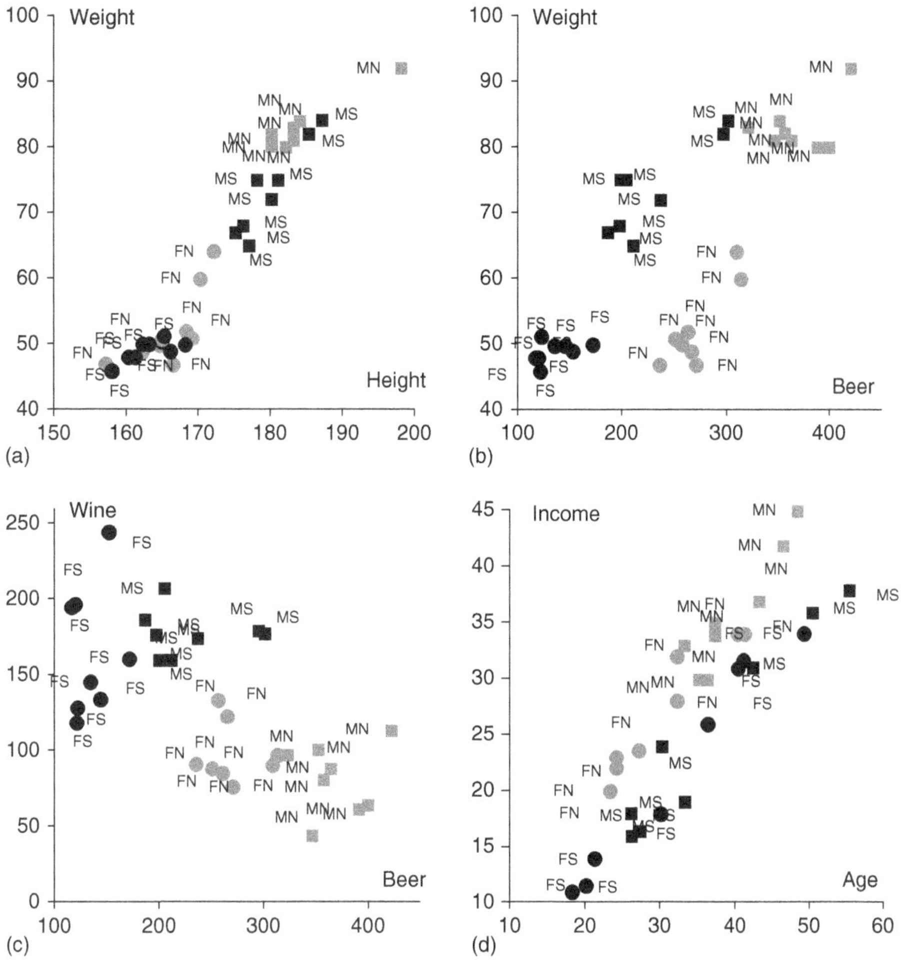

Figure 9.15 Relationships between variables in the "People" example. Women (F) are marked by the dots • and • and men (M) by the squares ■ and ■. The North (N) is represented in the gray ■ and the South (S) in the black • shading.

In principle, the data may not be explicitly prepared in the sheet but kept as they are. The special worksheet functions, collected in Chemometrics Add-In, can do the data centering and scaling automatically, in the course of PCA calculations (see Section 8.2.2). However, the autoscaled data matrix will be needed for the calculation of residuals in Section 9.3.8.

9.3.5 Scores and Loadings Calculation

The PCA decomposition is developed using special functions **`ScoresPCA`** and **`LoadingsPCA`**, available in Chemometrics Add-In (Sections 8.3.1 and 8.3.2). The first argument of the functions is Xraw, which is the name of the raw, nontransformed dataset. The last argument in both functions is 3, which means autoscaling. The second argument is missing, so all possible PCs are obtained, that is, $A = 12$.

The PCA calculations are done in worksheet PCA and shown in Figs 9.17 and 9.18.

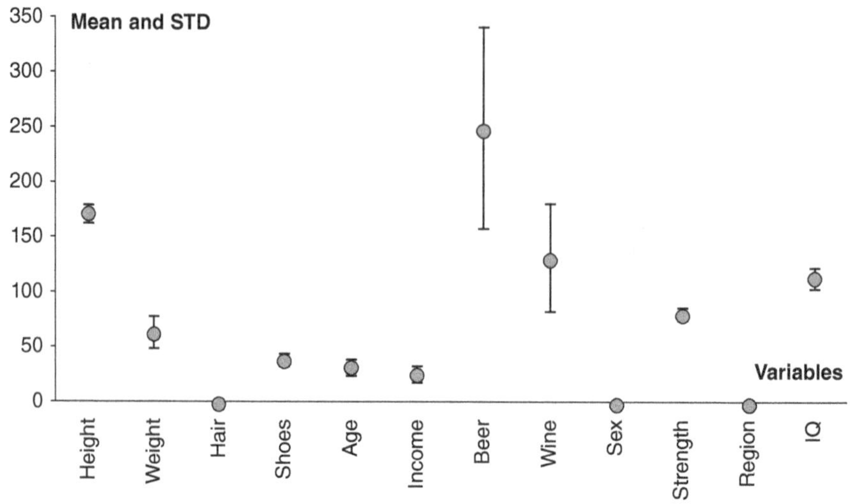

Figure 9.16 Mean and STD values for the "People" variables.

PCA

T Scores

	PC1	PC2	PC3	PC4	PC5	PC6	PC7	PC8	PC9	PC10	PC11	PC12
1	=ScoresPCA(Xraw,12,3)				1.060	-0.017	0.563	0.085	0.280	0.078	0.109	0.133
2	-3.114	-0.293	0.671	1.310	0.435	0.119	0.109	-0.456	-0.088	-0.034	0.051	-0.009
3	-2.997	-0.360	0.212	1.117	0.204	-0.017	0.120	-0.469	-0.149	-0.018	0.184	-0.208
4	-2.591	-0.928	0.863	2.321	-0.095	-0.076	-0.270	0.306	0.240	-0.192	-0.168	0.005
5	-2.588	-1.037	0.686	1.461	-0.347	-0.098	-0.447	0.345	0.083	0.021	-0.043	-0.005
6	-3.027	-1.400	0.284	-0.440	-0.633	-0.538	0.252	-0.343	0.042	-0.132	-0.247	0.020
7	-3.095	-1.069	-0.472	-0.144	-0.304	-0.059	-0.147	-0.025	-0.217	0.150	-0.077	-0.088
8	-3.113	-1.181	-1.068	0.242	-0.232	0.141	-0.211	-0.008	-0.038	0.203	-0.060	0.087
9	-2.130	2.356	1.503	-0.858	0.284	0.070	0.020	0.246	-0.171	0.095	-0.123	0.060
10	-2.510	2.525	1.542	-0.105	0.653	-0.087	0.222	0.228	-0.305	0.078	0.024	0.035
11	-0.463	1.992	1.090	0.206	-0.617	0.084	0.138	-0.186	-0.202	-0.421	0.069	0.174
12	-1.098	2.094	0.496	-0.198	-0.376	0.127	-0.126	0.299	-0.071	-0.122	0.112	-0.115
13	-1.438	1.527	-1.059	-0.788	-0.734	-0.124	0.015	-0.060	0.264	0.185	0.156	-0.057
14	-1.137	1.241	-2.200	-1.404	-0.953	0.200	0.033	0.063	-0.075	-0.002	-0.058	-0.009
15	-0.334	1.034	-2.931	-0.866	0.420	-0.654	-0.781	0.037	0.016	-0.283	0.060	-0.004
16	-0.652	2.360	0.125	0.055	-0.334	0.674	-0.398	-0.283	0.359	0.192	-0.007	-0.007
17	1.084	-1.845	0.409	0.123	-1.323	0.872	0.704	0.328	0.032	-0.029	0.078	-0.026
18	0.981	-1.434	1.645	-0.526	0.714	-0.035	-0.096	0.127	-0.005	0.003	-0.124	-0.053
19	0.567	-1.551	1.474	-1.154	0.677	-0.234	-0.047	0.031	0.068	0.213	-0.058	-0.052
20	1.663	-1.762	1.122	-1.394	0.018	-0.081	0.085	-0.218	0.224	-0.284	0.028	0.004
21	1.486	-1.813	0.904	-1.208	0.070	-0.147	-0.003	-0.023	0.043	-0.160	0.109	-0.034
22	2.464	-2.040	-0.222	1.202	-0.140	0.268	-0.491	-0.185	-0.091	0.166	0.191	0.233
23	1.396	-1.736	-1.130	-1.041	0.367	0.487	-0.170	0.133	-0.220	-0.007	0.123	-0.074
24	1.622	-1.894	-0.992	-0.419	0.287	0.463	-0.204	0.141	-0.225	-0.048	-0.033	0.015
25	2.005	0.344	-2.085	1.549	0.603	-0.377	0.509	0.471	0.105	-0.061	0.070	0.020
26	3.335	1.956	0.851	0.063	0.884	0.897	-0.141	-0.048	0.252	-0.128	-0.057	-0.046
27	3.711	0.147	0.195	0.279	-0.774	-0.624	-0.091	0.120	0.002	0.081	-0.114	-0.074
28	3.207	0.630	1.602	-1.358	-0.420	-0.480	-0.057	-0.047	-0.001	0.168	-0.026	0.112
29	3.423	1.021	1.482	1.036	0.016	-0.395	-0.084	-0.024	-0.035	0.124	0.177	-0.027
30	2.615	0.320	-0.600	0.784	-0.038	-0.789	0.455	-0.109	0.045	0.093	0.108	-0.007
31	2.958	0.688	-1.728	0.267	0.268	0.161	0.320	-0.246	-0.082	0.016	-0.249	-0.012
32	3.102	0.788	-1.603	0.988	0.359	0.249	0.220	-0.232	-0.078	0.055	-0.203	0.009

Figure 9.17 Scores calculation.

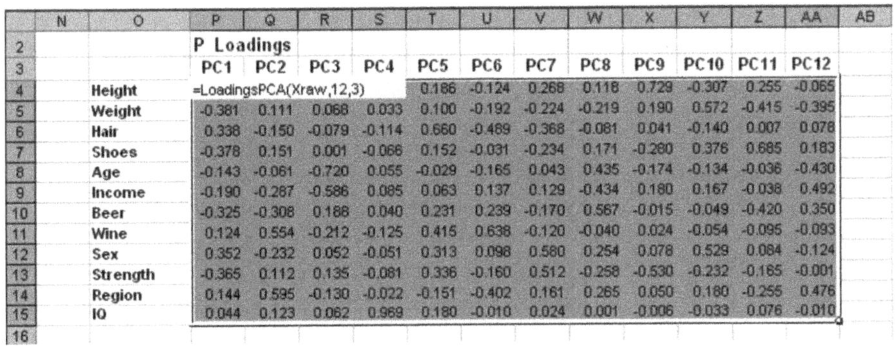

	N	O	P	Q	R	S	T	U	V	W	X	Y	Z	AA	AB
2			P Loadings												
3			PC1	PC2	PC3	PC4	PC5	PC6	PC7	PC8	PC9	PC10	PC11	PC12	
4		Height	=LoadingsPCA(Xraw,12,3)				0.186	-0.124	0.268	0.118	0.729	-0.307	0.255	-0.065	
5		Weight	-0.381	0.111	0.068	0.033	0.100	-0.192	-0.224	-0.219	0.190	0.572	-0.415	-0.395	
6		Hair	0.338	-0.150	-0.079	-0.114	0.660	-0.489	-0.368	-0.081	0.041	-0.140	0.007	0.078	
7		Shoes	-0.378	0.151	0.001	-0.066	0.152	-0.031	-0.234	0.171	-0.280	0.376	0.685	0.183	
8		Age	-0.143	-0.061	-0.720	0.055	-0.029	-0.165	0.043	0.435	-0.174	-0.134	-0.036	-0.430	
9		Income	-0.190	-0.287	-0.586	0.085	0.063	0.137	0.129	-0.434	0.180	0.167	-0.038	0.492	
10		Beer	-0.325	-0.308	0.188	0.040	0.231	0.239	-0.170	0.567	-0.015	-0.049	-0.420	0.350	
11		Wine	0.124	0.554	-0.212	-0.125	0.415	0.638	-0.120	-0.040	0.024	-0.054	-0.095	-0.093	
12		Sex	0.352	-0.232	0.052	-0.051	0.313	0.098	0.580	0.254	0.078	0.529	0.084	-0.124	
13		Strength	-0.365	0.112	0.135	-0.081	0.336	-0.160	0.512	-0.258	-0.530	-0.232	-0.165	-0.001	
14		Region	0.144	0.595	-0.130	-0.022	-0.151	-0.402	0.161	0.265	0.050	0.180	-0.255	0.476	
15		IQ	0.044	0.123	0.062	0.969	0.180	-0.010	0.024	0.001	-0.006	-0.033	0.076	-0.010	
16															

Figure 9.18 Loadings calculation.

9.3.6 Scores Plots

Let us look at the scores plots that depict the sample location in the projection space. In the first scores plot PC1–PC2 (worksheet Scores 1–2 and Fig. 9.19a), we see four distinct groups, arranged in four coordinate quadrants: females (F) are on the left, males (M) are on the right, the South (S) is above, and the North (N) is below. These findings immediately explain the sense of the first two components, PC1 and PC2. The first component discriminates people by gender and the second one separates them by location. These variables have the highest impact on the properties variation.

We continue our study analyzing the higher PC3–PC4 scores (worksheet Scores 3–4 and Fig. 9.19b). No distinct groups stand out here. However, after a careful examination of the plot together with the raw data table (Table 9.1), we can conclude that PC3 separates the elder/rich people from the younger/poor. To make this more evident, we have changed the marking in the plot. Now every person is presented by a dot, whose color and size varies according to income, that is, larger and brighter dots correspond to higher income. The age

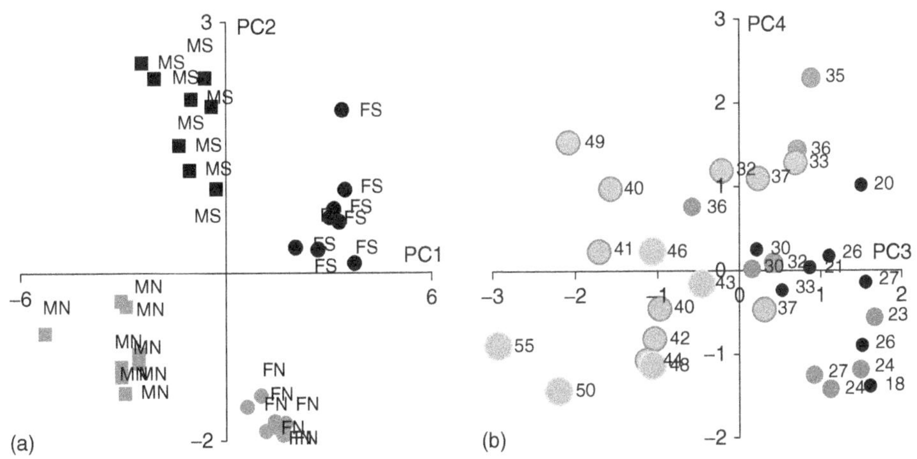

Figure 9.19 (a) PC1–PC2 scores plot with the marking used in Fig. 9.15. (b) PC3–PC4 scores plot with new marking. Larger and brighter dots correspond to higher income. The numbers represent age.

is shown next to each marker. In this plot, it can be seen that *Age* and *Income* is falling down from the left to right, that is, along PC3. But the meaning of PC4 is still unclear.

9.3.7 Loadings Plot

The loadings plots can be also useful in further data exploration. They will tell us what variables (see Table 9.1) are connected and what impacts these relations.

In the loadings plot for the first PC (worksheet Loadings and Fig. 9.20a), we see that variables *Height, Weight, Strength*, and *Shoes* form a compact group on the left side of the plot. Their marks are practically merged, which indicates a strong positive correlation. Variables *Hair* and *Sex* are in another group, which is located diagonally from the first group. This indicates a negative correlation between the variables of these groups, for example, between *Strength* and *Sex*. The variables *Wine* and *Region*, being closely related, have the maximum loadings on the second component. Variable *Income* is opposite to *Region*, and this reflects the differentiation in prosperity between the North and the South. One can also see the antithesis of *Beer* and *Region*/*Wine*.

In Fig. 9.19b, we see high loadings on the PC3 axis for variables *Age* and *Income*. This corresponds to the scores plot shown in Fig. 9.19a. Consider variables *Beer* and *IQ*. The former has large loadings on both PC1 and PC2, forming diagonal interaction with the other objects in the loadings plot. Variable *IQ* has no connections with other variables, as its value is close to zero for the first three PC loadings. It manifests itself only in the fourth component. We see that the *IQ* value does not depend on the place of residence, physiological characteristics, and preferences to drinks.

PCA was first applied in psychological research in the early 20th century, when people believed that factors like IQ or criminal behavior can be explained by an individual's physiological and social characteristics. If we compare the PCA results with the plots drawn above for the pairs of variables, it is clear that PCA gives a comprehensive overview of the data structure, which can be accessed at a glance. Therefore, an advantage of PCA is the conversion from a large number of paired variable plots into a very small number of scores and loadings plots.

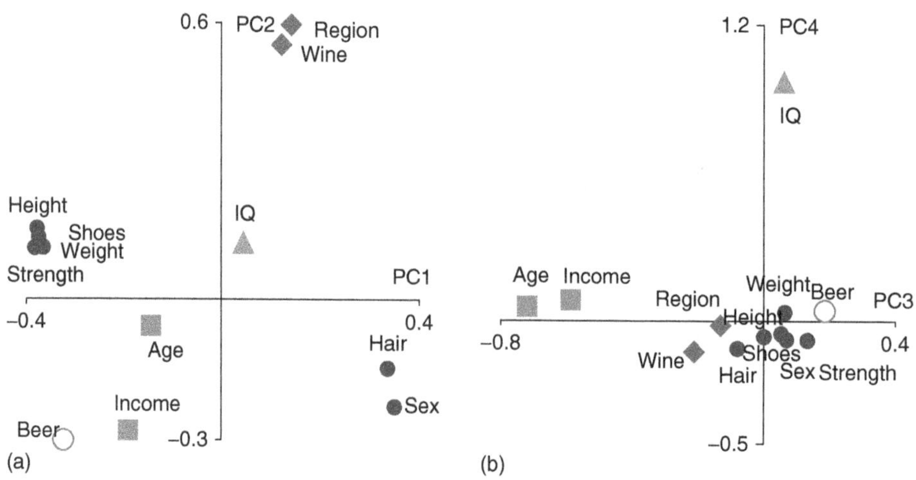

Figure 9.20 Loadings plots: PC1–PC2 (a) and PC3–PC4 (b).

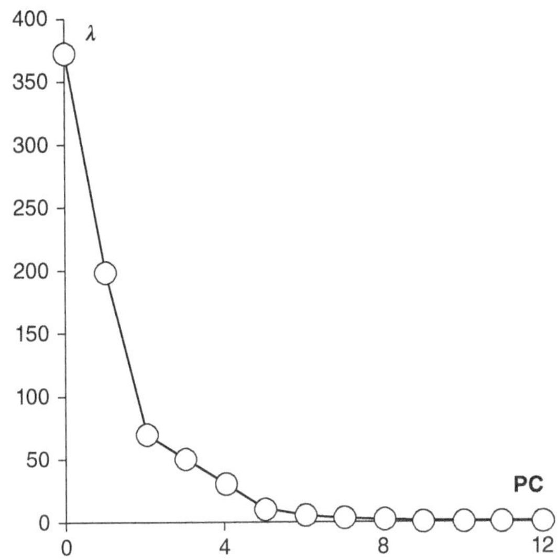

Figure 9.21 Eigenvalues plot.

9.3.8 Analysis of Residuals

How many PCs should be used in this example? To answer this question, we have to investigate how the model quality is changed with the number of PCs. Note that in this example, we will not perform validation. This is not necessary because the PCA model is applied for data exploration only. This model is not intended for prediction or classification.

Worksheet Residuals and Fig. 9.21 represent eigenvalues λ having dependence on the PCs number. A change in the curve behavior can be seen near the point PC = 5.

For calculation of the total residual variance (TRV) and ERV values, we can get the error matrix **E** for each number of PCs A and then compute residuals using formulas presented in Section 9.2.12. An example of calculations for $A = 4$ is shown in sheet *Residuals*. See worksheet Residuals and Fig. 9.22.

However, the same values can be obtained easily by using the following relations:

$$\mathrm{TRV}(A) = (1/IJ)\left(\lambda_0 - \sum_{a=1}^{A} \lambda_a\right), \quad \mathrm{ERV}(A) = 1 - (\mathrm{TRV}(A)/\mathrm{TRV}(0))$$

These values are shown in Fig. 9.23. The plot demonstrates that four PCs are sufficient for data description because at such complexity 94% of the data is explained, or, in other words, the noise remaining after projection onto the four-dimensional scores space constitutes only 6% of the original data.

CONCLUSION

The above example illustrates a small part of the advantages that are delivered by PCA. We have considered the data exploration problem, which does not intend to use the developed model for prediction or classification. Needless to say, abilities of PCA are much wider.

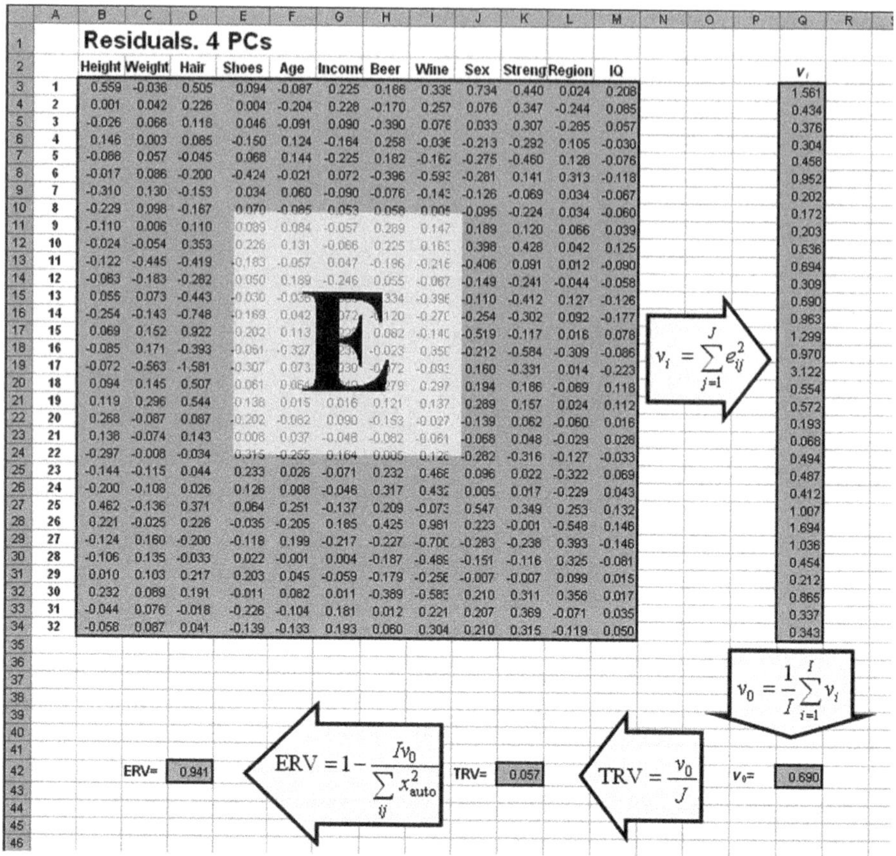

Figure 9.22 Analysis of residuals.

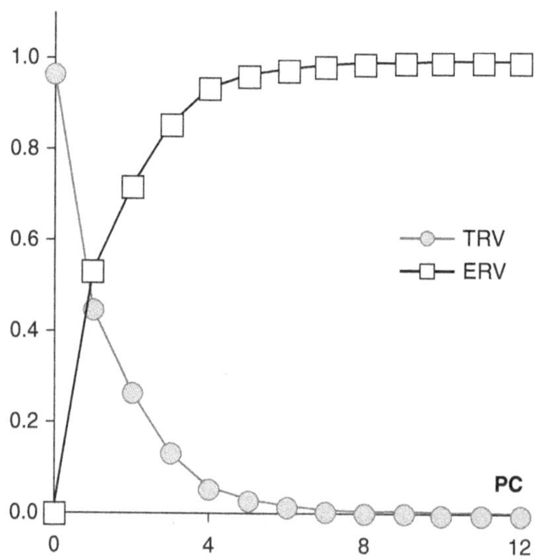

Figure 9.23 Total (TRV) and explained (ERV) residual variances.

PCA is the main method in chemometrics. It provides a basis for many other methods used in different areas. In calibration (Chapter 10), PCA transfers into principal components regression (PCR). In classification (Chapter 11), PCA serves as a method for preliminary data compression, which makes it possible to apply such classical techniques as linear discriminant analysis (LDA) in multivariate case. PCA also gives a background for such powerful method as SIMCA that can be considered a natural extension of the projection approach. In the multivariate curve resolution (Chapter 12), PCA plays the main role in providing a framework for almost all methods used in this area.

10

CALIBRATION

This chapter addresses the main methods used in calibration.

Preceding chapters	3–9
Dependent chapters	11–14
Matrix skills	Basic
Statistical skills	Basic
Excel skills	Basic
Chemometric skills	Basic
Chemometrics Add-In	Used
Accompanying workbook	Calibration.xls

10.1 THE BASICS

10.1.1 Problem Statement

The essence of *calibration* is as follows. Suppose there is a variable (property) y, the value of which should be determined. However, because of some circumstances (e.g., inaccessibility, high costs, duration), a direct measurement of y is impossible. At the same time, it is possible to easily measure (fast, cheap) other quantities $\mathbf{x} = (x_1, x_2, x_3, \ldots)$, which are related to y. The calibration problem is to establish a quantitative relationship between *variables* $\mathbf{x}$ and *response* y, that is,

$$y = f\left(x_1, x_2, x_3, \ldots \mid a_1, a_2, a_3, \ldots\right) + \varepsilon$$

Chemometrics in Excel, First Edition. Alexey L. Pomerantsev.

In practice, this means

1. a selection of an appropriate calibration function f and ...
2. an estimation of the unknown parameters $a_1, a_2, a_3, \ldots$ of the function.

Of course, calibration cannot be done precisely. We shall see below that it is not only impossible but also risky. A calibration model always contains errors (residuals) ε, which originate from sampling, measurement, modeling, etc.

The simplest example of calibration can be explained using a spring balance, which is no longer a well-known device used in a household. It consists of a spring with a hook used to weigh an object and a calibrated scale that follows the extension of the spring and indicates the weight of the object. The unknown response y is the object's weight and variable x is the spring elongation. According to Hooke's law,

$$y = ax + \varepsilon,$$

x and y are related through the spring's coefficient of stiffness, a constant a, which is unknown a priori. The calibration procedure is very simple. Firstly, we weigh a *standard sample* of 1 kg and mark the degree of spring elongation on the scale, then we use a 2-kg sample, etc. This calibration procedure results in a calibrated scale, which can be further used to determine the weight of a new, nonstandard sample.

This primitive example demonstrates the basic features of a calibration procedure. First of all, we need several calibration samples, for which values y are known in advance. Secondly, we suppose that the range of variation y must be fully covered by the calibration samples. Indeed, it is not possible to weigh samples heavier than 5 kg if all calibration samples were lighter than 5 kg.

Of course, in practice things are not as simple as in the simple example above. Sometimes, a set of responses $y_1, y_2, \ldots, y_K$, rather than a single response y, should be calibrated simultaneously. Various obstacle-related calibration procedures are discussed below. Also, we give a summary of the first outcomes as well as a general formulation of the calibration problem.

Let $\mathbf{Y}$ be the $(I \times K)$ matrix, where I is the number of samples used in calibration and K is the number of simultaneously calibrated responses. Matrix $\mathbf{Y}$ contains the response values, y, known from independent experiments using reference or standard samples. On the other hand, let $\mathbf{X}$ be the $(I \times J)$ matrix of variables, where I is also the number of samples and J is the number of independent variables (channels) used in calibration (see Fig. 10.1). Matrix $\mathbf{X}$ consists of alternative, usually multivariate ($J \gg 1$), measurements. It is required to develop a functional relationship between $\mathbf{Y}$ and $\mathbf{X}$ using the training dataset $(\mathbf{X}, \mathbf{Y})$.

The task of calibration is to construct a mathematical model connecting blocks $\mathbf{X}$ and $\mathbf{Y}$. This model can be used later on to predict the response values $y_1, y_2, \ldots, y_K$ using a row vector $\mathbf{x}$ of new measured variables.

10.1.2 Linear and Nonlinear Calibration

Theoretically, a functional relationship between variables $\mathbf{x}$ and a response y can be complex, for example,

$$y = b_0 \exp\left(b_1 x + b_2 x^2\right) + \varepsilon.$$

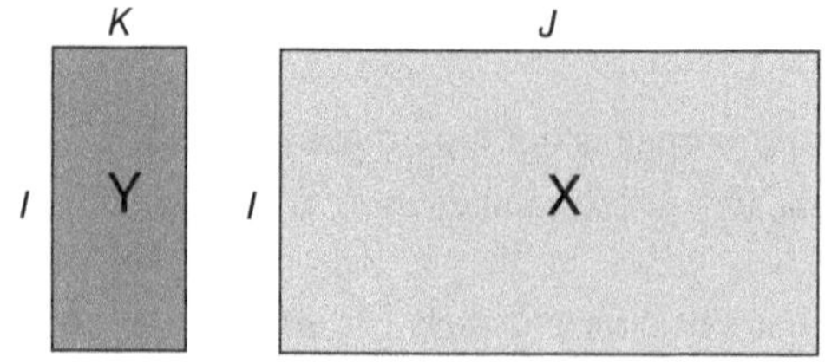

Figure 10.1 Training dataset.

However, in most cases, in practice, calibration functions are *linear*, that is, they have a form

$$y = b_0 + b_1 x_1 + b_2 x_2 + \cdots + b_J x_J + \varepsilon.$$

Note that in this book, the term *linear* is used instead of a more correct term *bilinear*, which implies linearity both with respect to variables **x** and coefficients **b**. Therefore, calibration

$$y = b_0 + b_1 x + b_2 x^2 + \varepsilon$$

is nonlinear, despite the fact that it can be easily "linearized" by introducing a new variable $x_2 = x^2$.

The main advantage of a bilinear model is its uniqueness, as for a given number of variables there exists only one linear model. All other calibration functions form an infinite set, which is difficult to select from. In this book, we consider only the linear calibration model, presented by equation

$$\mathbf{Y} = \mathbf{XB} + \mathbf{E} \tag{10.1}$$

Readers interested in nonlinear data analysis are invited to study the manual of Fitter software.[1]

A preference for (bi)linear soft models that are not burdened by an additional physical–chemical meaning is chemometric mainstream. This concept eliminates the influence of subjective factors that impact the selection of a calibration model. Nevertheless, we have to pay for linearity by a limited range of applicability of such models.

The issue is illustrated in Fig. 10.2, which shows four regions where a complex nonlinear dependence is approximated by linear models. Each model is valid only inside its own domain. Any attempt to use the model outside of a domain leads to a large error. The fundamental problem is setting a valid domain. In other words, how far the calibration model can be applied? The answer to this question is provided by *validation* methods.

10.1.3 Calibration and Validation

A properly developed calibration model is constructed using a dataset that consists of two independently derived and sufficiently representative subsets. The first subset, called the *training set* is used only for model development, that is, for model parameters estimation. The second subset, called *test set*, is used only for validation. For illustration see Fig. 10.3.

[1] http://polycert.chph.ras.ru/fitter.htm

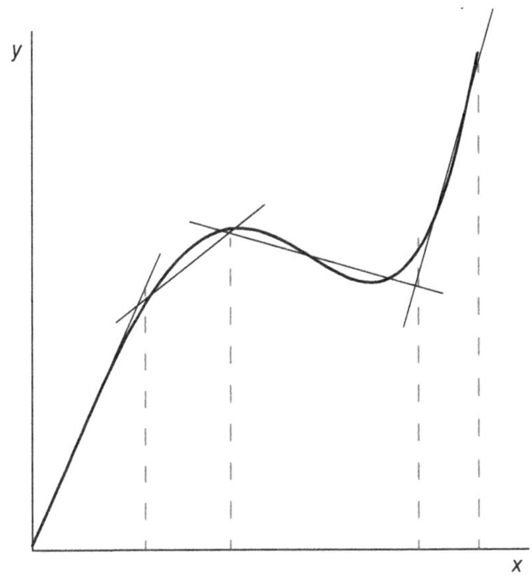

Figure 10.2 Linearization of calibration.

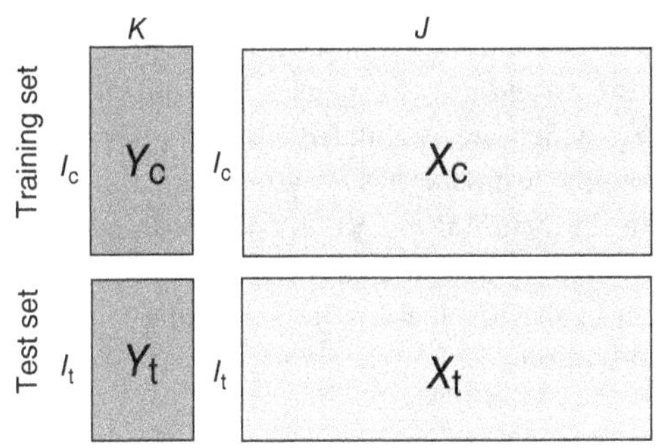

Figure 10.3 Training and test sets.

The developed model is applied to the test set data, and the obtained results are compared with the known validation values. Thus, the decision on the model correctness and accuracy is made by the *test set validation* method.

Sometimes, there is too little data to employ such a technique. In this case another approach, known as *cross-validation*, is applied. In this method, the test values are calculated using the following procedure.

A fixed fraction (e.g., the first 10%) of data is left out of the training set. A model based on the remaining 90% of the data is built and then applied to the excluded data subset. In the next iteration, the initially excluded data are returned and a new portion of data (next 10%) is excluded. Using the remaining data, a new model is constructed and then applied to the excluded 10%. This procedure is repeated until all data objects have been left out once (in our case, 10 times). The most popular (but unjustified) version of cross-validation method is the one in which the data points are eliminated one by one (*Leave-One-Out*, LOO), see Fig. 10.4.

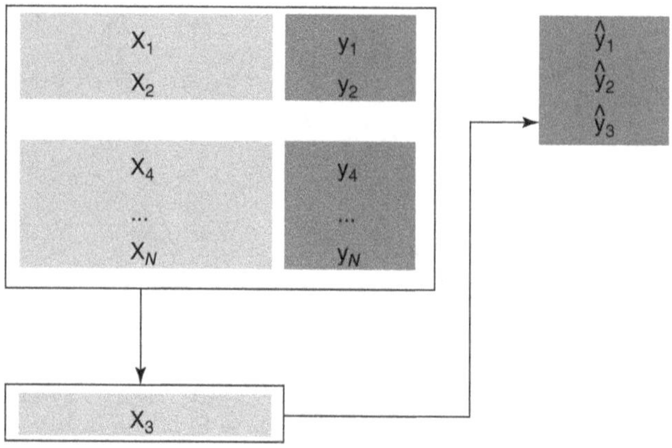

Figure 10.4 Cross-validation method.

Another validation method is leverage correction. This method is not discussed here and is recommended for self-study.

10.1.4 Calibration "Quality"

Calibration results are a set of values $\widehat{\mathbf{Y}}_c$, found with the help of the model based on the training set of data. $\widehat{\mathbf{Y}}_c$ are the *estimates* of responses $\mathbf{Y}_c$ (known a priori). Validation returns another set of values $\widehat{\mathbf{Y}}_t$, which are calculated using the same model and applied to the known responses $\mathbf{Y}_t$ from the test dataset. Deviations of estimates from their known counterparts are collected in the *residual matrix*: for the training set it is $\mathbf{E}_c = \mathbf{Y}_c - \widehat{\mathbf{Y}}_c$, for validation it is $\mathbf{E}_t = \mathbf{Y}_t - \widehat{\mathbf{Y}}_t$. For an illustration, see Fig. 10.5.

Let e_{ki} be an element of the residual matrix. The following indicators characterize an average "quality" of calibration.

- *Total residual variance* in training (TRVC) and in validation (TRVP) are defined by

$$\mathrm{TRVC} = (1/K/I_c)\sum_{k=1}^{K}\sum_{i=1}^{I_c} e_{ki}^2, \quad \mathrm{TRVP} = (1/K/I_t)\sum_{k=1}^{K}\sum_{i=1}^{I_t} e_{ki}^2 \tag{10.2}$$

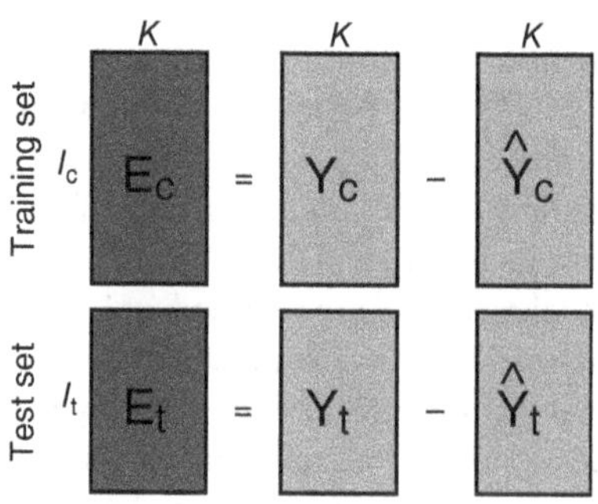

Figure 10.5 Residuals in training and in validation.

The smaller they are the better. These values correspond to formula (6.10) (Chapter 6) for $m = 0$.

- *Explained residual variance* in training (ERVC) and in validation (ERVP) are defined by

$$\mathrm{ERVC} = 1 - \sum_{k=1}^{K}\sum_{i=1}^{I_c} e_{ki}^2 \Big/ \sum_{k=1}^{K}\sum_{i=1}^{I_c} y_{ki}^2, \quad \mathrm{ERVP} = 1 - \sum_{k=1}^{K}\sum_{i=1}^{I_t} e_{ki}^2 \Big/ \sum_{k=1}^{K}\sum_{i=1}^{I_t} y_{ki}^2 \tag{10.3}$$

One would want to see values close to 1.

- *Root mean square error* in training (RMSEC) and in validation (RMSEP) are defined by

$$\mathrm{RMSEC}(k) = \sqrt{\left(1/I_c\right)\sum_{i=1}^{I_c} e_{ki}^2}, \quad \mathrm{RMSEP}(k) = \sqrt{\left(1/I_t\right)\sum_{i=1}^{I_t} e_{ki}^2} \tag{10.4}$$

The RMSEx values depend on k, which is a response index. They should be as small as possible.

- *Bias* in training (BIASC) and in validation (BIASP) are defined by

$$\mathrm{BIASC}(k) = \left(1/I_c\right)\sum_{i=1}^{I_c} e_{ki}, \quad \mathrm{BIASP}(k) = \left(1/I_t\right)\sum_{i=1}^{I_t} e_{ki} \tag{10.5}$$

The BIASx values depend on k, which is a response index. They should be close to zero.

- *Standard error* in training (SEC) and in validation (SEP) are defined by

$$\mathrm{SEC}(k) = \sqrt{\left(1/I_c\right)\sum_{i=1}^{I_c}\left(e_{ki} - \mathrm{BIASC}(k)\right)^2}, \quad \mathrm{SEP}(k) = \sqrt{\left(1/I_t\right)\sum_{i=1}^{I_t}\left(e_{ki} - \mathrm{BIASP}(k)\right)^2} \tag{10.6}$$

The standard error (SE) values depend on k, which is a response index. They are calculated by Eq. (6.9) from Chapter 6. The smaller they are the better.

- *Correlation coefficient* R^2 between the known y_{ki} and the estimated $\hat{y}_{ki}$ response values is defined by

$$R^2(k) = \left(\left(I\sum_i y_{ki}\hat{y}_{ki} - \sum_i y_{ki}\sum_i \hat{y}_{ki}\right) \Big/ \left(\sqrt{I\sum_i y_{ki}^2 - \left(\sum_i y_{ki}\right)^2}\sqrt{I\sum_i \hat{y}_{ki}^2 - \left(\sum_i \hat{y}_{ki}\right)^2}\right)\right) \tag{10.7}$$

The R^2 values depend on k, which is a response index. They are also calculated for the training set, that is, $R_c^2(k)$, and for the test set, that is, $R_t^2(k)$. These characteristics should be close to 1.

In all formulas above, e_{ij} values are the elements of matrix $\mathbf{E}_c$ or matrix $\mathbf{E}_t$. Characteristics with names ending on C (e.g., TRVC), use matrix $\mathbf{E}_c$ (training), and values with names ending on P (e.g., TRVP), employ matrix $\mathbf{E}_t$ (validation).

10.1.5 Uncertainty, Precision, and Accuracy

The large number of indicators that describe calibration quality originated because of historical reasons. More importantly, they reflect different features of uncertainty in training and in validation (prediction). To clarify the role of these indicators, it is useful to introduce two principal concepts such as precision and accuracy.

Precision determines the degree of closeness of independent repeated measurements.

Accuracy determines the degree of closeness to the true (standard) value of y.

Figure 10.6 demonstrates the essence of this in a nutshell. Plots (a) and (b) present estimates of good precision. In addition, plot (a) contains estimates of high accuracy. In plot (c) both precision and accuracy of the estimates are poor.

SEC and SEP indicators characterize calibration precision, while RMSEC and RMSEP describe its accuracy. BIASC and BIASP values determine an offset of calibration regarding the true value (Fig. 10.6b). Section 6.4.2 shows that

$$\mathrm{RMSE}^2 = \mathrm{SE}^2 + \mathrm{BIAS}^2. \tag{10.8}$$

Hence, RMSE indicators should be preferred over SE indicators at the stage of calibration development.

Indicators total residual variance (TRV) and explained residual variance (ERV) characterize the situation "in general," not distinguishing between particular responses, as

$$\mathrm{TRV} = (1/K)\sum_{k=1}^{K} \mathrm{RMSE}^2\,(k). \tag{10.9}$$

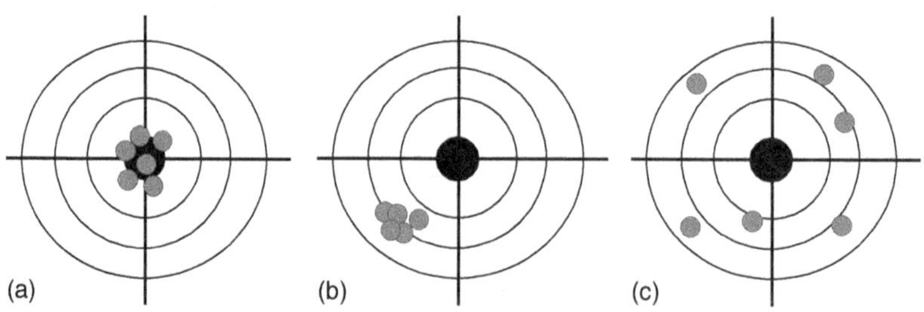

Figure 10.6 Precision and accuracy.

10.1.6 Underfitting and Overfitting

While calibrating, one can often increase the model's complexity. We provide a simple example (see worksheet Gun) to illustrate this process.

From a course of physics at school, we might know that a bullet fired at speed v from a gun tilted under an angle α with respect to the horizon covers a distance of L, which is given by equation

$$L = v^2 \sin(2\alpha)/g.$$

Suppose we have data $L(\alpha)$ presented in Fig. 10.7.

Let us forget the fact that the functional relationship between L (i.e., response y) and α (i.e., variable x) is known and try to develop a polynomial calibration

$$L = b_0 + b_1\alpha + b_2\alpha^2 + \cdots + b_n\alpha^n + \varepsilon.$$

using the first 14 observations as the training set and the last 5 points as the test set.

Figure 10.8a shows training (■) and validation (○) datasets and the calibrated values (curves) for $n = 0, 1, \ldots, 5$. It can be seen that at low polynomial complexity ($n = 0, 1$), the training data are far from the model. This is the case of *underfitting*. When the model complexity increases, the training data and the calibrated values move closer. However, at the redundant complexity level ($n = 4, 5$), the model again provides bad prediction for the test set. This is the case of *overfitting*.

Figure 10.8b shows the RMSEC and RMSEP values for the increasing model complexity. This is a typical plot in which RMSEC monotonously decreases, while RMSEP passes through a minimum. The point of the RMSEP minimum determines the optimal complexity of the model. In our example, it is 2. Now, we can look at the functional model above,

		α	L
calibration	1	0	0
	2	5	1.8525
	3	10	3.8964
	4	15	4.0166
	5	20	7.681
	6	25	6.0134
	7	30	8.8099
	8	35	10.068
	9	40	7.6277
	10	45	11.266
	11	50	8.1213
	12	55	7.4764
	13	60	9.2182
	14	65	7.5216
test	15	70	6.0098
	16	75	4.7233
	17	80	4.4673
	18	85	0.7638
	19	90	0.7246

Figure 10.7 The artillery example data.

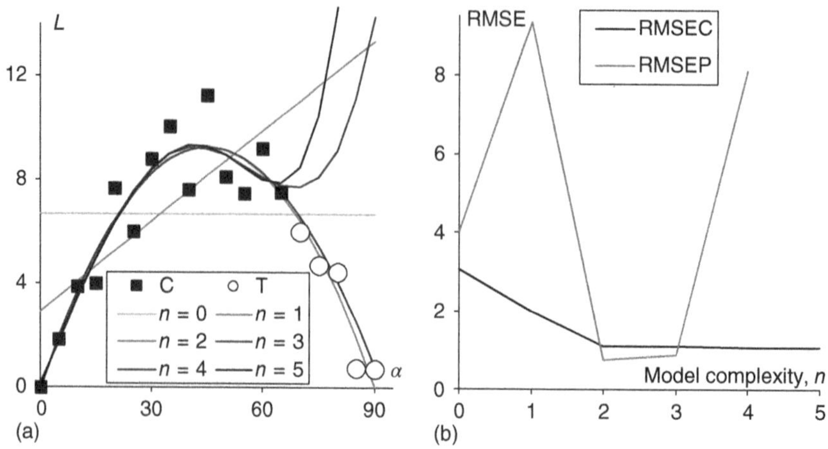

Figure 10.8 The artillery example. Calibration (a), the root mean square errors in calibration (RMSEC) and in validation (RMSEP) (b).

which is well approximated by a second-order polynomial, that is,

$$L = v^2/g \sin(2\alpha) \approx \text{Const}\, \alpha\, (\alpha - \pi/4)$$

In multivariate calibration, overfitting and underfitting are manifested through the choice of the number of latent (principal) components. When this number is small, the model does not fit the training set well, whereas when the model complexity increases, RMSEC monotonically decreases. However, the quality of the test set prediction can get worse (U-shaped RMSEP). The RMSEP minimum point, or rather the beginning of a plateau, corresponds to the optimal number of principal components (PCs). The problem of balanced data fitting has been addressed in the works by A. Höskuldsson, who introduced a new concept of modeling in 1988 called the *H-principle*.[2] According to this principle, the accuracy of modeling (RMSEC) and the accuracy of prediction (RMSEP) are related. Improving RMSEC inevitably entails deterioration of RMSEP, hence they must be considered together. For this reason alone, a multiple linear regression (MLR), which always involves a redundant number of parameters, certainly leads to unstable models that are unsuitable in practice.

10.1.7 Multicollinearity

Consider a multiple linear calibration

$$\mathbf{Y} = \mathbf{XB}. \tag{10.10}$$

In this case, we use only the training set. To simplify notation, we do not use the matrix indices meaning $\mathbf{Y} = \mathbf{Y}_c$ and $\mathbf{X} = \mathbf{X}_c$. A classic solution is given by the least squares method, presented in Section 6.7.2. The estimates of parameters **B** are found by minimizing the sum of squared residuals $(\mathbf{Y}-\mathbf{XB})^t(\mathbf{Y}-\mathbf{XB})$, so

$$\hat{\mathbf{B}} = \left(\mathbf{X}^t\mathbf{X}\right)^{-1}\mathbf{X}^t\mathbf{Y} \tag{10.11}$$

[2] A. Höskuldsson. *Prediction Methods in Science and Technology*, vol. 1, Thor Publishing, Copenhagen, 1996.

The response estimate is given by an equation

$$\hat{\mathbf{Y}} = \mathbf{X}\hat{\mathbf{B}}. \tag{10.12}$$

The main problem of this approach is the inversion of matrix $\mathbf{X}^t\mathbf{X}$. If the number of samples is less than the number of variables ($I < J$), the inverse matrix does not exist. Moreover, matrix $\mathbf{X}^t\mathbf{X}$ can be singular even at a sufficient number of samples ($I > J$). Consider a simple example that is presented in worksheet Multicollinearity.

Figure 10.9 presents the data that have only three samples ($I = 3$) and two variables ($J = 2$). Using an active element in worksheet Multicollinearity, the angle between vectors 1 and 2 can be adjusted between 90° and 0°. The smaller the angle is the smaller the determinant of $\mathbf{X}^t\mathbf{X}$ and the more singular the matrix will be. In extreme case, the angle is 0 and matrix $\mathbf{X}^t\mathbf{X}$ cannot be inverted. Vectors 1 and 2 are collinear, that is, they are located on the same line, which coincides with vector 3. Figure 10.10 illustrates the unfeasibility to develop classic calibration when the data are collinear.

The multivariate calibration (i.e., $J \gg 1$) is always burdened with *multicollinearity*, which is a situation when there exist multiple linkages between the **X** variables. This leads

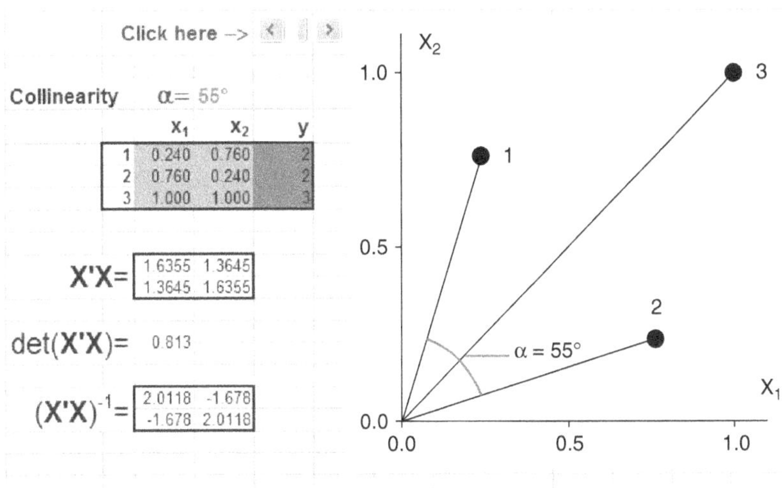

Figure 10.9 Example of collinear data.

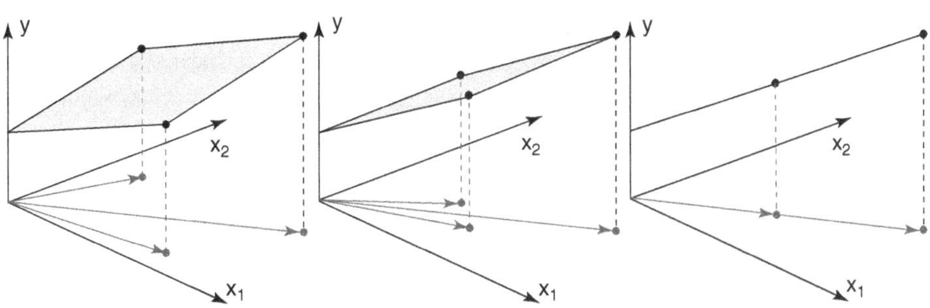

Figure 10.10 Example of collinear data.

to degeneration of matrix $\mathbf{X}^t\mathbf{X}$ that cannot be inverted. Below we will show what can be done in such a case.

10.1.8 Data Preprocessing

Preliminary data preparation is an important prerequisite of the correct modeling, and, therefore, a successful analysis. Preprocessing includes various transformations of the initial "raw" data. The simplest transformations are centering and scaling. They are described in Section 8.1.4.

10.2 SIMULATED DATA

10.2.1 The Principle of Linearity

To illustrate and compare different calibration methods, we will use a simulated example in which prototype is a popular problem of spectral data analysis. It is known that in such experiments the *principle of linearity* holds well. Suppose there are two substances A and B, mixed in concentrations c_A and c_B, then the spectrum of the mixture is

$$\mathrm{X} = c_A \cdot S_A + c_B \cdot S_B,$$

where S_A and S_B are the spectra of pure substances. Note that the same principle is applicable to chromatography. In this case, chromatographic profiles play the role of spectra.

10.2.2 "Pure" Spectra

To simulate "pure" spectra $S(\lambda)$, we use Gaussian peaks

$$S(\lambda) = S_0 \exp\left[-((\lambda - m)/v)^2\right],$$

where S_0, m, v are the spectrum parameters and $\lambda = 0, 1, 2, \ldots, 100$ are the conditional channel numbers (wave number, retention time, etc.). Figure 10.11 shows the spectra of three substances A, B, and C, used in the example. Substances A and B are the sought values (responses) and substance C is added to the system as noise, that is, an unexpected impurity.

It can be seen that the spectra strongly overlap, especially A and B, which have no regions where only one signal is present. This fact greatly hinders the classic approach to calibration. Pure spectra are calculated in worksheet "Pure Spectra." They can be changed by giving new values in cells B4:D6, which correspond to Gaussian peak parameters. For example, you may reduce the overlap by moving spectra A and B further apart and notice the changes in the entire workbook.

10.2.3 "Standard" Samples

In order to create simulated data, we should set the sample concentrations for all system components. It is assumed that we have 14 samples; the first nine samples comprise the training set and the last five comprise the test set. The concentrations are given in sheet Concentrations and shown in Fig. 10.12.

	S_0	m	v	λ-> 0	1	2	3	4	5	6
A	1	45	40	0.282	0.298	0.315	0.332	0.350	0.368	0.386
B	0.909	55	50	0.271	0.283	0.296	0.308	0.321	0.334	0.348
C	1.053	40	15	0.001	0.001	0.002	0.002	0.003	0.005	0.006

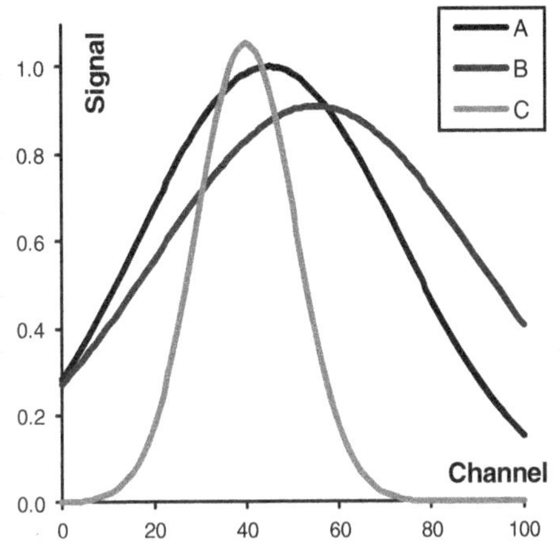

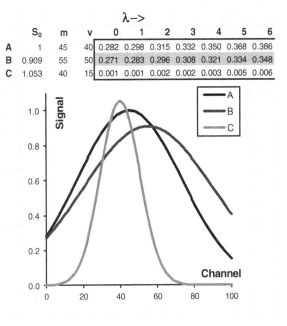

Figure 10.11 Pure spectra.

		A	B	C
calibration	1	0.90	0.85	0.15
	2	0.80	0.35	0.25
	3	0.70	0.25	0.10
	4	0.60	0.95	0.15
	5	0.50	0.45	0.20
	6	0.40	0.15	0.15
	7	0.20	0.55	0.10
	8	0.30	0.65	0.25
	9	0.10	0.75	0.20
test	10	0.55	0.70	0.25
	11	0.75	0.50	0.15
	12	0.25	0.30	0.20
	13	0.35	0.80	0.15
	14	1.45	0.20	0.20
mean		0.50	0.55	0.17

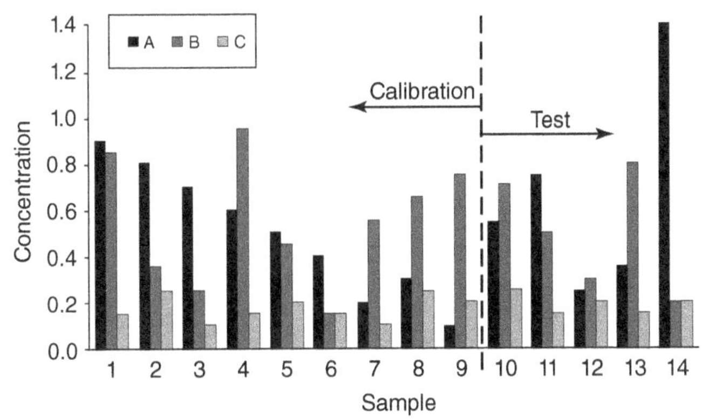

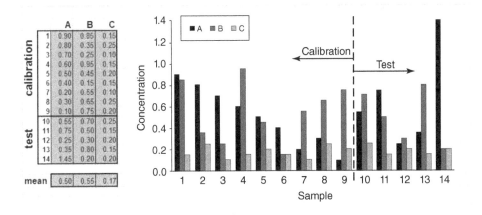

Figure 10.12 Concentrations in the training and test sets.

Note that the dataset contains no "pure" samples, that is, mixtures that contain only one (e.g., A) substance, while other concentrations are equal to zero. This is done to "prevent" the usage of conventional calibration methods.

10.2.4 X Data Creation

To obtain "experimental spectra" the concentrations' matrix **C** should be multiplied by the pure spectra matrix **S**, and the product has to be disturbed by random errors, that is,

$$\mathbf{X} = \mathbf{CS} + \mathbf{E}. \tag{10.13}$$

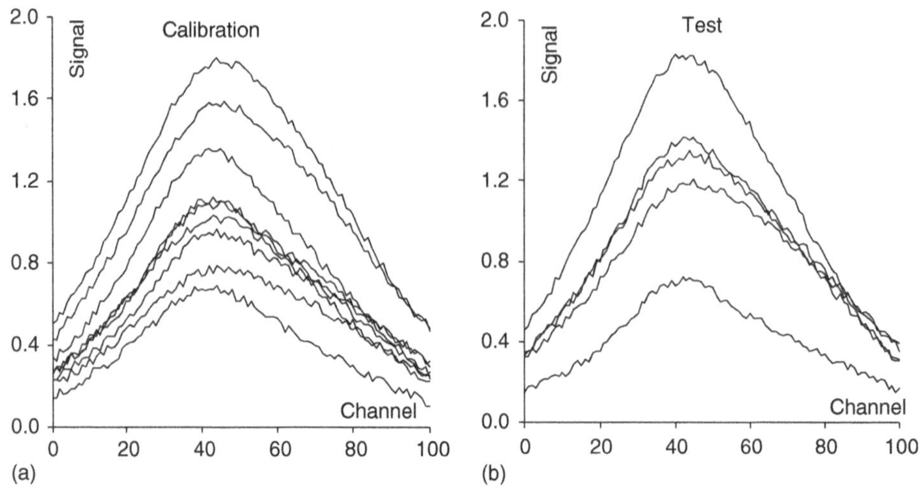

Figure 10.13 Simulated data.

Figure 10.13 presents the obtained training and test sets spectra (see worksheet Data). The modeled errors have a standard deviation of 0.015. The random errors are generated by a simple Visual Basic for Applications (VBA) macro (see Section 7.3.3), which initiates by pressing the button **Add New Error**. The desired value of the error (RMSE) is preset in cell E18, entitled Designed error. The error obtained in the result of random number generation is displayed in cell J18, entitled Obtained error.

Generating new data, one can learn a lot about methods of calibration.

10.2.5 Data Centering

In line with the concept presented in Section 10.1.8, raw data must be properly preprocessed. In this example, it is not necessary to scale the variables because all spectral channels have similar signal values. However, centering both spectra **X** and responses **Y** is necessary for the construction of calibration models.

To center **Y** concentrations, the mean values of each column of the training set are calculated. These averages are then subtracted from the corresponding **Y** values. The **X** data are centered in a similar way, that is, the mean values for each channel of the training set are calculated and then these values are subtracted from **X** data column-wise.

10.2.6 Data Overview

Thus, we have created a simulated dataset (**Y**, **X**). It contains the concentration matrix **Y** of size (14×2) and a spectral matrix **X** of size (14×101). Analyzing the data we will "forget" (i.e., we will not use in calculations) the fact that the system has one more "hidden" component C. It is interesting whether we can detect its presence. In addition, we will not use the spectra of pure substances A and B, which were used to construct the data. We will try to restore them and compare with the original spectra.

The data have been divided into two parts: the training (or calibration) set that includes nine samples and the test (or validation) set consisting of five objects. We will build calibration models employing different methods and using the training set only. The test set will be used for assessing the quality of the models.

The data and their analysis by various calibration methods are housed in Excel workbook Calibration.xls. This book includes the following sheets:

Intro: brief introduction
Layout: explanation of the names of arrays used in the example
Gun: illustration of underfitting and overfitting in calibration
Multicollinearity: illustration of multicollinearity in calibration
Pure Spectra: true pure spectra **S**
Concentrations: true concentration profiles **C**
Data: simulated data used in the example
UVR: univariate (single-channel) calibration
Vierordt: calibration by the Vierordt method
Indirect: indirect calibration
MLR: multiple linear regression
SWR: stepwise regression
PCR: principal component regression
PLS1: projection on latent structures 1
PLS2: projection on latent structures 2
Compare: comparison of different methods

10.3 CLASSIC CALIBRATION

Classic calibration is based on the same principle of linearity, which has been used to create simulated data,

$$\mathbf{X} = \mathbf{CS}. \tag{10.14}$$

In this equation, matrices $\mathbf{X} = \mathbf{X}_c$ and $\mathbf{C} = \mathbf{Y}_c$ represent the training subset of the raw (noncentered) data and $\mathbf{S}$ is the matrix of pure spectra, also called the *sensitivity matrix*. If matrix $\mathbf{S}$ is known a priori, the concentrations are calculated by an equation

$$\mathbf{C} = \mathbf{XS}^+$$

where $\mathbf{S}^+ = \mathbf{S}^t\left(\mathbf{SS}^t\right)^{-1}$ is a pseudo-inverse matrix (see Section 5.1.15). This is the case of *direct* calibration. However, in practice, matrix $\mathbf{S}$ is usually unknown and it should be reconstructed from the training data.

10.3.1 Univariate (Single Channel) Calibration

This is the simplest, naive approach to calibration. If for each substance (analyte) sought we select one channel (variable, wave number) from data $\mathbf{X}$, we get several column-vectors $\mathbf{x}$. Then, using the training part of a vector we develop a simple univariate regression

$$x = sc + b,$$

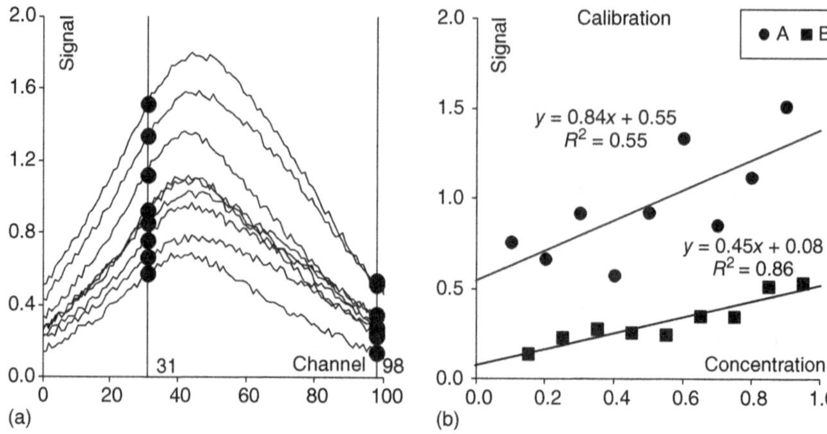

Figure 10.14 Univariate calibration.

for each substance; in this case, one regression for substance A $(c = c_A)$ and another regression for B $(c = c_B)$. The formulas for s (slope) and b (offset) coefficients are given in Section 6.7.1. In Excel, these can be easily obtained by using a standard function **LINEST**, as described in Section 7.2.7. Once coefficients c and b are found, the concentrations are calculated by equation

$$c = (x-b)/s.$$

Worksheet UVR and Fig. 10.14 present an example of two univariate regressions for channels 31 (A) and 98 (B). Figure 10.14b shows regression lines for substances A and B.

Note that each calibration equation contains a free term b, which contradicts the original equation (10.13). The offset term accounts for the influence of background, or, in this case, for the presence of an impurity (substance C). The coefficient s (slope) corresponds to the "pure" spectra reading. Figure 10.15 shows the pure spectra of substances A and B (curves) and the corresponding values of coefficients s (dots).

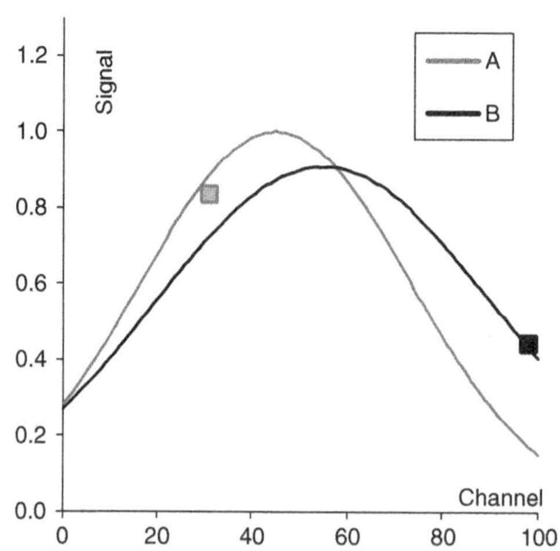

Figure 10.15 Pure spectra (curves) and their estimates (dots) for the selected channels.

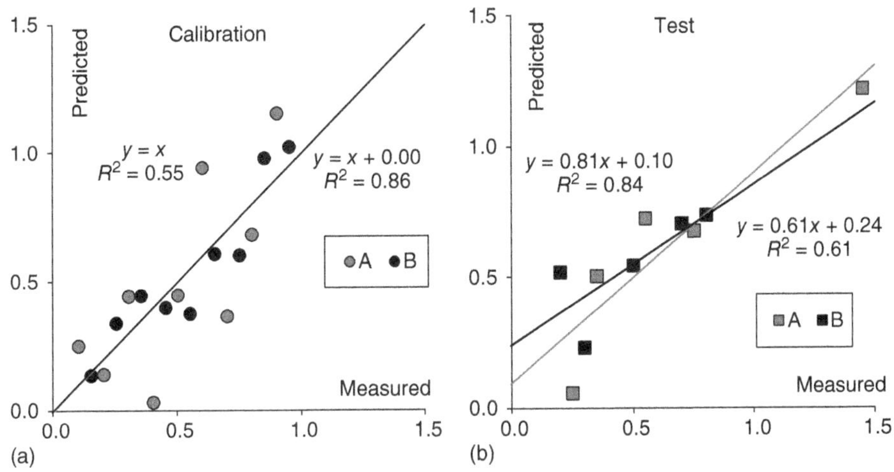

Figure 10.16 The "measured–predicted" plots for the univariate calibration. The training (a) and test (b) sets.

The plots "measured versus predicted" are traditionally used to visually assess the quality of calibration. In these plots, the horizontal axis shows the known (measured) concentration y, whereas the vertical axis represents the corresponding estimated values $\hat{y}$ determined (predicted) by the calibration model. These plots are shown in Fig. 10.16. Apart from the "measured–predicted" values (points), such graphs usually contain a regression line $ky + m$. Coefficients k (slope) and m (offset) characterize the "quality" of calibration. Ideally $k = 1$ and $m = 0$. The farther from this, the worse the calibration is. In addition, these plots usually present the coefficient of correlation R^2 between y and $\hat{y}$.

Table 10.1 shows the quality characteristics of univariate calibration calculated for substances A and B in accordance with the formulas provided in Section 10.1.4.

TABLE 10.1 Quality Characteristics of Univariate Calibration

A		B
$R_c^2 = 0.549$		$R_c^2 = 0.862$
RMSEC = 0.234		RMSEC = 0.103
BIASC = 0.000		BIASC = 0.000
SEC = 0.234		SEC = 0.103
$R_t^2 = 0.842$		$R_t^2 = 0.609$
RMSEP = 0.173		RMSEP = 0.151
BIASP = −0.032		BIASP = 0.050
SEP = 0.170		SEP = 0.151
	TRVC = 0.033	
	ERVC = 0.905	
	TRVP = 0.026	
	ERVP = 0.994	

Certainly, the univariate calibration can be carried out using any channel. The sheet UVR provides the flexibility to change the channel numbers for substances A and B, represented by the values in cells C2 and D2, and look at the variation in calibration quality. For example, the best RMSEP = 0.173 of substance A is obtained at the channel $\lambda = 31$ and the worst RMSEP = 0.702 turns up at $\lambda = 97$. Similarly, for substance B, the best value of RMSEP = 0.151 is reached at $\lambda = 98$ and the worst RMSEP = 0.611 is at $\lambda = 20$.

They say that it is better to build calibration using a point where the corresponding "pure spectrum" has a maximum. The example above clearly demonstrates that this is not true. At channel $\lambda = 45$ (the maximum of A) RMSEP = 0.205 and at channel $\lambda = 55$ (the maximum of B) RMSEP = 0.484.

10.3.2 The Vierordt Method

For the first time the problem of spectrophotometric data analysis was considered in 1873 by Karl von Vierordt. Exploring the changes of hemoglobin in the blood, he proposed a calibration method, which now, after 140 years, is still popular among a number of analysts. The Vierordt method is similar to univariate calibration. It also selects one channel per substance, but the regression equations are considered simultaneously

$$\tilde{\mathbf{X}}_c = \mathbf{Y}_c\mathbf{S}.$$

Here, matrix $\tilde{\mathbf{X}}_c$ has as many columns as the number of substances (we have 2), matrix $\mathbf{Y}_c$ contains known concentrations from the training set (we have 2), and matrix $\mathbf{S}$ is an unknown square (we have 2×2) sensitivity matrix that should be estimated. Note that the traditional Vierordt method does not use a free term, which is in full agreement with Eq. (10.13).

To assess the sensitivity matrix $\mathbf{S}$, we multiply this equation on the left by the transposed concentration matrix $\mathbf{Y}_c$

$$\mathbf{Y}_c^t\tilde{\mathbf{X}}_c = \mathbf{Y}_c^t\mathbf{Y}_c\mathbf{S}.$$

Then the matrix

$$\hat{\mathbf{S}} = \left(\mathbf{Y}_c^t\mathbf{Y}_c\right)^{-1}\mathbf{Y}_c^t\tilde{\mathbf{X}}_c$$

is an estimator of the sensitivity matrix.

Figure 10.17 shows the Vierordt method implementation for our example, where two analytic channels 29 and 100 are selected for the calibration. See also worksheet Vierordt.

The "measured–predicted" plot presents the signal (spectra $\mathbf{x}$) calibration results obtained for the selected channels. The estimated elements of the sensitivity matrix $\mathbf{S}$ that correspond to the elements of the pure spectra matrix are shown in Fig. 10.18.

For calibration and prediction, it is necessary to compute a matrix inverse to the sensitivity matrix $\mathbf{S}$. This can be easily done, as $\mathbf{S}$ is a square matrix. Matrix $\mathbf{V}$

$$\mathbf{V} = \hat{\mathbf{S}}^{-1}$$

is a linear estimator for concentration matrix $\mathbf{Y}$. In order to find the estimates of the training set concentrations, matrix $\mathbf{V}$ should be multiplied by the matrix of the corresponding spectra $\tilde{\mathbf{X}}_c$, that is,

$$\hat{\mathbf{Y}}_c = \tilde{\mathbf{X}}_c\mathbf{V}$$

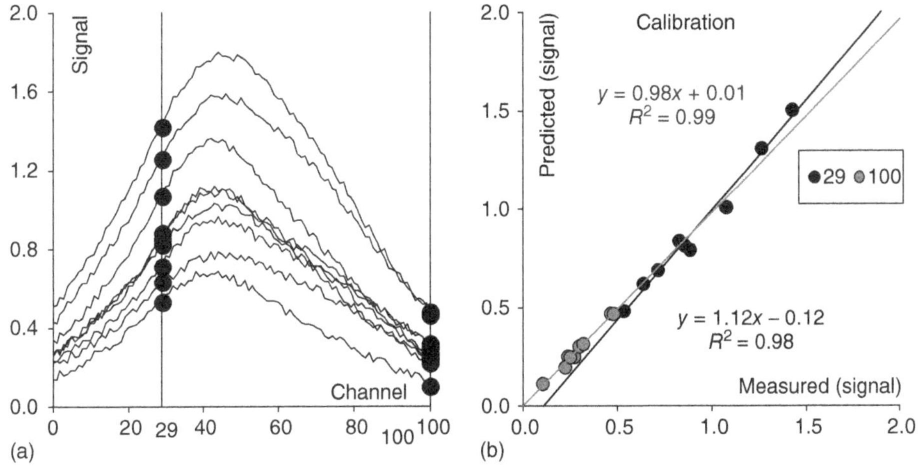

Figure 10.17 The Vierordt method application.

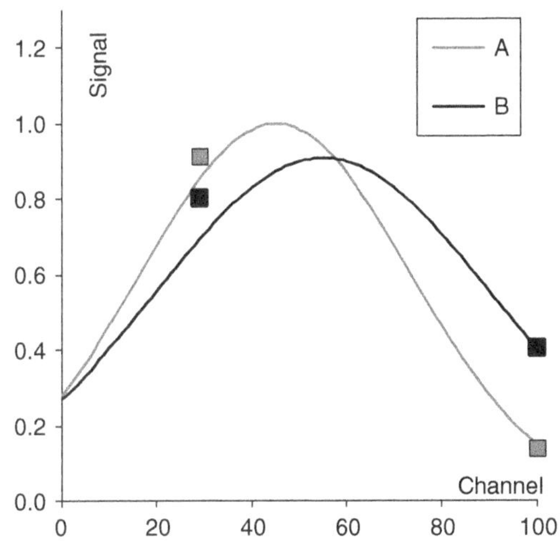

Figure 10.18 Elements of the sensitivity matrix (dots) in the Vierordt method and the pure spectra (curves).

Similarly, to estimate the test set concentrations, matrix $\mathbf{V}$ is multiplied by matrix $\widetilde{\mathbf{X}}_t$, that is,

$$\widehat{\mathbf{Y}}_t = \widetilde{\mathbf{X}}_t \mathbf{V}$$

The "measured–predicted" plots are shown in Fig. 10.19. Table 10.2 shows quality characteristics of the Vierordt calibration. They are calculated by the formulas introduced in Section 10.1.4.

The calibration quality greatly depends on the selection of analytic channels. Often the following method is recommended. Consider a plot of function $F(\lambda) = s_A/s_B$, where s_A and s_B are pure spectra. Using this plot two points λ_1 and λ_2 are found such that the difference

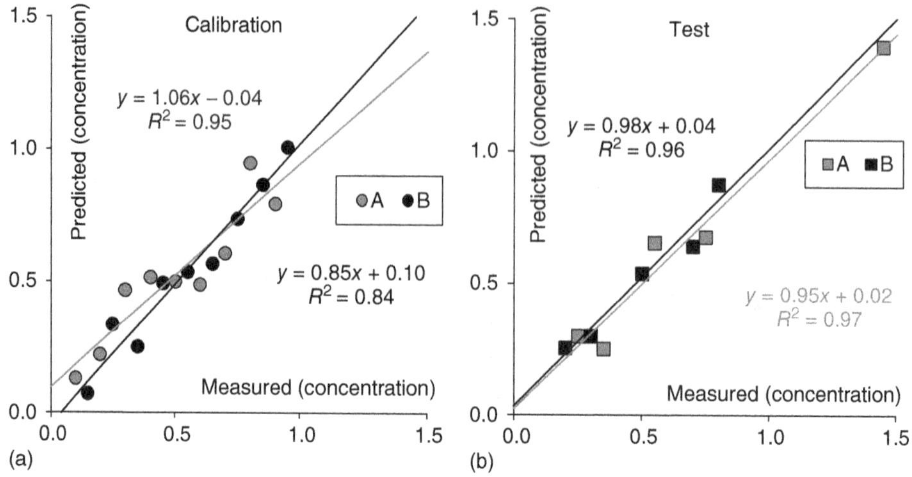

Figure 10.19 The "measured–predicted" plots for the Vierordt calibration. The training (a) and test (b) sets.

TABLE 10.2 Quality Characteristics of the Vierordt Calibration

A		B
$R_c^2 = 0.844$		$R_c^2 = 0.954$
RMSEC = 0.104		RMSEC = 0.062
BIASC = 0.020		BIASC = −0.007
SEC = 0.102		SEC = 0.062
$R_t^2 = 0.967$		$R_t^2 = 0.957$
RMSEP = 0.079		RMSEP = 0.053
BIASP = −0.012		BIASP = 0.023
SEP = 0.078		SEP = 0.047
	TRVC = 0.007	
	ERVC = 0.979	
	TRVP = 0.005	
	ERVP = 0.999	

$|F(\lambda_1)-F(\lambda_2)|$ is maximal. This approach is correct. However, it is impossible to plot $F(\lambda)$ without knowing the pure spectra.

10.3.3 Indirect Calibration

The Vierordt method uses as many channels as the number of substances to be determined. It is intuitive to generalize this approach utilizing all data available. Let us consider a general equation (10.14) of classic calibration

$$\mathbf{X} = \mathbf{CS}.$$

Similar to the Vierordt method, we multiply the equation by the transposed matrix $\mathbf{Y}_c$

$$\mathbf{Y}_c^t\mathbf{X}_c = \mathbf{Y}_c^t\mathbf{Y}_c\mathbf{S}.$$

Then the matrix

$$\hat{\mathbf{S}} = \left(\mathbf{Y}_c^t\mathbf{Y}_c\right)^{-1}\mathbf{Y}_c^t\mathbf{X}_c$$

is an estimator of the pure spectra matrix $\mathbf{S}$. Note that this is not a square matrix. Its dimension is $A \times J$ (in our example, it is 2×101). Let us compare the estimate with the matrix of pure spectra A and B. Figure 10.20 shows the true and the reconstructed spectra. The difference is mostly seen in the area where the impurity substance C has a large contribution. At the edge (at $\lambda < 20$ and $\lambda > 60$), where spectrum C is close to zero, spectra A and B are recovered almost perfectly. Unfortunately, having no idea about the presence of substance C, we cannot notice this fact.

Continuing with the calibration development, we set

$$\mathbf{S} = \hat{\mathbf{S}} = \mathbf{U}$$

and multiply Eq. (10.14) on the right by matrix $\mathbf{U}^t$, that is,

$$\mathbf{X}_c\mathbf{U}^t = \mathbf{Y}_c\mathbf{U}\mathbf{U}^t.$$

Then the (101×2) matrix

$$\mathbf{V}^t = \mathbf{U}^t\left(\mathbf{U}\mathbf{U}^t\right)^{-1}$$

is a linear estimator for the concentration matrix $\mathbf{Y}$.

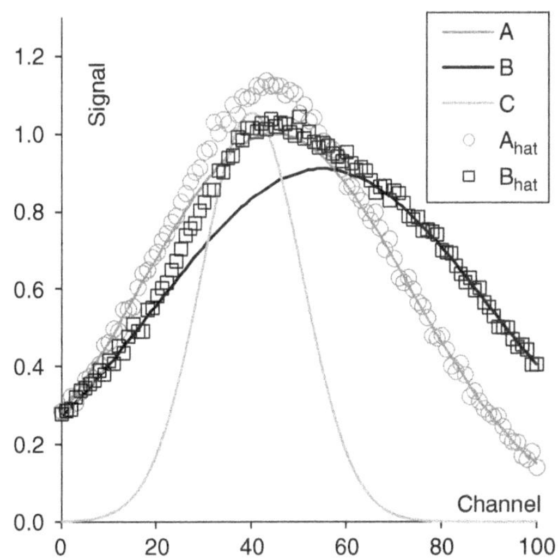

Figure 10.20 Estimation of the pure spectra matrix by indirect calibration.

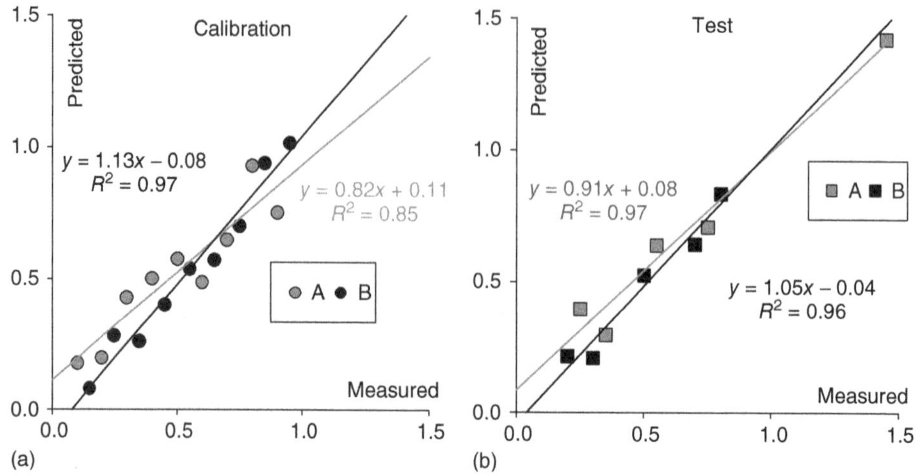

Figure 10.21 The "measured–predicted" plots for the indirect calibration. The training (a) and the test (b) sets.

In order to find the concentrations in the training set, matrix **V** should be multiplied by the matrix of the corresponding spectra $\mathbf{X}_c$, that is,

$$\hat{\mathbf{Y}}_c = \mathbf{X}_c\mathbf{V}.$$

Similarly, to predict the test set concentrations, matrix **V** is multiplied by matrix $\mathbf{X}_t$, that is,

$$\hat{\mathbf{Y}}_t = \mathbf{X}_t\mathbf{V}.$$

The "measured–predicted" plots are shown in Fig. 10.21. Table 10.3 shows the quality characteristics of indirect calibration calculated by formulas introduced in Section 10.1.4.

There is a strong opinion that classic calibration is better than inverse calibration. To explain justification of the benefits, it is claimed that the classic calibration is based on fundamental physical principles, such as the Lambert–Beer law. From a mathematical point of view, it is the opposite. Firstly, classic calibration is inverted regarding the desired **Y** values. Secondly, the procedure used to estimate the responses is unstable. Indeed, if there is a linear relationship between concentrations A and B, for example, $c_B \sim 0.5c_A$, matrix $\mathbf{Y}^t\mathbf{Y}$ is either singular or its inverse is calculated with a huge error. The linear interdependences between the concentrations often manifest themselves during the environmental analysis and are caused by natural reasons. Thirdly, classic calibration does not allow deviations from an ideal system, for example, the impurities (substance C) cannot be detected. Finally, we shall see that the quality of classic calibration (accuracy, precision) is inferior to many modern methods.

10.4 INVERSE CALIBRATION

A basic model of inverse calibration is given by Eq. (10.10)

$$\mathbf{Y} = \mathbf{XB}$$

TABLE 10.3 Quality Characteristics of the Indirect Calibration

A		B
$R_c^2 = 0.854$		$R_c^2 = 0.968$
RMSEC = 0.102		RMSEC = 0.064
BIASC = 0.024		BIASC = −0.015
SEC = 0.099		SEC = 0.062
$R_t^2 = 0.968$		$R_t^2 = 0.960$
RMSEP = 0.084		RMSEP = 0.052
BIASP = 0.024		BIASP = −0.013
SEP = 0.080		SEP = 0.050
	TRVC = 0.007	
	ERVC = 0.979	
	TRVP = 0.005	
	ERVP = 0.999	

where the targeted value **Y** (concentration) is expressed directly in terms of the known spectra matrix **X**. Despite the fact that such a representation of calibration equation contradicts the fundamental relation given in Eq. (10.14), this approach provides a better quality of modeling.

10.4.1 Multiple Linear Calibration

MLR (see Section 6.7.3) is the simplest version of the inverse calibration. In Section 10.1.7, we have already discussed the MLR properties in relation to multicollinearity problem. In particular, it is noted that the number of variables in multiple regression must be less than the number of training samples. In our example, the number of training samples is 9; therefore, we should select 8 channels of 101 and then build the MLR calibration for these selected variables only. Certainly, we can use fewer variables but not more. Worksheet MLR contains an active element, which can be used to change the selection of the first channel; the remaining 7 channels are calculated automatically.

Figure 10.22 shows the channels selected with a regular step of 13. The first channel can be selected randomly between 0 and 9, all subsequent variables are uniquely defined. The result of this selection is a (14 × 8) matrix of independent variables **X**, consisting of two blocks, the training set (9 × 8) and the test set (5 × 8). With the help of the training set, we construct a multiple regression between **X** and **Y** given by Eq. (10.10). This can be obtained using Eqs (10.11) and (10.12), but the easier way is to use an Excel standard function **TREND** described in Section 7.2.7.

Figure 10.23 shows the "measured–predicted" plots for the multiple calibration. It is evident that the training set model is overfitted, while the test set prediction is unsatisfactory. Both the offset and low correlation are shown in Fig. 10.23b. Table 10.4 shows the MLR calibration quality. These characteristics are calculated by formulas from Section 10.1.4.

We are dealing with a typical overfitted model (see Section 10.1.6), that is, the number of selected variables is too large. The attempts to change the set of variables will not improve

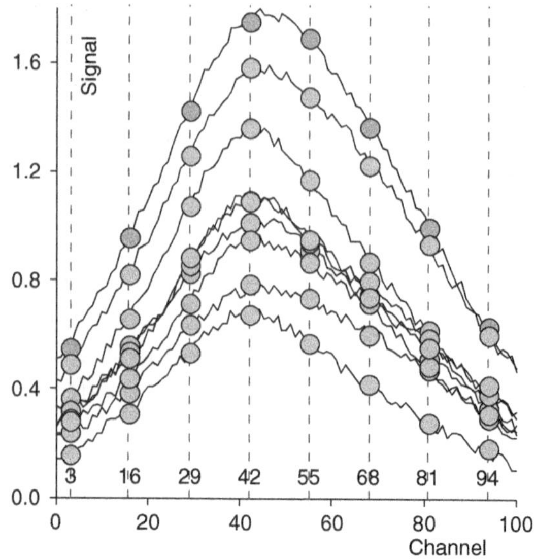

Figure 10.22 Multiple linear regression.

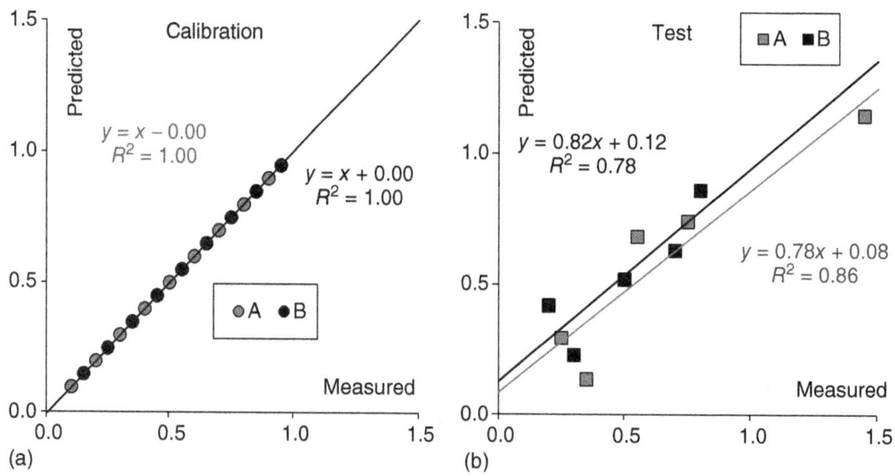

Figure 10.23 The "measured–predicted" plots for the MLR calibration. The training (a) and the test (b) sets.

the situation, making multiple calibration an unacceptable method. It leads to model overfitting and provides unsatisfactory results when it is applied to a new (test) set of samples.

10.4.2 Stepwise Calibration

We have just seen that the multiple linear calibration is not satisfactory and presents a clear example of overfitting. In this section, we discuss a *stepwise regression* (SWR), in which the variable selection method introduces a way to cope with overfitting. The idea of the method is as follows.

TABLE 10.4 Quality Characteristics of Multiple Calibration

A		B
$R_c^2 = 1.000$		$R_c^2 = 1.000$
RMSEC = 0.000		RMSEC = 0.000
BIASC = 0.000		BIASC = 0.000
SEC = 0.000		SEC = 0.000
$R_t^2 = 0.863$		$R_t^2 = 0.782$
RMSEP = 0.176		RMSEP = 0.112
BIASP = −0.068		BIASP = 0.033
SEP = 0.163		SEP = 0.107
	TRVC = 0.000	
	ERVC = 1.000	
	TRVP = 0.022	
	ERVP = 0.995	

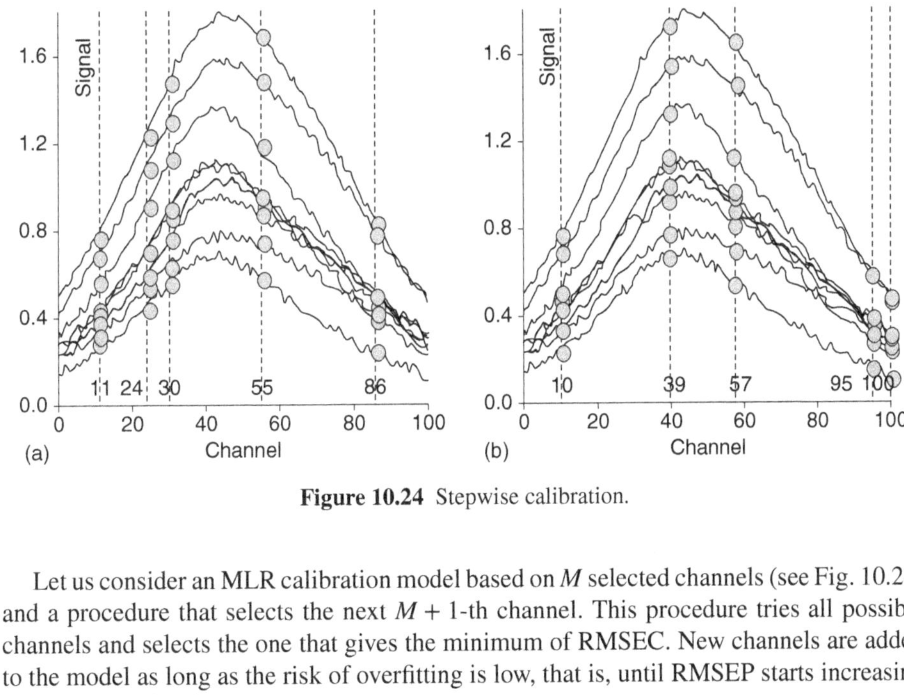

Figure 10.24 Stepwise calibration.

Let us consider an MLR calibration model based on M selected channels (see Fig. 10.24) and a procedure that selects the next $M + 1$-th channel. This procedure tries all possible channels and selects the one that gives the minimum of RMSEC. New channels are added to the model as long as the risk of overfitting is low, that is, until RMSEP starts increasing (see Section 10.1.6).

Obviously, in our example (see worksheet SWR), the best result for substances A and B is achieved at different channels. For A, these channels are 24, 86, 11, 30, … , and for B, these are 100, 10, 95, 39, 57, … All channels are presented in the order of selection. Channel selection is a simple but a time-consuming procedure, which can be easily automated through a small Excel macro. There are many ways to select the "optimal" variables in

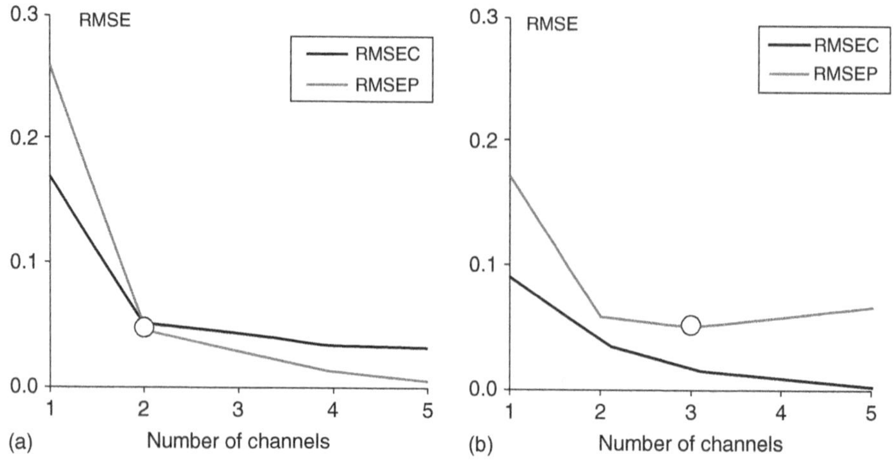

Figure 10.25 (a,b) Root mean square error of training (RMSEC) and of validation (RMSEP) in stepwise calibration.

SWR. The method presented here is a very simple one. One should just choose the channel at which the minimum of the root mean square error in training, RMSEC, is achieved.

Figure 10.25 shows how the root mean square error in training (RMSEC) and in validation (RMSEP) change when the number of the SWR channels grows. Following the principle of RMSEP minimum, the optimal number of channels for substance B is three. This is clearly seen in Fig. 10.25b. However, the choice of the channel number for substance A is difficult. The RMSEP curve in Fig. 10.25a has no minimum. This often happens during the analysis of complicated data. In our example, the optimal channels for substance A are located on the edges of the spectral area where the impact of the hidden impurity C is immaterial. Compare Figs 10.11 and 10.24. SWR calibration for substance A cannot "notice" the presence of substance C. In case of doubt, one should choose a break point on the RMSEP curve. Hence, we choose only two channels for substance A.

Figure 10.26 presents the "measured–predicted" plots for the stepwise calibration. It can be seen that the fitting is much better balanced here, that is, the difference in the accuracy of training and validation is not as drastic as that in MLR calibration.

Table 10.5 shows the SWR calibration quality. These characteristics are calculated by formulas given in Section 10.1.4.

Summarizing, we can conclude that the SWR gave the best results among all calibration methods that have been investigated. However, there are methods that are even more accurate.

10.5 LATENT VARIABLES CALIBRATION

10.5.1 Projection Methods

In stepwise calibration, we used only two or three most informative channels from the 101 channels available. Other channels were rejected as "excessive," that is, unnecessary. The method gives an acceptable result; however, the approach is somewhat wasteful because the discarded data may contain useful information. In particular, using SWR, we did not

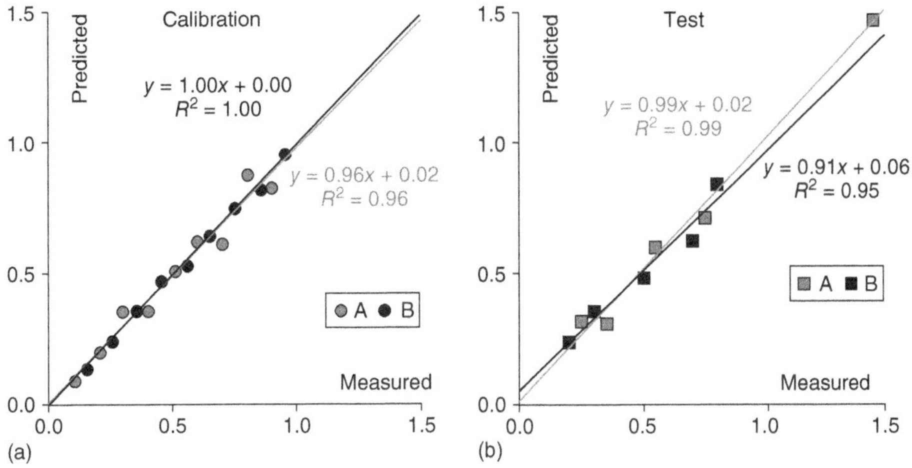

Figure 10.26 The "measured–predicted" plots for the SWR calibration. The training (a) and the test (b) sets.

TABLE 10.5 Quality Characteristics of the SWR Calibration

A		B
$R_c^2 = 0.961$		$R_c^2 = 0.996$
RMSEC = 0.051		RMSEC = 0.015
BIASC = 0.000		BIASC = 0.000
SEC = 0.051		SEC = 0.015
$R_t^2 = 0.989$		$R_t^2 = 0.954$
RMSEP = 0.031		RMSEP = 0.052
BIASP = 0.013		BIASP = 0.013
SEP = 0.046		SEP = 0.050
	TRVC = 0.001	
	ERVC = 0.996	
	TRVP = 0.002	
	ERVP = 0.995	

detect the presence of the third substance, being impurity C, which manifests itself at the excluded channels. We consider the problem of balancing the desire to employ all available data (channels) below, and, at the same time, to avoid overfitting and multicollinearity, which are inevitable in case of a large number of regression variables.

The solution to this dilemma was proposed by Karl Pearson in the year 1901. We should use new latent variables $\mathbf{t}_a$, $(a = 1, \ldots, A)$ that are linear combinations of the original variables $\mathbf{x}_j$ $(j = 1, \ldots, J)$, that is,

$$\mathbf{t}_a = \mathbf{p}_{a1}\mathbf{x}_1 + \cdots + \mathbf{p}_{aJ}\mathbf{x}_J,$$

or in a matrix form

$$\mathbf{X} = \mathbf{T}\mathbf{P}^{t} + \mathbf{E} = \sum_{a=1}^{A} \mathbf{t}_a \mathbf{p}_a^{t} + \mathbf{E}. \tag{10.15}$$

In this equation, **T** is the *scores matrix*. It is of dimension $(I \times A)$. Matrix **P** is the *loadings matrix* and is of dimension $(A \times J)$. **E** is a $(I \times J)$ matrix of residuals.

New latent variables $\mathbf{t}_a$ are called the *principal components* (PCs) and therefore the method is known as the *principal component analysis* (PCA). This method is described in detail in Chapter 9. The number of columns $\mathbf{t}_a$ in matrix **T** and $\mathbf{p}_a$ in matrix **P** equals to an effective (chemical) rank of matrix **X**. This value is denoted by A and is called the *number of PCs*. It is certainly less than the number of variables J or the number of samples I.

To illustrate PCA, we get back to the method used for the simulated data construction (Section 10.2.4). Matrix **X** is a product of the concentration matrix **C** and the matrix of pure spectra **S**; see Eq. (10.13). The number of rows in **X** is equal to the number of samples (I), that is, each row corresponds to a sample spectrum recorded for J wavelengths. The number of rows in matrix **C** is also equal to I, and the number of its columns corresponds to the number of the mixture components ($A = 3$). The number of rows in matrix **S** is equal to the number of channels (wavelengths) J and the number of columns equals A. In the analysis of the experimental data **X** burdened with the errors represented by matrix **E** in Eq. (10.13), the effective rank A may not correspond to the actual number of the components in the mixture.

The PCA method is often used for the explorative analysis of chemical data. In general, the scores matrix **T** and the loadings matrix **P** cannot be interpreted as concentrations and spectra. Similarly, the number of PCs A is not the number of chemical components presented in the system. However, even a formal analysis of the scores and loadings can be very useful for better understanding of the data structure. We give a simple two-dimensional illustration of PCA.

Figure 10.27a presents the data consisting of two strongly correlated variables x_1 and x_2. In Fig. 10.27b, the same data are presented in new coordinates. Vector $\mathbf{p}_1$ of the first principal component (PC1) loading determines a new axis direction along which the largest data

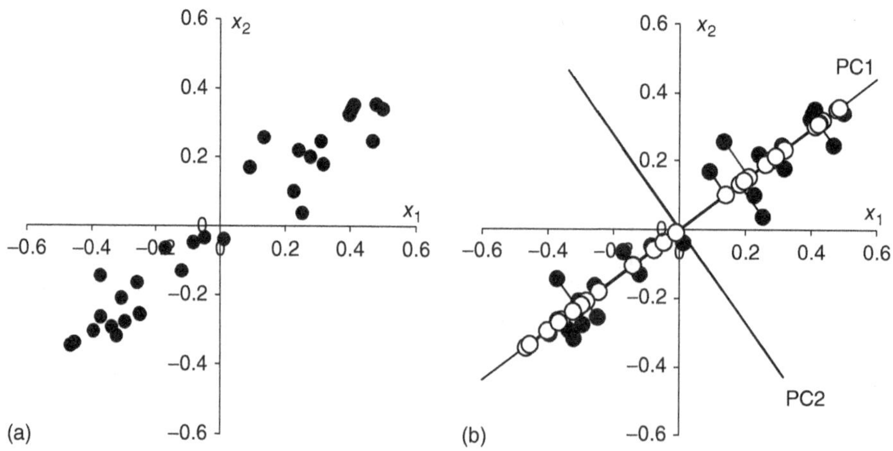

Figure 10.27 Graphic illustration of the principal components. (a) Data in the original coordinates and (b) data in the principal components coordinates.

variation is observed. The projections of all initial data points onto this axis compose vector $\mathbf{t}_1$. The second principal component is orthogonal to the first one, and its direction (PC2) corresponds to the largest variation in the residuals shown by the segments perpendicular to the first axis. This trivial example shows that the method of PCs is carried out in steps. At each step, the residuals $\mathbf{E}_a$ are explored. As the direction (axis) of the largest variation is chosen, the data are projected onto this axis and new residuals are calculated, etc. This algorithm is called *non-linear iterative partial least squares* (NIPALS). It is presented in Section 15.5.3.

PCA can be interpreted as a projection of the data onto a subspace of a lower dimension A. The emerging residuals $\mathbf{E}$ are considered as noise, which does not carry significant chemical information. Figure 10.28 provides a graphical illustration of this interpretation.

In case PCA is used to solve a calibration (regression) problem, it is called the *principal component regression* (PCR). In PCA, the projection is based on data $\mathbf{X}$ only and the response matrix $\mathbf{Y}$ is not used. A PCA generalization is the *method of projection on latent structures* (PLS), which is the most popular method of multivariate calibration now. This method is described in Section 8.1.3. PLS is a lot like PCR, with an important distinction that PLS conducts a simultaneous decomposition of matrices $\mathbf{X}$ and $\mathbf{Y}$.

$$\mathbf{X} = \mathbf{TP}^{\mathrm{t}} + \mathbf{E}, \quad \mathbf{Y} = \mathbf{UQ}^{\mathrm{t}} + \mathbf{F}, \quad \mathbf{T} = \mathbf{XW} + \mathbf{G} \tag{10.16}$$

Projections are constructed coherently in order to maximize the correlation between vector $\mathbf{t}_a$ of X-scores ($\mathbf{T}$) and vector $\mathbf{u}_a$ of Y-scores ($\mathbf{U}$). Therefore, the PLS decomposition better accounts for the complex relationships, using a smaller number of PCs.

When there are multiple responses $\mathbf{Y}$ (i.e., $K > 1$), one can construct two projections of the original data, PLS1 and PLS2. In PLS1, a particular projection subspace is built for each response $\mathbf{y}_k$ separately. Both scores $\mathbf{T}$ ($\mathbf{U}$) and loadings $\mathbf{P}$ ($\mathbf{W}$, $\mathbf{Q}$) depend on the response in use. PLS2 method uses a joint projection subspace, common for all responses. See Fig. 10.29.

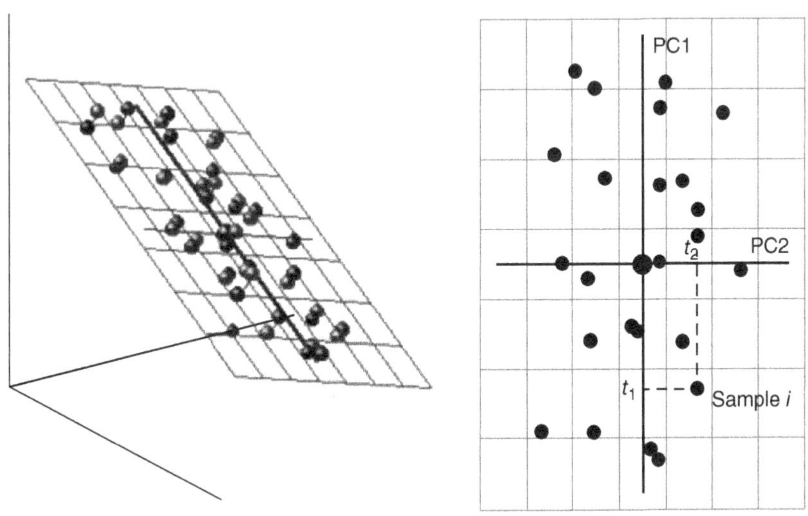

Figure 10.28 Principal component analysis as a subspace projection.

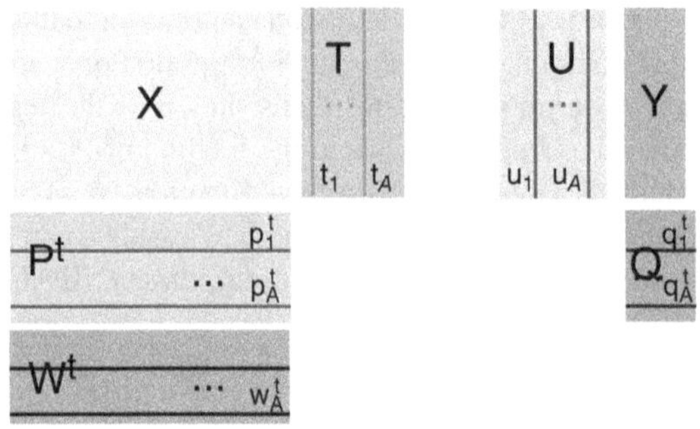

Figure 10.29 Graphical presentation of PLS method.

From Eqs (10.15) and (10.16), it is clear that both PCA and PLS methods do not take into account the constant term in the decomposition of matrix **X**. It is initially assumed that all columns of matrices **X** and **Y** have a zero mean, that is,

$$\sum_{i=1}^{I} x_{ij} = 0 \text{ and } \sum_{i=1}^{I} y_{ik} = 0, \quad \text{for each } j = 1, \ldots, J \text{ and } k = 1, \ldots, K.$$

This condition can be easily satisfied by data centering (Sections 10.1.8 and 10.2.5). However, after calculating a calibration model based on the centered data, we should not forget to recalculate the obtained estimates $\hat{\mathbf{y}}_{\text{center}}$ adding back the mean values vector

$$\hat{\mathbf{y}} = \hat{\mathbf{y}}_{\text{center}} + m_k \mathbf{1}, \quad m_k = \frac{1}{I}\sum_{i=1}^{I} y_{ik}$$

We list some useful formulas below, the proofs of which can be found in numerous textbooks on multivariate analysis.

In all methods

$$\mathbf{T}^{\mathrm{t}}\mathbf{T} = \operatorname{diag}\left(\lambda_1, \ldots, \lambda_A\right)$$

In PCR

$$\mathbf{P}^{\mathrm{t}}\mathbf{P} = \mathbf{I} = \operatorname{diag}(1, \ldots, 1)$$

In PLS

$$\mathbf{W}^{\mathrm{t}}\mathbf{W} = \mathbf{I} = \operatorname{diag}(1, \ldots, 1)$$

10.5.2 Latent Variables Regression

Once data **X** are projected onto the A-dimensional subspace, the initial calibration problem given by Eq. (10.10) transforms into a series of regressions on the latent variables **T**, that is,

$$\mathbf{T}_k \mathbf{b}_k = \mathbf{y}_k, \quad k = 1, \ldots, K \tag{10.17}$$

Here, the index k numbers the responses in matrix **Y**. Note that in both PCR and PLS2 methods, there is only one matrix of latent variables, that is, $\mathbf{T} = \mathbf{T}_1 = \cdots = \mathbf{T}_K$. Vectors $\mathbf{b}_k$

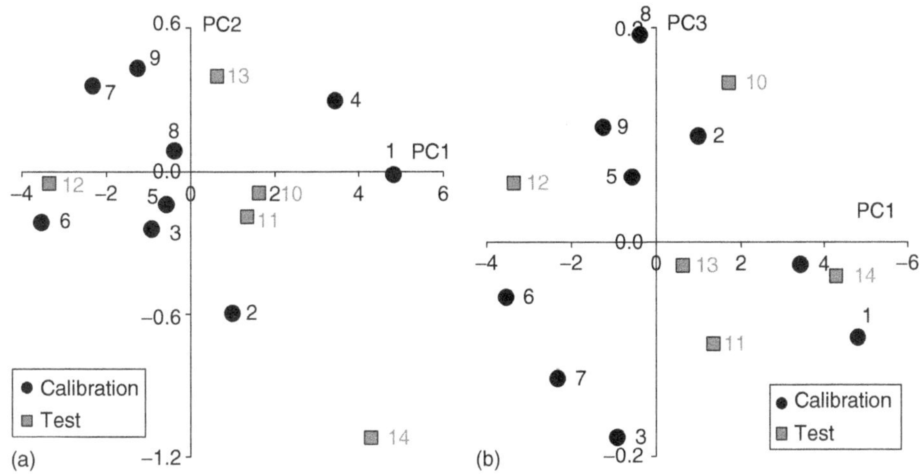

Figure 10.30 **T**-scores plots in RCR.

represent the unknown coefficients. The scores matrices $\mathbf{T}_k$ are of dimension $(I \times A)$, where A is much smaller than the number of variables J. The latter implies that matrix $\mathbf{T}_k^t\mathbf{T}_k$ can be steadily inverted, and multicollinearity (Section 10.1.7) is not an issue. However, overfitting is still a problem (Section 10.1.6) and must be resolved.

The complexity of the projection model is completely determined by the number of PCs A, which is an effective dimension of multivariate data. The choice of A is a major challenge in projection methods. There are many ways to determine A. We have already noted that it often relates to the chemical rank of the system, that is, the number of components present. However, the most universal way to assess A is through the analysis of RMSE plots as it has already been done in Sections 10.1.6 and 10.4.2.

While constructing a multivariate calibration, a great attention is paid to the scores and loadings plots. They contain information useful for understanding the data structure. In the scores plots (Fig. 10.30) each sample is represented by coordinates $(\mathbf{t}_i, \mathbf{t}_j)$, often by $(\mathbf{t}_1, \mathbf{t}_2)$. A proximity of two points (e.g., samples 6 and 7) means that the samples are similar. The points that are close to the origin (e.g., samples 5 and 11) represent the most typical samples. In contrast, the points that are far away (e.g., samples 1 and 14) are suspicious extremes, or maybe even outliers, which are discussed in Section 6.3.2.

While the scores plot is used to analyze the samples relationship, the loadings plot is applied to assess the role of variables. Figure 10.31 shows the loadings coefficients depending on the channel number j. The following pattern can be observed. The greater the number of PCs the noisier the loadings. The first component loading is almost smooth, and its shape is similar to the spectra of the pure substances A and B (see Fig. 10.11). The second and third PCs have more complex forms, but some trends still can be traced there. As of component four, only noise is visible. Such analysis of the loadings plots helps to understand that in this example only two or three PCs should be used.

10.5.3 Implementation of Latent Variable Calibration

The latent variables calibration is usually based on a centered training set $(\mathbf{X}_c, \mathbf{Y}_c)$. In order to predict a test or a new dataset $(\mathbf{X}_t, \mathbf{Y}_t)$, they should be preprocessed too, and then projected

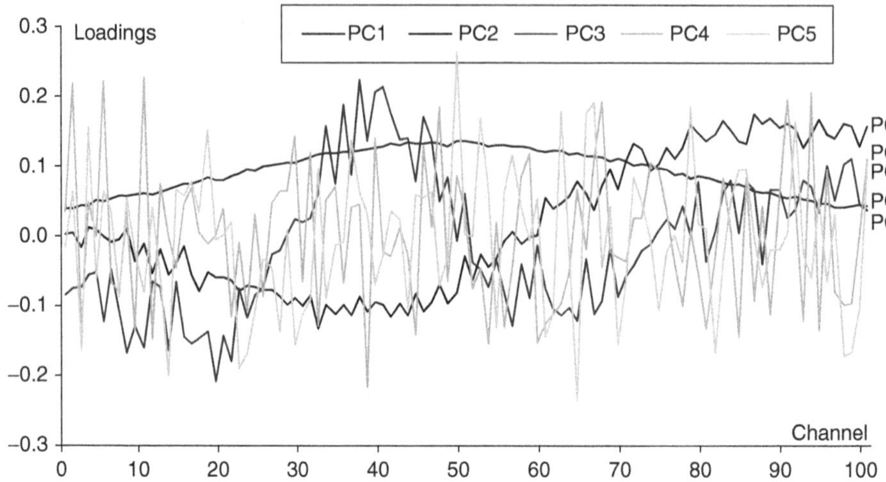

Figure 10.31 Loadings **P** in PCR.

onto the developed subspace. In other words, the scores values $\mathbf{T}_t$ have to be found. In PCR this is very simple, as

$$\mathbf{T}_t = \mathbf{X}_t\mathbf{P},$$

where **P** is the loadings matrix found with the help of the training set.

For PLS, the situation is much more complicated. Computational aspects of multivariate calibration including all relevant algorithms and related MATLAB® codes for PCA, PLS1, and PLS2 are presented in Chapter 15.

There are numerous programs (see Section 4.4) developed to work with projection methods. Among them are huge commercial packages, such as the Unscrambler and Simca-P, which present a special environment for multivariate data manipulation. In particular, all the calibration methods discussed above are implemented in these packages. Multivariate Analysis Add-in deserves a prominent spot among free packages. It is an Excel add-in developed at Chemometrics Center of the University of Bristol under the direction of Prof. R. Brereton.

In this book, all projection calculations (PCA/PLS) are performed using a special Excel add-in called Chemometrics Add-In. It supplements the list of standard Excel functions and enables calculation of projection decompositions directly in a workbook. The setup instructions are provided in Chapter 3, and a detailed user tutorial can be found in Chapter 8.

10.5.4 Principal Component Regression (PCR)

Let us consider the PCR application for our simulated example. All calculations are shown on worksheet PCR.

Firstly, using the training dataset $\mathbf{X}_c$, the PCA scores matrix **T** is found. It is of dimension (9×5), that is, the matrix consists of five columns, which are the PCs $\mathbf{t}_a$, $a = 1, 2, \ldots, 5$. From the regression point of view, **T** is a matrix of predictors, that is, of independent variables. Two responses are the centered concentrations of substances A and B. With the help of a standard **TREND** function, five regressions are developed for each response, using one, two, … , five PCA PCs, respectively. In this way, the PCR calibration is conducted for the training set. There is a fine detail to these simple calculations. The regressions are built using centered concentrations values, A_c and B_c, simply because Excel 2003 has a critical error corrected in the later versions of Excel starting with Excel 2007. This bug is discussed in Section 7.2.8.

The test set data are also projected onto the PCA subspace with their scores $\mathbf{t}_1, \ldots, \mathbf{t}_5$ calculated. These values are pasted into the developed calibration models, and the corresponding estimates of the centered concentrations A_c^{hat} are found. The final predicted concentrations values A^{hat} are obtained by correcting for centering.

The values of the root mean square errors in calibration (RMSEC) and in validation (RMSEP) are presented in Fig. 10.32.

Figure 10.32 shows that for both substances the RMSEP minimum is obtained at three PC ($A = 3$). Thus, using PCR, we can easily reveal that the system has not two but three components.

Figure 10.33 shows the TRV and the ERV. These plots also confirm that the effective dimension is three. At PC = 3, the explained variance is almost equal to 1, while the total variance is close to 0. At the same time, it is clear that the RMSE plots (Fig. 10.32) are more visible than the variance plots (Fig. 10.33).

Figure 10.34 presents the "measured–predicted" plots for PCR with three PCs. The result looks perfect as it has high correlation coefficients and small biases both in the training and in the test sets. Comparing these plots with their counterparts in other methods, that are, Figs 10.16, 10.19, 10.21, 10.23, and 10.26, one can see the obvious benefits of the PCR.

Table 10.6 shows the PCR calibration quality. These characteristics are calculated by formulas given in Section 10.1.4.

We have discovered that PCR has evident advantages over classic calibration methods. This method of calibration is more accurate and has a smaller bias. This is explained by the fact that in a multivariate calibration, all available data are used. At the same time, the data are modified to avoid both multicollinearity and overfitting.

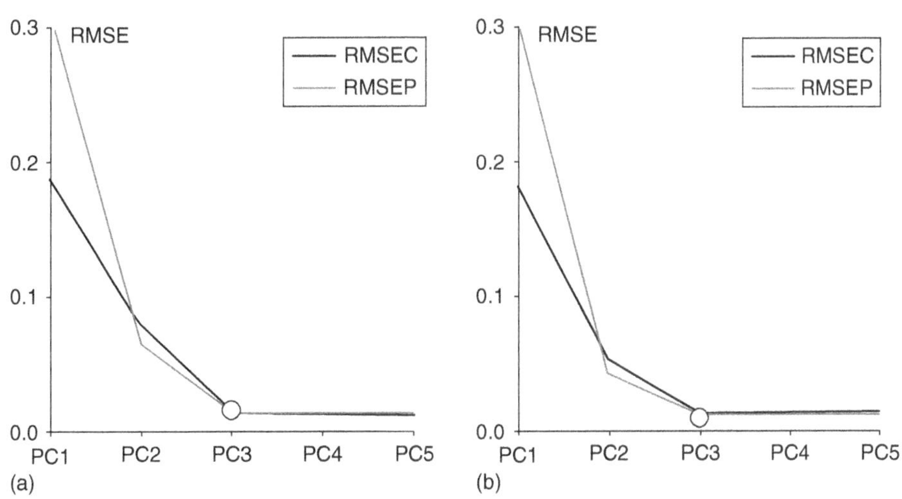

Figure 10.32 The root mean square errors in calibration (RMSEC) and validation (RMSEP) for PCR.

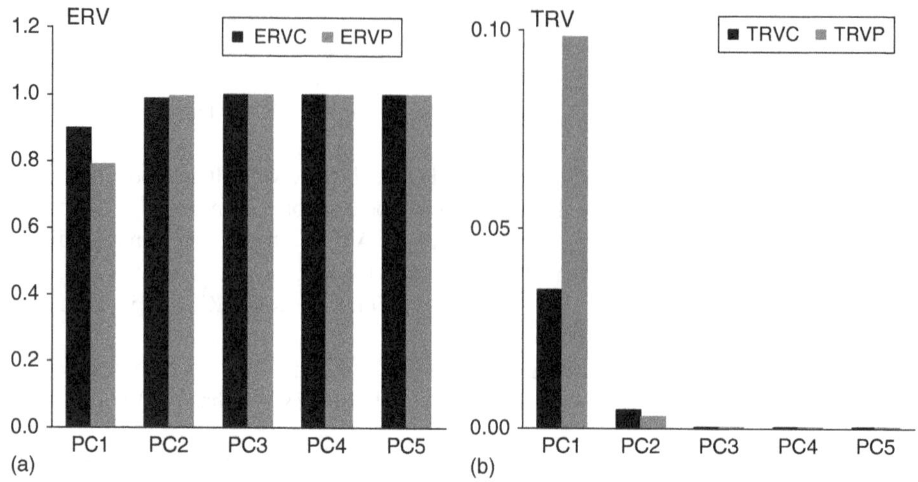

Figure 10.33 Total (TRV) and explained (ERV) residual variances in PCR.

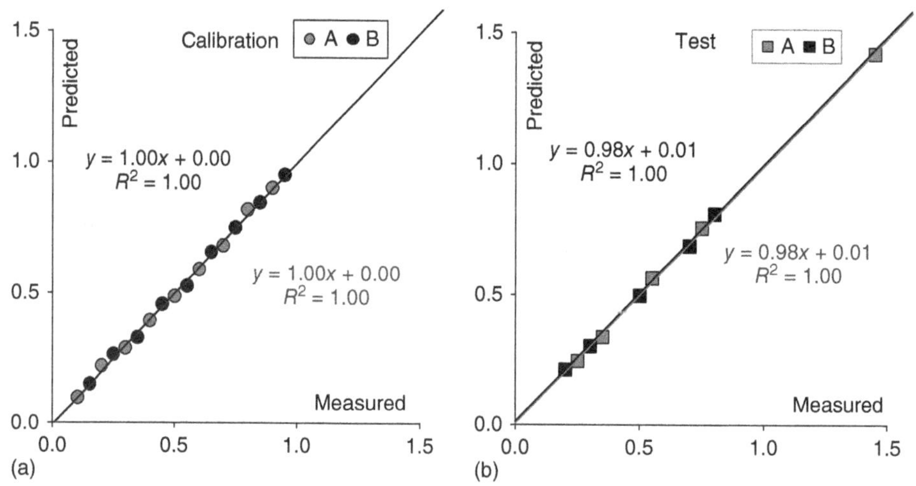

Figure 10.34 The "measured–predicted" plots for the PCR calibration. The training (a) and the test (b) sets.

10.5.5 Projection on the Latent Structures-1 (PLS1)

Regression on the latent structures (PLS1) is very similar to PCR. The main difference is that when constructing the projection space not only predictors **X** but also responses **Y** are taken into account. As a result, we obtain two particular scores matrices **T**, one for each of the responses (A and B).

Worksheet PLS1 and Fig. 10.35 present the first two PLS1 scores plots for responses A and B. Despite the seeming differences, they are similar to each other and to the PCA plot shown in Fig. 10.30. To understand this, it is enough to change the signs of the first PLS PC in response A. Such a transformation of the PCs is legitimate. As introduced in Section

TABLE 10.6 Quality Characteristics of the PCR Calibration

A	B
$R_c^2 = 0.997$	$R_c^2 = 0.998$
RMSEC = 0.013	RMSEC = 0.012
BIASC = 0.000	BIASC = 0.000
SEC = 0.013	SEC = 0.012
$R_t^2 = 0.999$	$R_t^2 = 0.999$
RMSEP = 0.015	RMSEP = 0.010
BIASP = −0.003	BIASP = 0.004
SEP = 0.015	SEP = 0.009
TRVC = 0.002	
ERVC = 1.000	
TRVP = 0.000	
ERVP = 1.000	

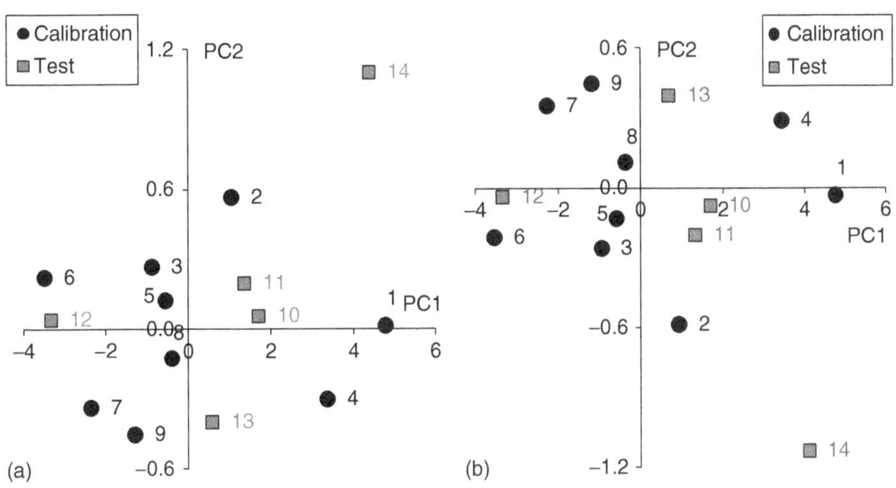

Figure 10.35 PLS1 scores **T**.

9.2.11, both PCA and PLS decompositions are not unique. It is always possible to paste an arbitrary matrix **O**, such that $\mathbf{OO}^t = \mathbf{I}$, in those decompositions. Indeed,

$$\mathbf{X} = \mathbf{TP}^t = \mathbf{TIP}^t = \mathbf{T}\left(\mathbf{OO}^t\right)\mathbf{P}^t = (\mathbf{TO})(\mathbf{PO})^t = \left(\mathbf{T}_{\text{new}}\right)\left(\mathbf{P}_{\text{new}}\right)^t$$

The PLS loadings display a similar behavior. Figure 10.36 shows the weighted loadings **W** obtained for response A. We use the weighted loadings **W**, but not loadings **P**, because the former are essentially closer to PCA loadings **P**.

The development of PLS1 calibration is quite similar to that in PCR. The only distinction is that in the former method a particular scores matrix is used for each response.

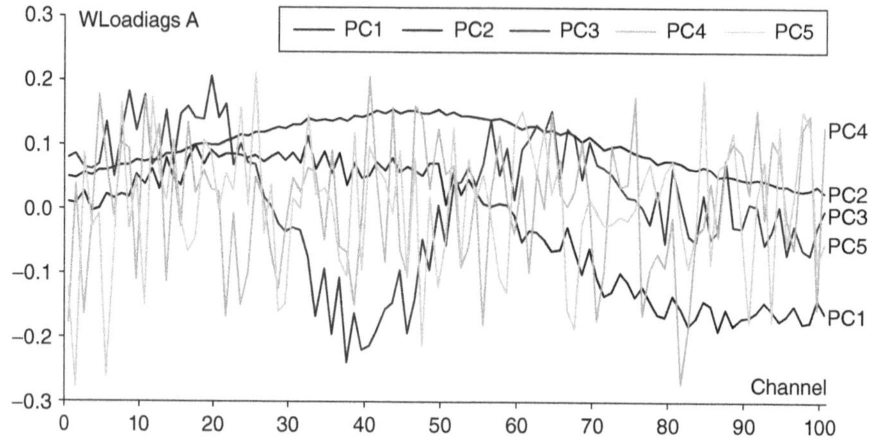

Figure 10.36 Weighted PLS1 loadings **W** for response A.

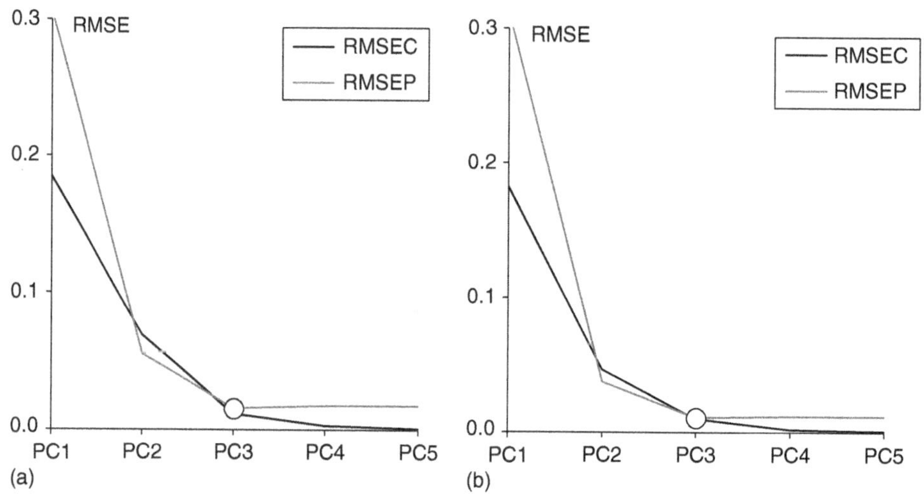

Figure 10.37 The root mean square errors in calibration (RMSEC) and validation (RMSEP) in PLS1 calibration.

Figure 10.37 shows the root mean square residuals in training (RMSEC) and in validation (RMSEP) for the PLS1 calibration. This plot is similar to its counterpart obtained with the PCR method (Fig. 10.32). Here, it is also obvious that the minimum of RMSEP is obtained for three PC ($A = 3$). Some distinction can be seen only in the RMSEC curves, which do not stop at PC = 3 but continue decreasing for the elder PCs. This reflects a PLS feature that focuses on maintaining a maximum correlation between **T** and **U** scores.

Naturally, the "measured–predicted" plots for calibration by the PLS1 method (see Fig. 10.38) are similar to those for PCR calibration (Fig. 10.34). The same can be said about the calibration quality characteristics, which are given in Table 10.7.

Thus, in our example, the PLS1 calibration is very similar to the PCR calibration.

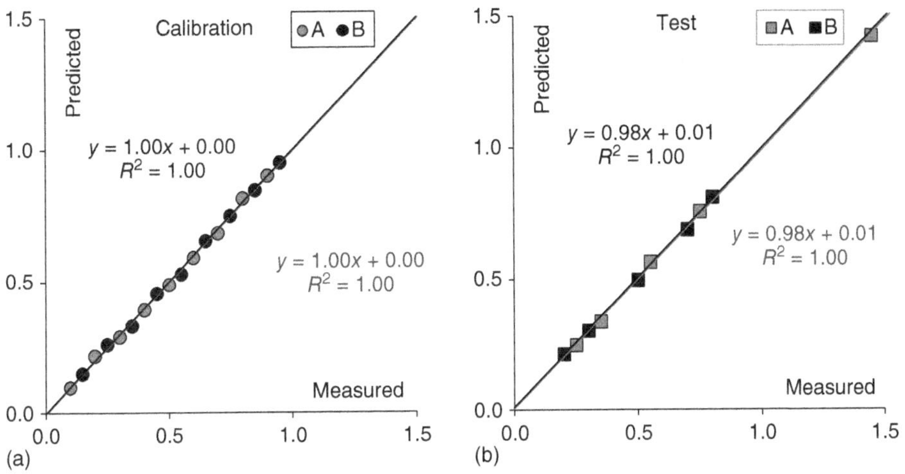

Figure 10.38 The "measured–predicted" plots for the PLS1 calibration. The training (a) and the test (b) sets.

TABLE 10.7 Quality Characteristics of the PLS1 Calibration

A		B
$R_c^2 = 0.998$		$R_c^2 = 0.998$
RMSEC = 0.011		RMSEC = 0.010
BIASC = 0.000		BIASC = 0.000
SEC = 0.011		SEC = 0.010
$R_t^2 = 0.999$		$R_t^2 = 0.998$
RMSEP = 0.015		RMSEP = 0.010
BIASP = −0.003		BIASP = 0.003
SEP = 0.015		SEP = 0.010
	TRVC = 0.001	
	ERVC = 1.000	
	TRVP = 0.000	
	ERVP = 1.000	

10.5.6 Projection on the Latent Structures-2 (PLS2)

Although the PLS1 method is very similar to PCR calibration, the distinctions between PLS1 and PLS2 methods are even less pronounced.

Both **X** and **Y** values are used in the construction of the PLS2 projection. However, in PLS2 all responses **Y** are considered together unlike in PLS1, where each response generates its own projection subspace. Therefore, the PLS2 method develops one common projection subspace (as in PCR). In the result, we obtain a pair of the scores matrices **T** and **U** and three loadings matrices **P**, **W**, and **Q**.

Worksheet PLS2 and Fig. 10.39 present **T** scores plots obtained by the PLS2 method in our example. Comparing them with the plots given in Figs 10.30 and 10.35, we can

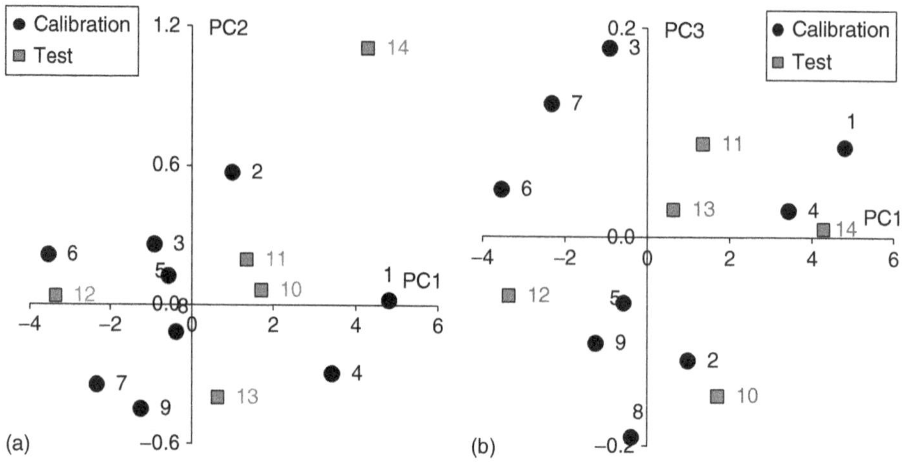

Figure 10.39 PLS2 scores **T**.

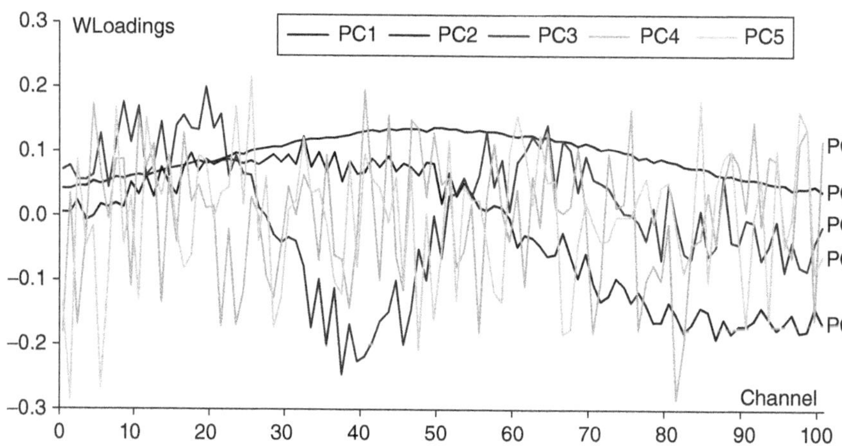

Figure 10.40 Weighted PLS2 loadings **W**.

easily notice the similarities. The similarities can also be seen for the weighted loadings **W** presented in Fig. 10.40 and those shown in Figs 10.31 and 10.36.

Once the PLS2 decomposition is constructed, the calibration is built in a way explained in the PCR method.

Analyzing Fig. 10.41, which shows the root mean square errors in PLS2 calibration (RMSEC) and validation (RMSEP), it is easy to determine the number of PCs. It is $A = 3$. Dissimilarities between these plots and their counterparts in Fig. 10.37 (PLS1) are imperceptible to the eye.

The "measured–predicted" plots for PLS1 (Fig. 10.38) and for PLS2 (Fig. 10.42) are also indistinguishable. The quality characteristics of PLS2 calibration, which are listed in Table 10.8, are identical to those for PLS1 (Table 10.7).

It seems that the PLS2 method adds nothing to the PLS1 method. Indeed, in most cases the methods are very similar. Moreover, if the responses in **Y** matrix are not correlated, the

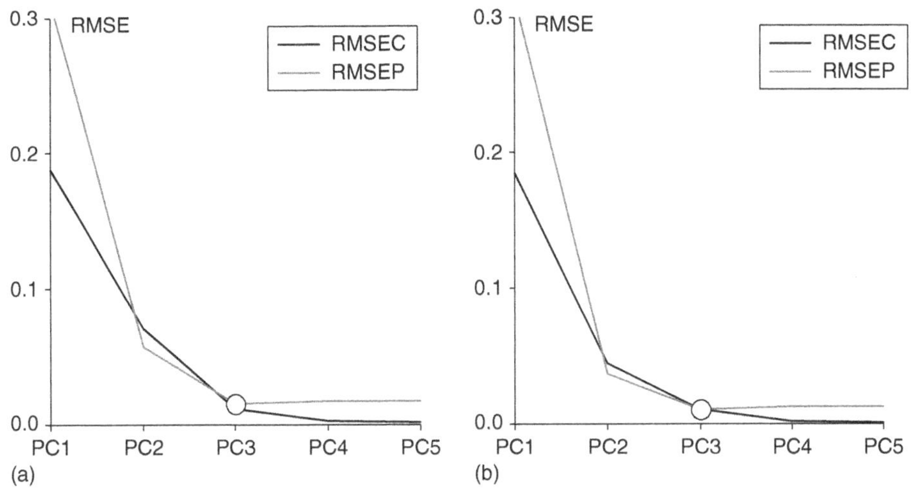

Figure 10.41 The root mean square errors in calibration (RMSEC) and validation (RMSEP) in PLS2 calibration.

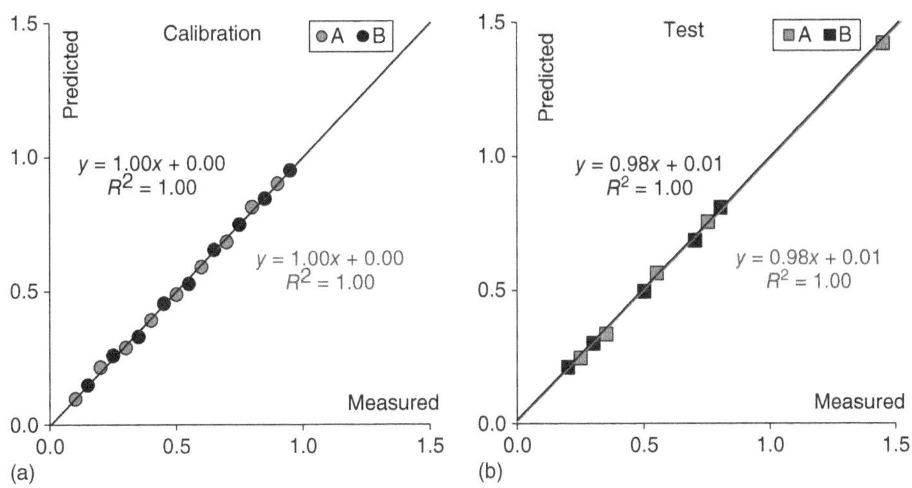

Figure 10.42 The "measured–predicted" plots for the PLS2 calibration. The training (a) and the test (b) sets.

PLS1 results are usually better than the PLS2 ones. To clarify this, it is instrumental to make a PCA decomposition of matrix **Y**. If we detect the dependences between the responses, the PLS2 method could be useful. Further discussion of this interesting issue is beyond the scope of this book.

10.6 METHODS COMPARISON

We start the comparison (see worksheet Compare) with RMSEC and RMSEP characteristics (Section 10.1.4) that better reflect the calibration quality.

TABLE 10.8 Quality Characteristics of the PLS2 Calibration

A	B
$R_c^2 = 0.998$	$R_c^2 = 0.998$
RMSEC = 0.011	RMSEC = 0.010
BIASC = 0.000	BIASC = 0.000
SEC = 0.011	SEC = 0.010
$R_t^2 = 0.999$	$R_t^2 = 0.998$
RMSEP = 0.015	RMSEP = 0.010
BIASP = −0.003	BIASP = 0.004
SEP = 0.015	SEP = 0.010
TRVC = 0.001	
ERVC = 1.000	
TRVP = 0.000	
ERVP = 1.000	

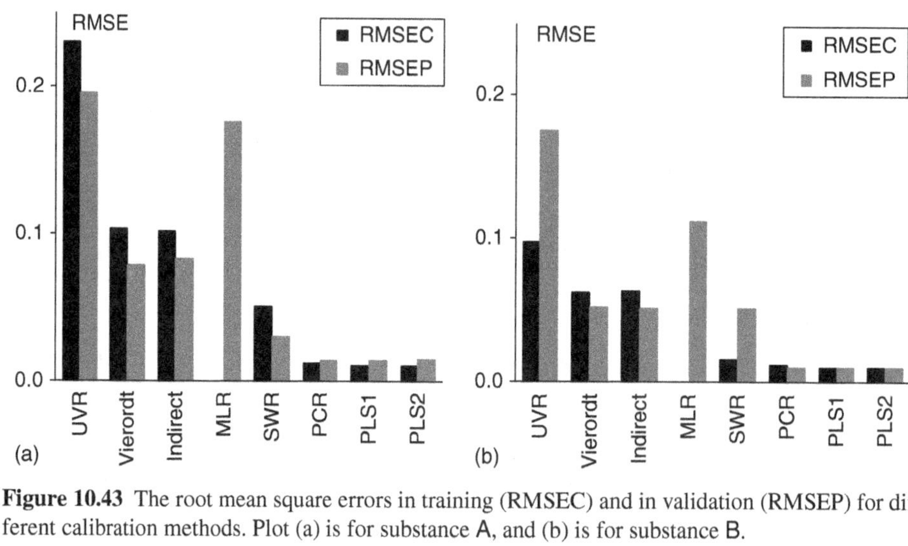

Figure 10.43 The root mean square errors in training (RMSEC) and in validation (RMSEP) for different calibration methods. Plot (a) is for substance A, and (b) is for substance B.

Figure 10.43 shows the RMSEC (training) and RMSEP (validation) values obtained by the calibration methods considered above. The first evident finding is an obvious superiority of the latent variables calibration, that is, PCR, PLS, and PLS2, over the conventional approaches. Another interesting feature is that very often RMSEP is larger than RMSEC. This is typical for a proper calibration method, unless the distinction between RMSEP and RMSEC is as striking as in the case of substance B calibration by the MLR (Section 10.4.1) or the SWR (Section 10.4.2).

Figure 10.44 shows the bias values (10.5) obtained in training (BIASC) and in validation (BIASP). The plots demonstrate the superiority of the projection methods, which yield

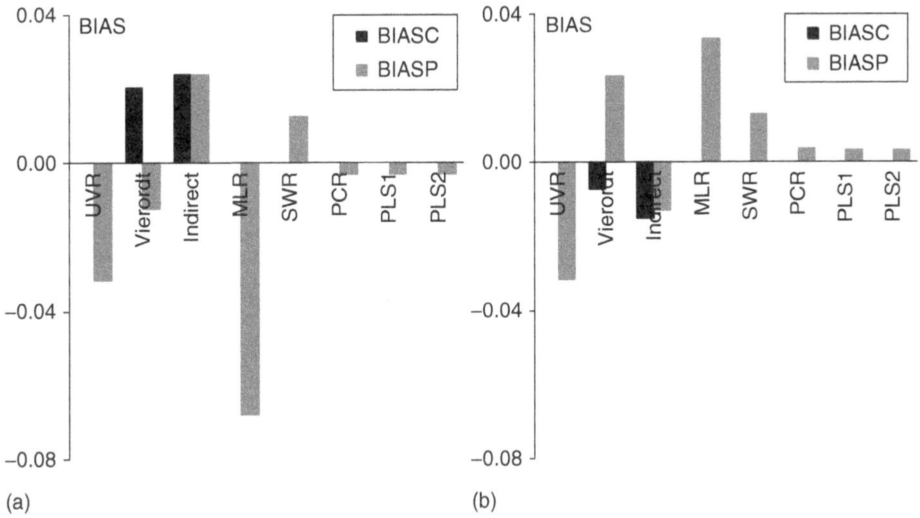

Figure 10.44 Bias in training (BIASC) and in validation (BIASP) for different calibration methods. Plot (a) is for substance A, and (b) is for substance B.

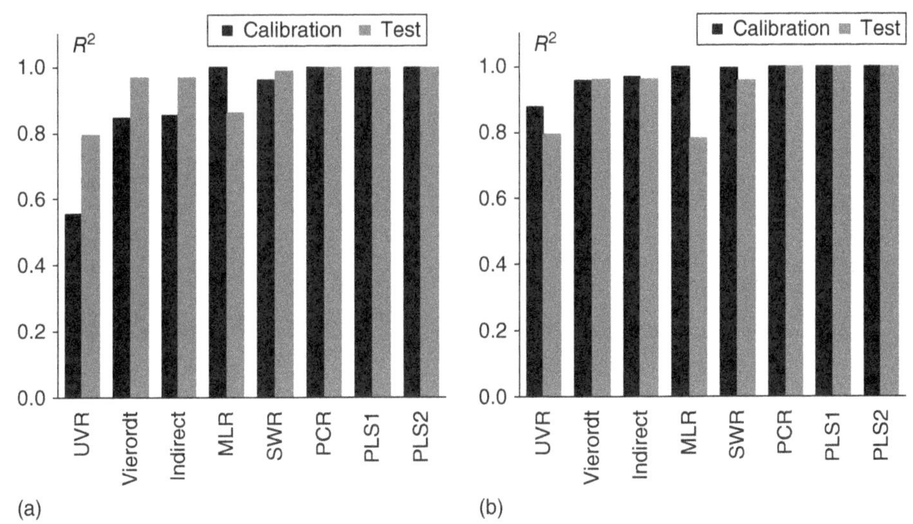

Figure 10.45 Correlation coefficients in training and in validation for different calibration methods. Plot (a) is for substance A, and (b) is for substance B.

significantly lower values of systematic deviation. Similarly, BIASC is as a rule smaller than BIASP, and for UVR (Section 10.3.1) and MLR (Section 10.4.1) this difference is particularly noticeable.

Figure 10.45 shows correlation coefficients between the known and evaluated response values, introduced in Eq. (10.7). We see that the projection methods not only give the best R^2 values (closer to one) but also provide a proper balance between training and validation stages.

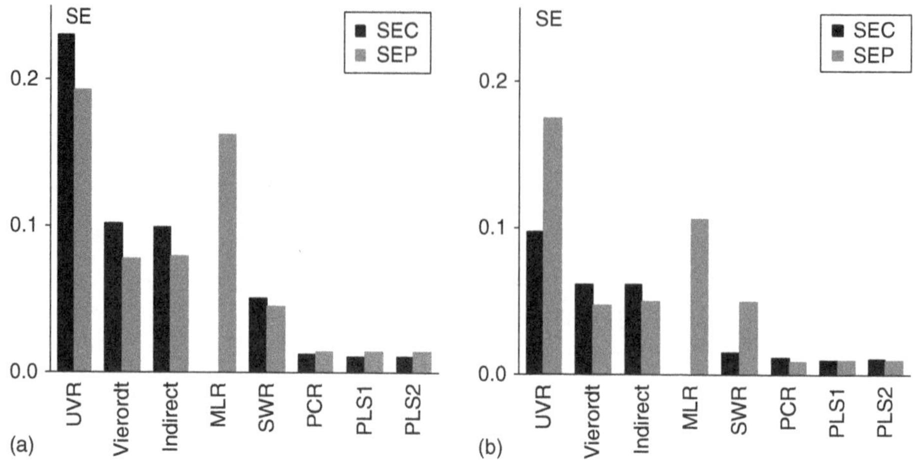

Figure 10.46 Standard errors in training (SEC) and in validation (SEP) for different calibration methods. Plot (a) is for substance A, and (b) is for substance B.

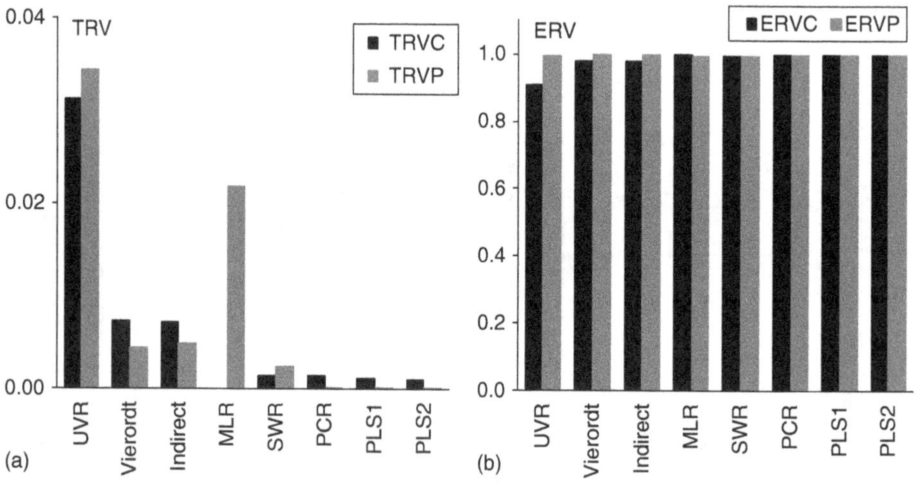

Figure 10.47 Total residual variance in training (TRVC) and in validation (TRVP). Explained residual variance in training (ERVC) and in validation (ERVP) for different calibration methods. Plot (a) is for substance A, and (b) is for substance B.

Figure 10.46 shows the SEs defined by Eq. (10.6) in training (SEC) and in validation (SEP). As in our example, all bias values are small compared with RMSE, these plots add nothing new to the features already explained when discussing Fig. 10.43.

Figure 10.47a shows the total residual variance (given by Eq. (10.2)) in calibration (TRVC) and in validation (TRVP). Figure 10.47b demonstrates the explained residual variance (Eq. (10.3)) in training (ERVC) and in validation (ERVP). These plots again represent the same pattern as before, that is, the latent variables calibration methods are better than the traditional ones. It can be also noted that the ERV plots are inconvenient for the analysis of the model quality as they provide a poor overview of various methods' pros and cons.

CONCLUSION

Summarizing the study of the different calibration methods, we can conclude that

- a single channel calibration leads to underfitting, that is, the RMSEC and RMSEP values are too large;
- the multiple calibration, in contrast, leads to overfitting. that is, the RMSEC value is considerably smaller than RMSEP;
- the best results are obtained in the latent variables method (PCR and PLS) where a proper balance between the RMSEC and RMSEP values is achieved.

11

CLASSIFICATION

This chapter describes the most popular classification methods used in chemometrics.

Preceding chapters	5–10
Dependent chapters	
Matrix skills	Basic
Statistical skills	Basic
Excel skills	Basic
Chemometric skills	Basic
Chemometrics Add-In	Used
Accompanying workbook	Iris.xls

11.1 THE BASICS

11.1.1 Problem Statement

Classification is a procedure used to sort *objects* into groups (classes) in accordance with numerical values of the variables (features) that characterize the properties of these objects. The initial data for classification is contained in matrix **X**, where each row represents an object and each column corresponds to a variable. The number of objects (rows of **X**) is denoted by the letter I and the number of variables (columns of **X**) by the letter J. The number of classes is denoted by K.

The term *classification* is applied not only to the allocation procedure itself but also to its result. The term *pattern recognition* can be considered a synonym. In mathematical statistics, classification is often called *discrimination*. The method (algorithm) that conducts classification is called a *classifier*. The classifier maps a row-vector, which characterizes the

Chemometrics in Excel, First Edition. Alexey L. Pomerantsev.

features of the objects onto an integer (labeled 1, 2, …) that corresponds with the number of the class to which the object is allocated.

11.1.2 Types of Classes

Classification can be done for different numbers of classes. A *one-class* classification is an allocation to a single selected group. As an example, consider the problem of separating apples from other fruits in a basket. A *two-class classification* is the most simple, base case, which is often referred as *discrimination*. An example is the separation of apples and pears provided that there are no other fruits in the basket. A *multiclass classification* is often reduced to a sequence of either one-class (soft independent modeling of class analogy, SIMCA) or two-class (linear discriminate analysis, LDA) problems. This is the most complex problem.

In most cases, the classes are isolated and they do not overlap so that each object belongs to one class only. However, there may be problems with overlapping classes when an object can belong simultaneously to more than one class.

11.1.3 Hypothesis Testing

In mathematical statistics (Section 6.6), we considered the problem of *hypotheses testing*, which, in fact, is very close to classification. Let us illustrate this with a simple example.

Suppose we have a mix of plums and apples, which should be sorted automatically. Typically, a plum is smaller than an apple so the problem can be solved by using an appropriate sieve. The analysis of the objects in the mix shows that their sizes are well described by the normal distribution with the following parameters. For plums: mean = 3, variance = 1.4. For apples: mean = 8, variance = 2.1. Hence, it is wise to choose a mesh size equal to 5 (see Fig. 11.1).

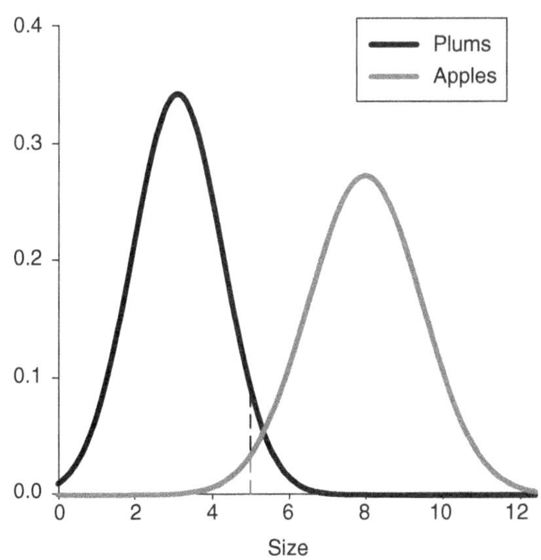

Figure 11.1 Distributions of the sizes.

From a statistical point of view (Section 6.6.2), we are testing the hypothesis that the mean of the normal distribution is 3 (a plum), against the alternative that the mean is 8 (an apple). The test is performed using a single observation x. A critical level (cutoff) is 5. If $x \le 5$ (the acceptance area), the hypothesis is accepted, that is, the object is a plum. If $x > 5$, the alternative is accepted, that is, the object is an apple.

11.1.4 Errors in Classification

Obviously, in the example above, the classification is not ideal because small apples will fall into the class of plums, whereas large plums will remain with the apples. Using the size distributions of the objects, we can calculate the probabilities of these events

$$\alpha = 1 - \Phi(5|3, 1.4) = 0.05, \quad \beta = \Phi(5|8, 2.1) = 0.01,$$

where Φ is the cumulative normal distribution function (Section 6.2.3). To calculate these values in Excel, one can use the standard worksheet function **`NORMDIST`**.

The value α (the rate of incorrect rejection of the null hypothesis) is called the *type I error* and value β (the rate of false acceptance of the alternative hypothesis) is named the *type II error* (Section 6.6.3). If the hypothesis and the alternative are swapped, the type I error becomes the type II error and vice versa.

Thus, at the critical level 5, 5% of plums remain with the apples and 1% of apples are mixed with the plums. If the critical level is reduced to 4, there will be no apples among the plums, but the "loss" of plums will rise to 20%. If the cutoff is increased to 6, we will miss only 1% of plums, but 5% of apples will mix with plums. It is clear from this problem that no sieve can perfectly separate plums from apples, that is, errors will always occur.

During hypothesis testing (classification), it is important to decide what type of error should be minimized. Consider two classic examples. In the field of law, testing the hypothesis "not guilty," we should minimize the type I error, that is, the risk of a false judgment. In medicine, testing the hypothesis "healthy," we should aim at minimizing of the type II error, that is, the chance that a disease will not be detected.

Is it possible to reduce both errors simultaneously? In principle it is possible. In order to do so, one should change the very decision-making procedure by making it more efficient. A mainstream approach would be to increase the number of variables that characterize the objects to be classified. In our example, a new useful variable could be the color–blue for plums and green for apples. Therefore, classification methods used in chemometrics are always based on multivariate data.

11.1.5 One-Class Classification

In one-class classification case, the type I error α is called the *significance level*. The type II error for such a classification is equal to $1-\alpha$. The intuition behind this is very simple. An alternative to the one-class classification includes all conceivable objects, which fall outside the class. Therefore, whatever classifier we develop, there will always be an object that does not belong to the class but is very similar to its members. Imagine that we are selecting plums by distinguishing them from all other existing objects. After careful examination of the classification method, one can develop an artificial object (e.g., a plastic dummy), which meets all the test criteria.

11.1.6 Training and Validation

In addition to the vector of variables **x**, a classifier depends on the unknown parameters, which should be evaluated in order to minimize the classification error. This is called classifier *training*. It is performed using the *training set* $\mathbf{X}_c$. Another important stage is classifier *validation*. This is done using a new *test set* $\mathbf{X}_t$. An alternative to the test set validation is cross-validation (Section 10.1.3).

Classification and calibration (Chapter 10) have a lot in common. Theoretically, classification can be viewed as a regression, the response of which is an integer 1, 2, ..., that is, the class number. This approach is used in classification by partial least squares (PLS) discrimination method, which is presented in Section 11.3.3. Similar to calibration, classification also has a large risk of overfitting, called *overtraining*, in this context. To avoid this risk, it is necessary to analyze how the errors in training and validation change when a classification method becomes more complex.

11.1.7 Supervised and Unsupervised Training

There are two types of classifier training. When it is known which class each object from the training set belongs to, the procedure is called a *supervised* classification (training). An *unsupervised* training is the opposite case when this membership is unknown in advance.

11.1.8 The Curse of Dimensionality

Classification involves a substantial problem, poetically referred as the *curse of dimensionality*. In a nutshell, as the number of variables J increases, the complexity of classification problem grows exponentially. Therefore, even a relatively modest number J (about 10) can cause trouble. Note that in chemical applications (e.g., in spectral data analysis) one may deal with $J = 1000$ and even $J = 10{,}000$ variables.

In classical classification methods, a large dimension results in *multicollinearity* that manifests itself as the singularity of $\mathbf{X}^t\mathbf{X}$ matrix, which should be inverted in linear and quadratic discriminate analysis (Sections 11.3.1 and 11.3.2). In object distance methods (such as k-nearest neighbor (*kNN*), Section 11.4.2), the large dimension leads to the alignment of all distances. A usual way to address this problem is to employ a method of dimensionality reduction, such as the *principal component analysis* (PCA) (Chapter 9).

11.1.9 Data Preprocessing

When solving classification problems, the raw data should be properly prepared in the same way as it is done in the method of principal components (PCs) (Section 8.1.4). For example, data centering is automatically embedded in the algorithms of many classification methods.

Data transformation plays a separate role in classification.

Let us illustrate this with an example. Figure 11.2 shows data that belong to two classes. In Fig. 11.2a, they are presented in their "natural" form. In Fig. 11.2b, the same data are shown in polar coordinates.

$$x_1 = r\cos(\varphi), \quad x_2 = r\sin(\varphi).$$

It is clear that it is very difficult to separate the classes in original coordinates. However, in polar coordinates the discrimination is obvious. Figure 11.3 presents such a classification

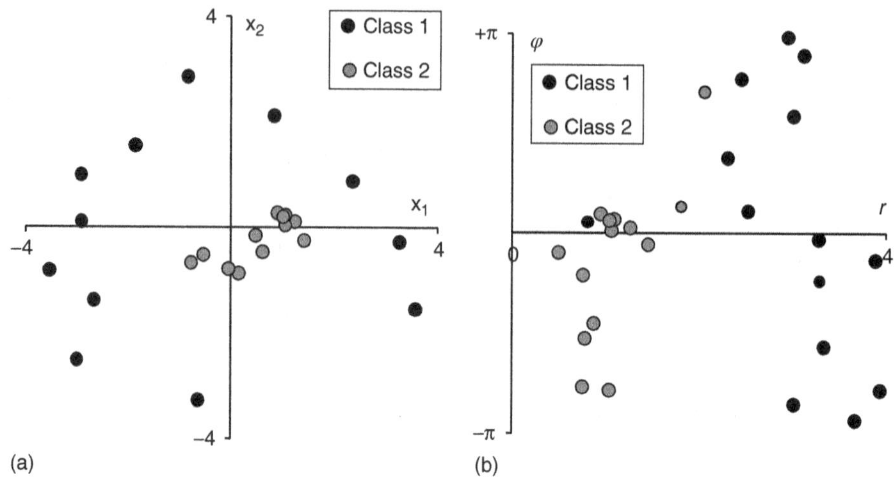

Figure 11.2 Objects in Cartesian (a) and in polar (b) coordinates.

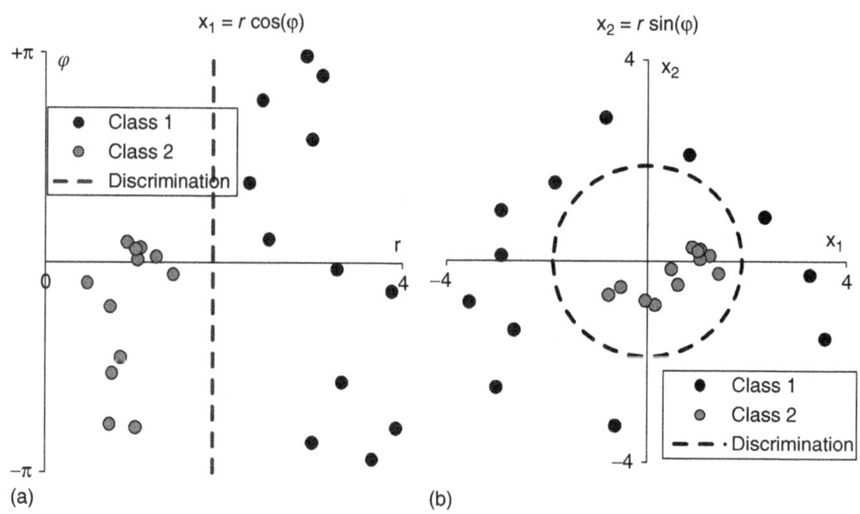

Figure 11.3 Classification in polar coordinates (a) and its image in Cartesian coordinates (b).

(a) and its image in the original Cartesian coordinates (b). This simple principle serves as a base for nonlinear methods of classification, for example, the *Support Vector Machine*.

11.2 DATA

11.2.1 Example

To illustrate various classification methods, we use a famous example being Fisher's Iris dataset. It became popular since the seminal work[1] by Sir Ronald Fisher, in which he

[1]R.A. Fisher (1936). The use of multiple measurements in taxonomic problems. Annals of Eugenics 7 (2): 179–188.

(a) (b) (c)

Figure 11.4 Fisher's Irises (a–c): *Setosa*, *Versicolor*, and *Virginica*.

proposed the LDA method. The dataset includes three classes of 50 samples. Each class corresponds to one of Iris species: *Iris setosa* (Class 1), *Iris versicolor* (Class 2), and *Iris virginica* (Class 3) (see Fig. 11.4).

In the above-mentioned paper, Ronald Fisher used the data collected by the American botanist Edgar Anderson, who had studied 150 samples and measured the following characteristics of the flowers:

- sepal length;
- sepal width;
- petal length;
- petal width.

These values (in centimeter) are provided in the table in worksheet Data. In the attempt to understand what sepals and petals are, one could take a look at Wikipedia.[2] It says the following:

> *"The inflorescences are fan-shaped and contain one or more symmetrical six-lobed flowers. These grow on a pedicel or lack a footstalk. The three sepals, which are spreading or droop downwards, are referred to as "falls"....... The three, sometimes reduced, petals stand upright, partly behind the sepal bases."*

After reading the description, my urge is to leave the exact distinction between the two to the discretion of botanists.

11.2.2 Data Subsets

The original array of data (three classes containing 50 samples each) is divided into two parts: the *training* and the *test* sets. The first set $\mathbf{X}_c$ includes the first 40 samples from each class (120 samples in total). The second set $\mathbf{X}_t$ contains 10 remaining samples of each class (30 samples in total). Obviously, the first set is used to train the classifier, whereas the second set is employed for method validation.

The classes are labeled according to their Latin names *Setosa*, *Versicolor*, and *Virginica*. Variables are labeled by two letters abbreviating objects' characteristics. *SL* is the sepal length, *SW* is the sepal width, *PL* is the petal length, and *PW* is the petal width.

[2]http://en.wikipedia.org/wiki/Iris_(plant).

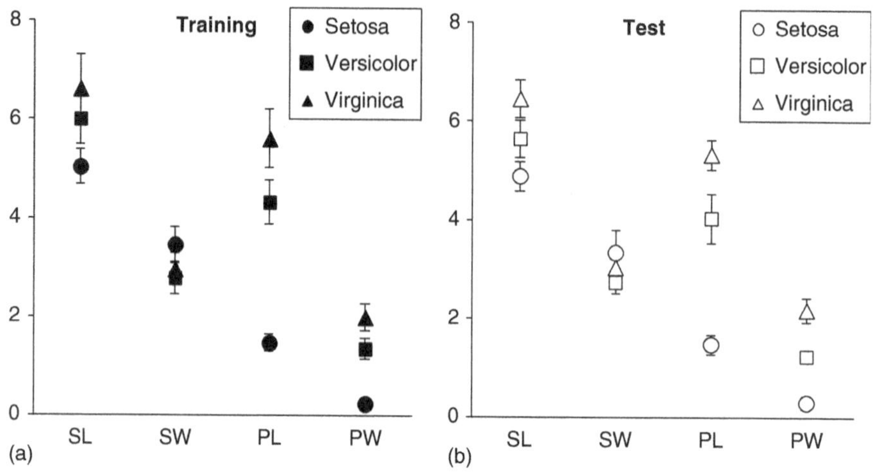

Figure 11.5 Statistical characteristics of the training and test sets.

Figure 11.5 shows the main statistical characteristics of the training and test sets. Mean values (m) of each variable (*SL*, *SW*, *PL*, and *PW*) are shown by the dots, and their standard deviations (s) are presented by the bars. Shapes of the markers correspond to a class: circles represent *Setosa*, boxes stand for *Versicolor*, and triangles mark *Virginica*. Filled marks correspond to the training set and the open ones (○, □, Δ) represent the test set. The same marking is also used later on in this chapter. Figure 11.5 demonstrates that m and s differ for all classes. In addition, we can conclude that the partition into the training and the test sets has been done correctly as the corresponding plots (a,b) are similar.

11.2.3 Workbook Iris.xls

This chapter is accompanied by file Iris.xls. This Excel workbook includes the following sheets:

Intro: a brief introduction
Data: the dataset
PCA: principal component analysis (PCA)
PCA-LDA: linear discriminant analysis (LDA) of the compressed data
LDA: LDA of the original data
QDA: quadratic discriminant analysis of the original data
PLSDA: partial least squares (PLS) discriminant analysis
PLSDA-PCA-LDA: LDA after PLSDA
SIMCA_1: SIMCA classification based on the first class
SIMCA_2: SIMCA classification based on the second class
SIMCA_3: SIMCA classification based on the third class
KNN: k nearest neighbors' method
PCA-Explore: clustering by PCA method
KMeans: clustering by the K-means method

11.2.4 Principal Component Analysis

PCA is the main tool used in chemometrics (Chapter 9). In classification, it is used for two purposes. First of all, PCA reduces data dimensionality, replacing multiple variables by a small set (usually 2–5) of PCs. Secondly, PCA serves as a basis for many classification methods, such as SIMCA, which is described in Section 11.3.4.

In our example, the number of variables is rather small, just four, so the first goal is not so important. However, we will construct a PCA model to see how the dimension can be reduced. PCA is performed using special worksheet functions **`ScoresPCA`** and **`LoadingsPCA`** (Sections 8.3.1 and 8.3.2). The PCA model is developed using the training set $\mathbf{X}_c$ and then applied to the test set $\mathbf{X}_t$. Figure 11.5 implies that the data should be centered but not scaled (Section 8.1.4). The plots of the first PCs are shown in Fig. 11.6. The higher-order PCs (PC3–PC4) plots are presented in sheet PCA.

In order to determine what number of PCs is sufficient for data modeling (Section 9.2.10), we should explore a plot in which the explained residual variance (ERV, Section 9.2.9) for the training and the test sets is depicted as a function of the number of PC. Figure 11.7 shows that PC = 2 is enough because this model explains 98% of the variation in both the training and test sets.

11.3 SUPERVISED CLASSIFICATION

11.3.1 Linear Discriminant Analysis (LDA)

LDA is the oldest classification method, developed by Ronald Fisher, and published in the paper mentioned in Section 11.2.1. The method is designed to enable separation into two classes. The training set consists of two matrices $\mathbf{X}_1$ and $\mathbf{X}_2$, which have I_1 and I_2 rows (samples), respectively. The number of variables (columns) in each of the matrices is equal to J. The initial assumptions are as follows:

1. Each class ($k = 1$ or 2) is modeled by the normal distribution $N(\boldsymbol{\mu}_k, \boldsymbol{\Sigma}_k)$
2. Covariance matrices of the two classes are identical $\boldsymbol{\Sigma}_1 = \boldsymbol{\Sigma}_2 = \boldsymbol{\Sigma}$ (11.1)

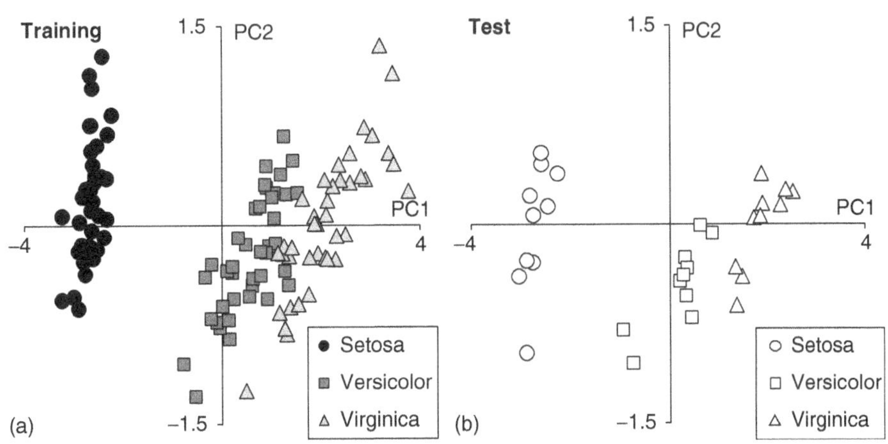

Figure 11.6 PCA scores of the data.

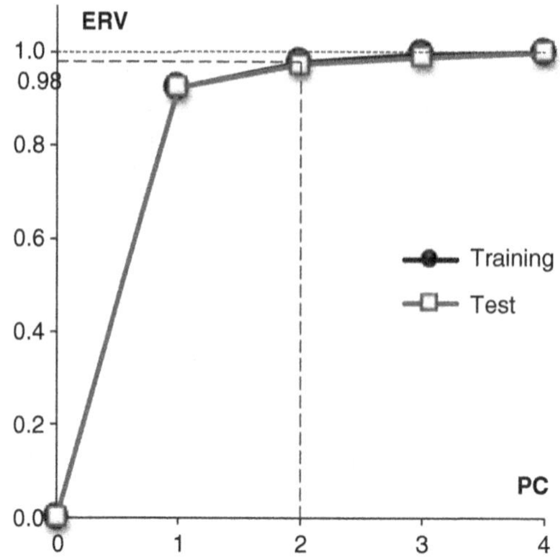

Figure 11.7 Explained residual variance (ERV) for the training and test sets.

The LDA classification rule is very simple and states that a new sample **x** belongs to the class which is closer by the Mahalanobis metric (Section 6.2.7)

$$d_k = (\mathbf{x} - \boldsymbol{\mu}_k)\,\boldsymbol{\Sigma}^{-1}(\mathbf{x} - \boldsymbol{\mu}_k)^{\mathrm{t}}, \quad k = 1,2 \tag{11.2}$$

In practice, the unknown expectations $\boldsymbol{\mu}_k$ and the covariance matrix $\boldsymbol{\Sigma}$ are replaced by their estimators (see Section 6.3.5)

$$\mathbf{m}_k = (1/I_k)\sum_{i=1}^{I_k}\mathbf{x}_i, \quad \mathbf{S} = (1/(I_1 + I_2 - 2))\left(\widetilde{\mathbf{X}}_1^{\mathrm{t}}\widetilde{\mathbf{X}}_1 + \widetilde{\mathbf{X}}_2^{\mathrm{t}}\widetilde{\mathbf{X}}_2\right) \tag{11.3}$$

In this formula, $\widetilde{\mathbf{X}}_k$ is the centered matrix $\mathbf{X}_k$ (see Section 8.1.4).

Setting $d_1 = d_2$ in formula 11.2, we obtain an equation of a curve, which separates the classes. Here, the quadratic terms $\mathbf{x}\mathbf{S}^{-1}\mathbf{x}^{\mathrm{t}}$ are reduced and the equation becomes linear.

$$\mathbf{x}\mathbf{w}_1^{\mathrm{t}} - v_1 = \mathbf{x}\mathbf{w}_2^{\mathrm{t}} - v_2 \tag{11.4}$$

where

$$\mathbf{w}_k = \mathbf{m}_k\mathbf{S}^{-1}, \quad v_k = 0.5\mathbf{m}_k\mathbf{S}^{-1}\mathbf{m}_k^{\mathrm{t}} \tag{11.5}$$

The terms at the two sides of Eq. 11.4 are called the *LDA-scores*, $f_1 = \mathbf{x}\mathbf{w}_1^{\mathrm{t}} - v_1$ and $f_2 = \mathbf{x}\mathbf{w}_2^{\mathrm{t}} - v_2$. An object belongs to Class 1, if $f_1 > f_2$ and, vice versa, to Class 2, if $f_1 < f_2$.

The main LDA difficulty is the inversion of matrix **S**. If **S** is degenerate, the method cannot be used. Therefore, before using LDA, the original data matrix **X** is often replaced by the PCA-scores matrix **T**, which is not singular.

We demonstrate the application of LDA to the Iris dataset. First of all, we will apply PCA to reduce the number of variables and then move on with LDA. As demonstrated in Section 11.2.4, a two-PCs model is sufficient. As LDA is a two-class discriminator, we will

	C	D	E	F
153				
154	**m1c**	-2.685	0.189	
155	**m23c**	1.342	-0.094	
156	**m2**	0.574	-0.224	
157	**m3**	2.111	0.035	
158				

Figure 11.8 Mean values.

perform classification in two steps. We start with developing a classifier, which separates Class 1 (*Setosa*) from all other Irises united in Class 23 (*Versicolor* + *Virginica*). Next, we construct a second classifier, which separates Class 2 (*Versicolor*) from Class 3 (*Virginica*). All calculations are presented in sheet PCA-LDA.

We begin with the calculation of mean values for all classes in the training set. We should calculate the averages for Class 1 ($I_1 = 40$), the combined Class 23 ($I_{23} = 80$), Class 2 ($I_2 = 40$), and Class 3 ($I_3 = 40$). The values are presented in arrays, which have local names m1c, m23c, m2c, and m3c (see Fig. 11.8 and worksheet PCA-LDA).

Then, we compute two covariance matrices (one for Classes 1 and 23 and another one for Classes 2 and 3) and invert them. The results are located in sheet PCA-LDA in the arrays with local names Sinv123 and Sinv23. After that, using formulas in Eq. 11.5, we calculate all other necessary values (see Fig. 11.9 and worksheet PCA-LDA).

	I	J	K	L
2		S_{123}		
3		0.718	0.258	
4		0.258	0.239	
5				
6		S_{123}^{-1}		
7		2.274	-2.450	
8		-2.450	6.819	
9				
10		$w_1=m_1S_{123}^{-1}$		
11		-6.567	7.864	
12				
13		$w_{23}=m_{23}S_{123}^{-1}$		
14		3.284	-3.932	
15				
16		$v_1=0.5m_1S_{123}^{-1}m_1^t$		
17		9.559		
18				
19		$v_{23}=0.5m_{23}S_{123}^{-1}m_{23}^t$		
20		2.390		
21				

	P	Q	R	S
42		S_{23}		
43		0.455	0.262	
44		0.262	0.238	
45				
46		S_{23}^{-1}		
47		6.001	-6.602	
48		-6.602	11.461	
49				
50		$w_2=m_2S_{23}^{-1}$		
51		4.921	-6.354	
52				
53		$w_3=m_3S_{23}^{-1}$		
54		12.44	-13.54	
55				
56		$v_2=0.5m_2S_{23}^{-1}m_2^t$		
57		2.1229		
58				
59		$v_3=0.5m_3S_{23}^{-1}m_3^t$		
60		12.894		
61				

Figure 11.9 Calculation of the covariance matrices and other the LDA values.

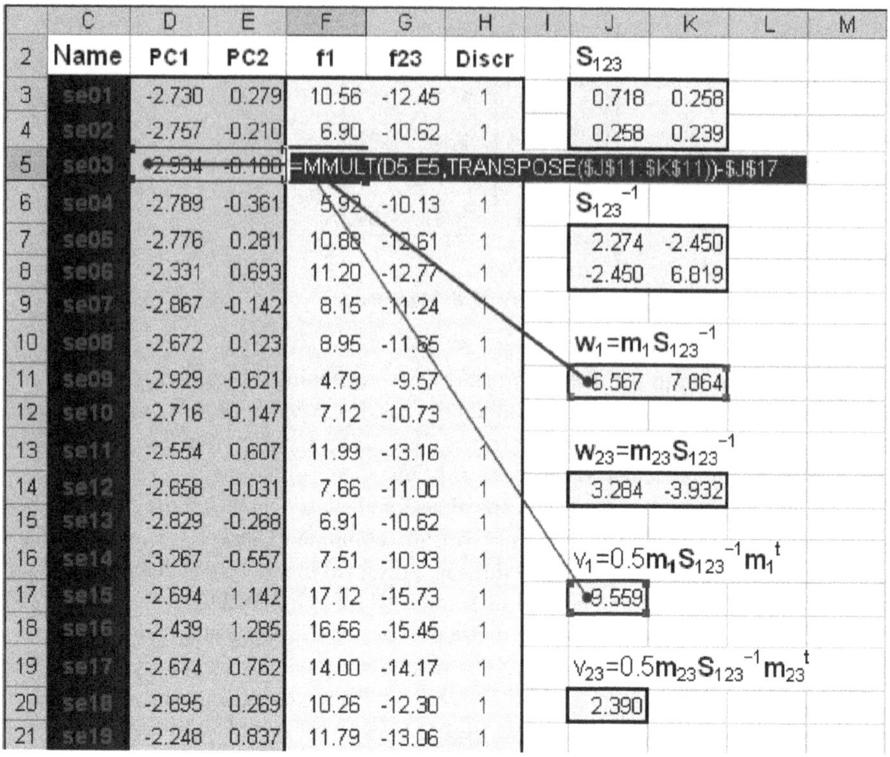

	C	D	E	F	G	H	I	J	K
2	Name	PC1	PC2	f1	f23	Discr		S123	
3	se01	-2.730	0.279	10.56	-12.45	1		0.718	0.258
4	se02	-2.757	-0.210	6.90	-10.62	1		0.258	0.239
5	se03	-2.934	-0.188	=MMULT(D5:E5,TRANSPOSE($J$11:$K$11))-$J$17					
6	se04	-2.789	-0.361	5.92	-10.13	1		S123^-1	
7	se05	-2.776	0.281	10.88	-12.61	1		2.274	-2.450
8	se06	-2.331	0.693	11.20	-12.77	1		-2.450	6.819
9	se07	-2.867	-0.142	8.15	-11.24				
10	se08	-2.672	0.123	8.95	-11.85	1		w1=m1 S123^-1	
11	se09	-2.929	-0.621	4.79	-9.57	1		6.567	7.864
12	se10	-2.716	-0.147	7.12	-10.73	1			
13	se11	-2.554	0.607	11.99	-13.16	1		w23=m23 S123^-1	
14	se12	-2.658	-0.031	7.66	-11.00	1		3.284	-3.932
15	se13	-2.829	-0.268	6.91	-10.62	1			
16	se14	-3.267	-0.557	7.51	-10.93	1		v1=0.5m1 S123^-1 m1^t	
17	se15	-2.694	1.142	17.12	-15.73	1		9.559	
18	se16	-2.439	1.285	16.56	-15.45	1			
19	se17	-2.674	0.762	14.00	-14.17	1		v23=0.5m23 S123^-1 m23^t	
20	se18	-2.695	0.269	10.26	-12.30	1		2.390	
21	se19	-2.248	0.837	11.79	-13.06	1			

Figure 11.10 Calculation of the LDA scores for Classes 1 and 23.

Now, using Eq. 11.4, we calculate the LDA-scores f_1 and f_{23} and determine the sample's membership in Classes 1 and 23 (see Fig. 11.10 and worksheet PCA-LDA).

Separation of Classes 2 and 3 is carried out in a similar way (see Fig. 11.11 and worksheet PCA-LDA).

The result of the discrimination between Classes 1 and 23 is shown in Fig. 11.12. The separating line (Eq. 11.4) correctly divides the classes in both the training and in test sets.

The results of the second round of discrimination between Classes 2 and 3 are shown in Fig. 11.13. One can notice that there are several classification errors that accrued in the training set. Two samples from Class 2 are erroneously classified as belonging to Class 3, and two samples from Class 3 are assigned to Class 2. These points are shown by enlarged markers in Fig. 11.13. There are no errors in the test set.

The Iris dataset has few variables, the covariance matrices are not degenerate and therefore the LDA method can be applied to the original data directly, without the use of PCA. These calculations are presented in sheet PCA-LDA. Figures 11.14 and 11.15 show the results of the direct LDA classification. As there are four variables (not two), the plots illustrating the results can only be presented in the coordinates of the LDA-scores (f_1, f_2), and the discriminating line is the bisector $f_1 = f_2$ of the first quadrant. The second discrimination reveals misclassifications in the training set: two samples from Class 2 are classified as those belonging to Class 3 and one sample from Class 3 is assigned to Class 2. These points are shown by enlarged markers in Fig. 11.14. There are no errors in the test set.

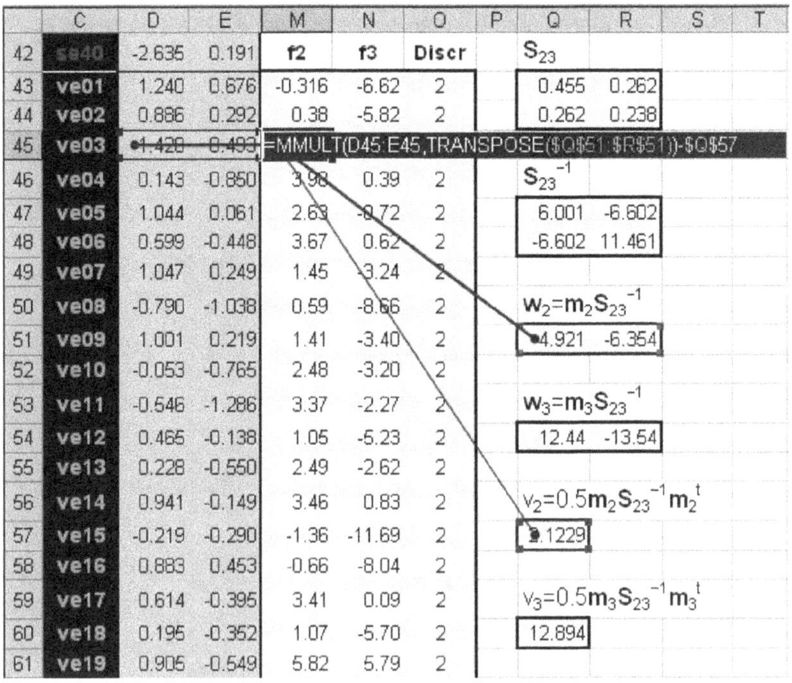

	C	D	E	M	N	O	P	Q	R	S	T
42	se40	-2.635	0.191	f2	f3	Discr		S_{23}			
43	ve01	1.240	0.676	-0.316	-6.62	2		0.455	0.262		
44	ve02	0.886	0.292	0.38	-5.82	2		0.262	0.238		
45	ve03	-1.428	-0.493	=MMULT(D45:E45,TRANSPOSE($Q$51:$R$51))-$Q$57							
46	ve04	0.143	-0.850	3.98	0.39	2		S_{23}^{-1}			
47	ve05	1.044	0.061	2.63	-0.72	2		6.001	-6.602		
48	ve06	0.599	-0.448	3.67	0.62	2		-6.602	11.461		
49	ve07	1.047	0.249	1.45	-3.24	2					
50	ve08	-0.790	-1.038	0.59	-8.66	2		$w_2=m_2S_{23}^{-1}$			
51	ve09	1.001	0.219	1.41	-3.40	2		4.921	-6.354		
52	ve10	-0.053	-0.765	2.48	-3.20	2					
53	ve11	-0.546	-1.286	3.37	-2.27	2		$w_3=m_3S_{23}^{-1}$			
54	ve12	0.465	-0.138	1.05	-6.23	2		12.44	-13.54		
55	ve13	0.228	-0.550	2.49	-2.62	2					
56	ve14	0.941	-0.149	3.46	0.83	2		$v_2=0.5m_2S_{23}^{-1}m_2^t$			
57	ve15	-0.219	-0.290	-1.36	-11.69	2		[illegible].1229			
58	ve16	0.883	0.453	-0.66	-8.04	2					
59	ve17	0.614	-0.395	3.41	0.09	2		$v_3=0.5m_3S_{23}^{-1}m_3^t$			
60	ve18	0.195	-0.352	1.07	-5.70	2		12.894			
61	ve19	0.905	-0.549	5.82	5.79	2					

Figure 11.11 Calculation of the LDA scores for Classes 2 and 3.

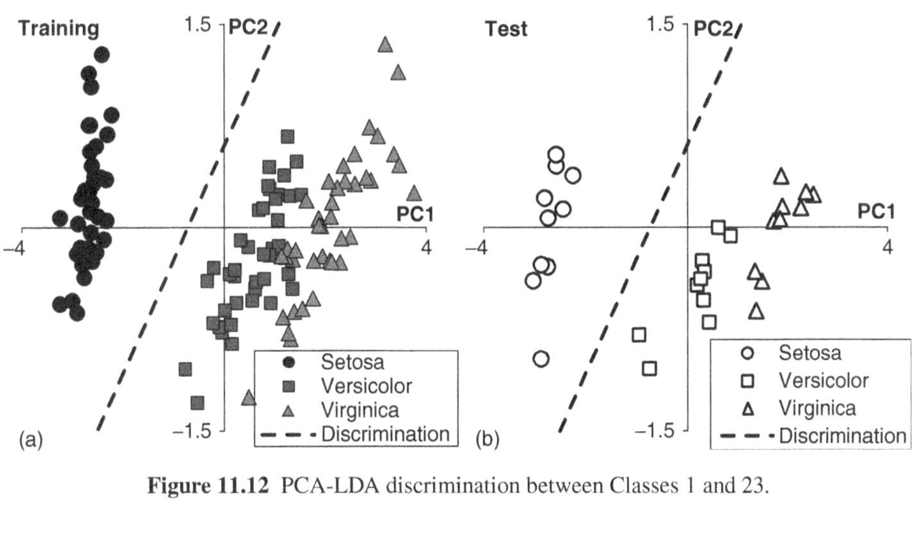

Figure 11.12 PCA-LDA discrimination between Classes 1 and 23.

Disadvantages of LDA

1. LDA does not work when the covariance matrix is singular, for example, for a large number of variables. It requires regularization, for example, by PCA.
2. LDA method is not suitable if the covariance matrices of the classes are different.

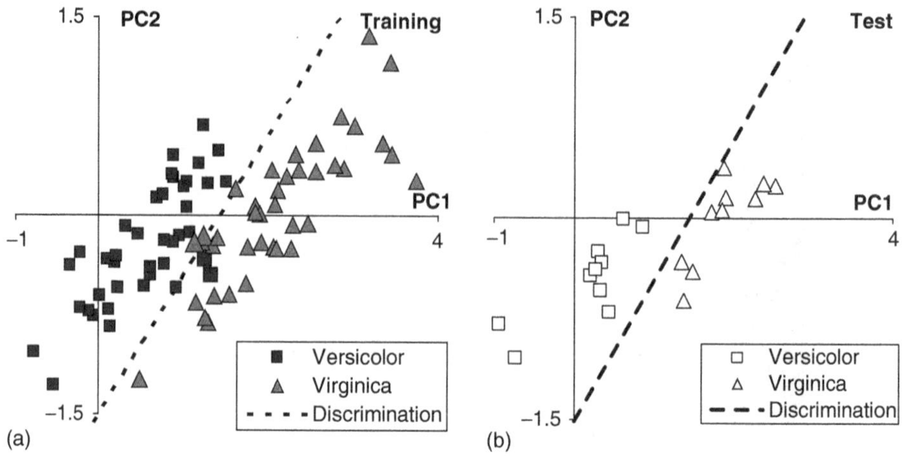

Figure 11.13 PCA-LDA discrimination between Classes 2 and 3.

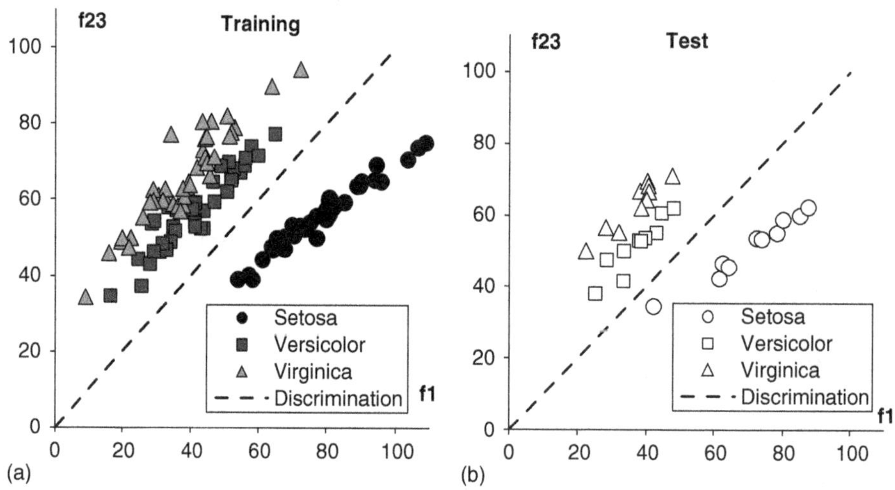

Figure 11.14 Direct LDA discrimination between Classes 1 and 23.

3. LDA implicitly uses the assumption of normality.
4. LDA does not allow interchanging the errors of the first and the second kind.

Advantages of LDA

1. LDA is easy to apply.

11.3.2 Quadratic Discriminant Analysis (QDA)

Quadratic discriminant analysis (*QDA*) is a generalization of LDA. QDA is a multiclass method that can be used for simultaneous classification of several classes.

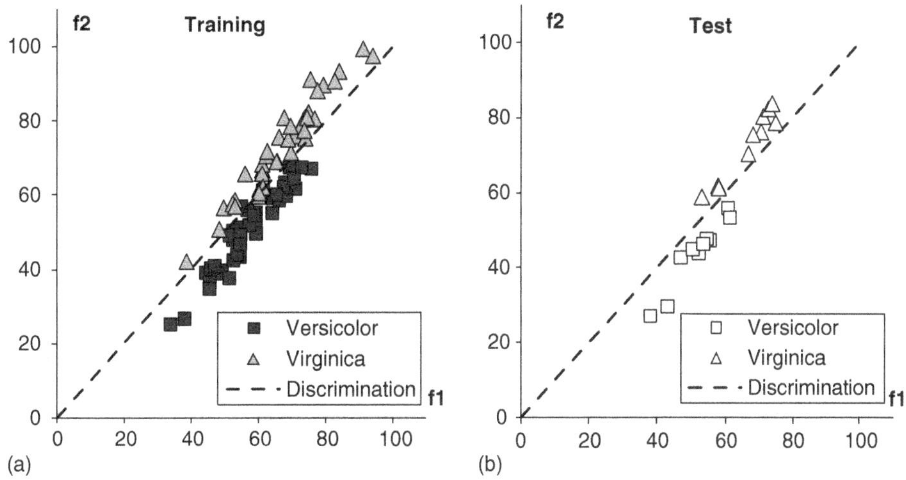

Figure 11.15 Direct LDA discrimination between Classes 2 and 3.

A training set consists of K matrices $\mathbf{X}_1, \ldots, \mathbf{X}_K$, which have $I_1, \ldots, I_K$ rows (samples) correspondingly. The number of variables (columns) is the same for all matrices and equals J. Preserving the first assumption of LDA given by Eq. 11.1, we relax the second one, that is, we presume that the covariance matrices $\boldsymbol{\Sigma}_k$ are different. Then the *QDA-scores* are calculated by

$$f_k = (\mathbf{x} - \boldsymbol{\mu}_k)\, \boldsymbol{\Sigma}_k^{-1} (\mathbf{x} - \boldsymbol{\mu}_k)^{\mathrm{t}} + \log\left(\det\left(\boldsymbol{\Sigma}_k^{-1}\right)\right), \qquad k = 1, \ldots, K \tag{11.6}$$

The QDA classification rule is as follows. A new sample $\mathbf{x}$ belongs to the class that provides the smallest QDA-score. In practice, similar to LDA, the unknown expectations $\boldsymbol{\mu}_k$ and the covariance matrices $\boldsymbol{\Sigma}_k$ are replaced by their estimates

$$\mathbf{m}_k = (1/I_k) \sum_{i=1}^{I_k} \mathbf{x}_i, \quad \mathbf{S}_k = (1/I_k)\, \widetilde{\mathbf{X}}_k^{\mathrm{t}} \widetilde{\mathbf{X}}_k \tag{11.7}$$

In these formulas, $\widetilde{\mathbf{X}}_k$ is the centered matrix $\mathbf{X}_k$, (see Section 8.1.4). A surface that separates classes k and l is determined by a quadratic equation

$$f_k = f_l.$$

That is why the method is called *quadratic*.

Consider the application of the QDA method to the Iris dataset. The calculations are shown in sheet QDA. The training set consists of three classes (with local names X1c, X2c, and X3c) containing 40 samples each. Variable-wise mean values of these arrays are computed in sheet QDA under the local names m1c, m2c, and m3c (see Fig. 11.16).

Then we calculate the covariance matrices and invert them (local names Sinv1, Sinv2, and Sinv3; see Fig. 11.17 and sheet QDA).

At last, we can calculate the QDA scores and determine the class membership as shown in Fig. 11.18.

	C	D	E	F	G	H
153						
154	m1c	5.038	3.453	1.460	0.235	
155	m2c	6.010	2.780	4.318	1.350	
156	m3c	6.623	2.960	5.608	1.990	
157						

Figure 11.16 Mean values in QDA.

	L	M	N	O	P	Q	R	S	T	U	V	W	X	Y	Z	AA
2		S_1					S_2					S_3				
3		0.131	0.097	0.013	0.013		0.274	0.087	0.172	0.052		0.468	0.110	0.358	0.051	
4		0.097	0.130	0.002	0.012		0.087	0.111	0.081	0.045		0.110	0.113	0.081	0.046	
5		0.013	0.002	0.030	0.005		0.172	0.081	0.204	0.074		0.358	0.081	0.345	0.059	
6		0.013	0.012	0.005	0.010		0.052	0.045	0.074	0.043		0.051	0.046	0.059	0.074	
7																
8		S_1^{-1}					S_2^{-1}					S_3^{-1}				
9		18.81	-13.38	-6.47	-5.99		9.13	-4.08	-9.58	9.60		12.07	-4.89	-12.11	4.39	
10		-13.38	18.25	5.95	-7.42		-4.08	17.71	3.62	-19.90		-4.89	14.82	3.00	-8.26	
11		-6.47	5.95	39.49	-19.32		-9.58	3.62	23.00	-31.54		-12.11	3.00	15.79	-6.12	
12		-5.99	-7.42	-19.32	133.02		9.60	-19.90	-31.54	86.51		4.39	-8.26	-6.12	20.47	
13																
14		log(det(S_1^{-1}))					log(det(S_2^{-1}))					log(det(S_3^{-1}))				
15		5.806					4.639					3.845				
16																

Figure 11.17 Covariance matrices and their inverses in QDA.

	D	E	F	G	H	I	J	K	L	M	N	O	P	Q
2	SL	SW	PL	PW	f1	f2	f3	Discr		S_1				
3	5.100	3.500	1.400	0.200	1.8133	4.7207	5.1247	1		0.131	0.097	0.013	0.013	
4	~~4.900~~	~~3.000~~	~~1.400~~	~~0.200~~	=LN(MMULT((D4:G4-m1c),MMULT(S1inv,TRANSPOSE(D4:G4-m1c)))+$M$15)									
5	4.700	3.200	1.300	0.200	1.9876	4.5653	5.0003	1		0.013	0.002	0.030	0.005	
6	4.600	3.100	1.500	0.200	2.0174	4.4311	4.8633	1		0.013	0.012	0.005	0.010	
7	5.000	3.600	1.400	0.200	1.8761	4.764	5.1215	1						
8	5.400	3.900	1.700	0.400	2.3317	4.7561	5.1176	1		S_1^{-1}				
9	4.600	3.400	1.400	0.300	2.281	4.5754	4.9468	1		18.81	-13.38	-6.47	-5.99	
10	5.000	3.400	1.500	0.200	1.802	4.6248	5.0345	1		-13.38	18.25	5.95	-7.42	
11	4.400	2.900	1.400	0.200	2.2152	4.3462	4.8142	1		-6.47	5.95	39.49	-19.32	
12	4.900	3.100	1.500	0.100	2.175	4.5251	4.9758	1		-5.99	-7.42	-19.32	133.02	
13	5.400	3.700	1.500	0.200	2.013	4.8211	5.2039	1						
14	4.800	3.400	1.600	0.200	2.0702	4.5845	4.9424	1		log(det(S_1^{-1}))				
15	4.800	3.000	1.400	0.100	2.178	4.4957	4.9744	1		5.806				
16	4.300	3.000	1.100	0.100	2.5579	4.5673	4.974	1						

Figure 11.18 Calculation of the QDA-scores and class memberships.

The classification results are presented in the QDA-scores plots shown in Fig. 11.19. These plots (as well as the QDA-scores analysis) demonstrate that the training set classification is done with errors: three samples from the second class (*Versicolor*) are assigned to the third class (*Virginica*). There are no errors in the test set.

QDA retains most of the LDA shortcomings.

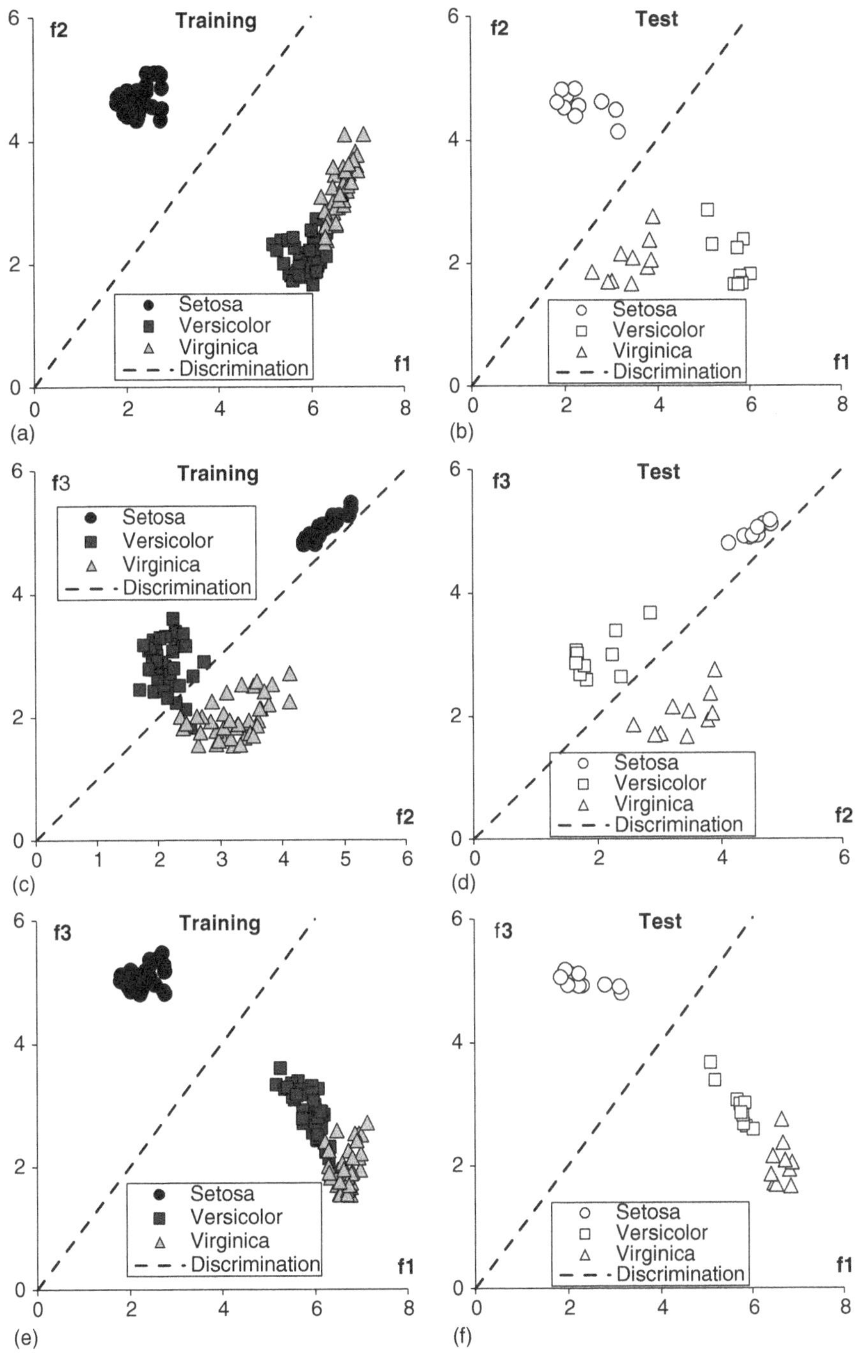

Figure 11.19 The results of QDA.

Disadvantages of QDA

1. QDA does not work when the covariance matrices are degenerate, for example, for a large number of variables. It requires regularization, for example, by PCA.
2. QDA implicitly uses the assumption of normality.
3. QDA does not allow interchanging the errors of the first and second kind.

Advantages of QDA

1. It is suitable when the covariance matrices are different.

11.3.3 PLS Discriminant Analysis (PLSDA)

In Section 11.1.6, we noted that classification and calibration problems have a lot in common. The *partial least squares-discriminant analysis* (PLSDA) method is based on this notion. Classification data are treated as a set of predictors (the training set $\mathbf{X}_c$ and the test set $\mathbf{X}_t$). These data are supplied with new *dummy* (fictitious) response matrices, $\mathbf{Y}_c$ for the training set, and $\mathbf{Y}_t$ for the test set. These matrices are formed as follows.

The number of responses (columns) in dummy matrices must be equal to the number of classes (three in our case). Each row matrices $\mathbf{Y}$ should be filled with zeros (0), or ones (1), depending on the class membership. If the row (object) belongs to the k-th class, the element in the k-th column must be 1, whereas the other elements in this row should be zeros. An example of such a dummy matrix $\mathbf{Y}$ is presented in sheet PLSDA (see areas with local names Yc and Yt).

The predictor block $\mathbf{X}_c$ and the response block $\mathbf{Y}_c$ are connected by a PLS2 regression (Section 8.1.3). This regression is employed to calculate the predicted dummy responses $\mathbf{Y}^{\text{hat}}$ for both training and test sets. In the result, a set of new responses $y_1, \dots, y_K$ is obtained for each sample $\mathbf{x}$. The PLSDA classification rule goes as follows. A sample belongs to the class the prediction y_k for which is closer to 1. In other words, we select an index k, which corresponds to

$$\min \{ |1 - y_1|, \dots, |1 - y_K| \} \tag{11.8}$$

The application of PLSDA to the Iris dataset is presented in sheet PLSDA. The procedure starts with the formation of the dummy response matrix and PLS2-scores calculation by the special function **ScoresPLS2** (Section 8.5.1).

Figure 11.20 shows the relevant parts of this sheet. Note that PLS2-scores for the test set are obtained using a slightly different formula.

A standard worksheet function **TREND** is applied to calculate predicted response values $\mathbf{Y}^{\text{hat}}$. In Excel 2003, this function gives an incorrect result sometimes, as explained in Section 7.2.8. To avoid this error, we use a formula

= TREND (Y1c-Y1m, Tc) + Y1m

that employs centered dummy responses (see Fig. 11.21).

PLSDA classification results are shown in Fig. 11.22. In the training set, 15 samples from the second class (*Versicolor*) are erroneously attributed to the third class (*Virginica*), and four samples from the third class (*Virginica*) are erroneously attributed to the second class (*Versicolor*). The test set classification also has errors. One sample from the first class

	C	D	E	F	G	H	I	J
2	Name	PC1	PC2	PC3	Y1	Y2	Y3	
3	se01	=ScoresPLS2(Xc,Yc,3,1,1)				0	0	
4	se02	-2.746	0.288	0.188	1	0	0	
5	se03	-2.921	0.175	-0.040	1	0	0	
6	se04	-2.769	0.329	-0.070	1	0	0	
7	se05	-2.785	-0.292	-0.064	1	0	0	
8	se06	-2.359	-0.711	-0.107	1	0	0	
9	se07	-2.855	0.034	-0.281	1	0	0	
10	se08	-2.675	-0.100	0.033	1	0	0	
11	se09	-2.897	0.577	-0.087	1	0	0	
12	se10	-2.708	0.223	0.187	1	0	0	

	C	D	E	F	G	H	I	J
123	se41	=ScoresPLS2(Xc,Yc,3,1,1,Xt)				0	0	
124	se42	-2.843	1.052	0.244	1	0	0	
125	se43	-3.019	0.294	-0.237	1	0	0	
126	se44	-2.455	-0.230	-0.270	1	0	0	
127	se45	-2.271	-0.477	-0.268	1	0	0	
128	se46	-2.742	0.314	0.062	1	0	0	
129	se47	-2.603	-0.487	-0.121	1	0	0	
130	se48	-2.867	0.227	-0.114	1	0	0	
131	se49	-2.613	-0.489	0.071	1	0	0	
132	se50	-2.749	-0.022	0.095	1	0	0	
133	ve41	0.455	0.646	-0.066	0	1	0	
134	ve42	0.848	0.046	-0.023	0	1	0	

Figure 11.20 Developing of PLS2 regression.

	D	E	F	G	H	I	J	K	L	M	N
2	PC1	PC2	PC3	Y1	Y2	Y3		Y_1^{hat}	Y_2^{hat}	Y_3^{hat}	
3	-2.741	-0.242	0.054	1	0	0		=TREND(Y1c-Y1m,Tc)+Y1m			
4	-2.746	0.288	0.188	1	0	0		0.841	0.378	-0.219	
5	-2.921	0.175	-0.040	1	0	0		0.896	0.254	-0.150	
6	-2.769	0.329	-0.070	1	0	0		0.824	0.309	-0.132	
7	-2.785	-0.292	-0.064	1	0	0		0.986	0.078	-0.063	
8	-2.359	-0.711	-0.107	1	0	0		1.001	-0.073	0.072	
9	-2.855	0.034	-0.281	1	0	0		0.908	0.126	-0.034	
10	-2.675	-0.100	0.033	1	0	0		0.918	0.186	-0.104	
11	-2.897	0.577	-0.087	1	0	0		0.787	0.390	-0.176	
12	-2.708	0.223	0.187	1	0	0		0.850	0.355	-0.205	

Figure 11.21 Prediction of dummy responses.

is erroneously attributed to the second one, and two samples from the second class are erroneously attributed to the third class. Thus, we can conclude that PLSDA classification did not provide satisfactory results. However, the situation could be greatly improved if we give up "bad" classification rule presented in Eq. 11.8 and continue with calculations further.

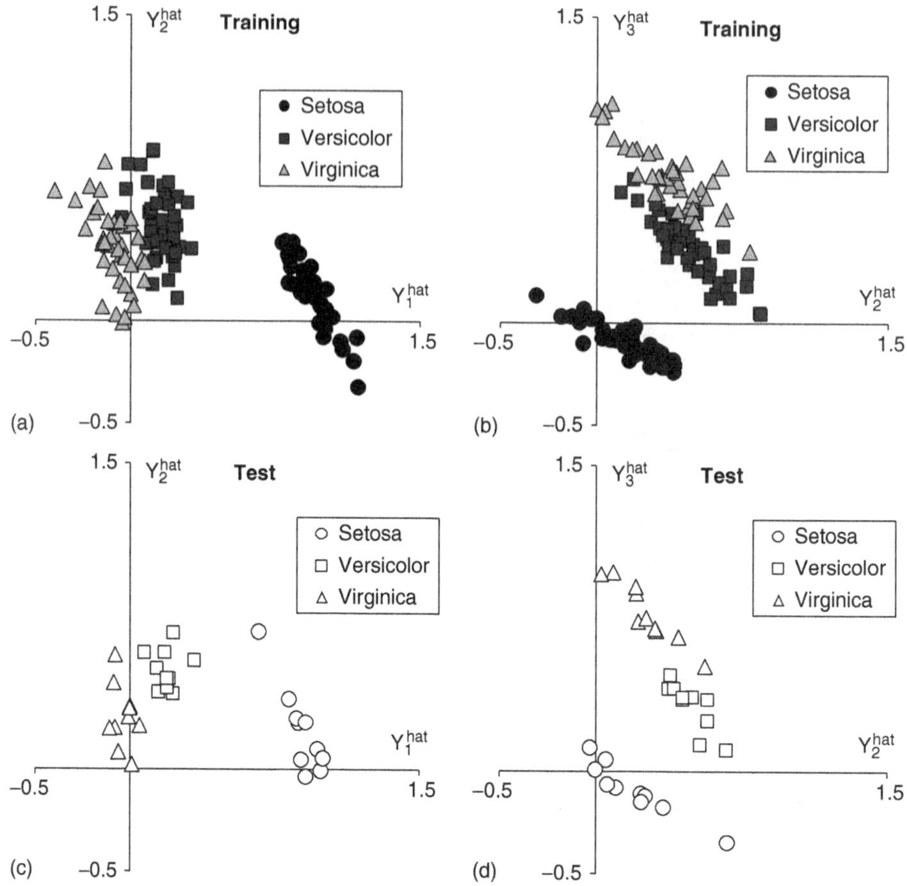

Figure 11.22 Results of PLSDA classification.

We will consider the predicted responses $\mathbf{Y}^{\mathrm{hat}}$ not as the ultimate but as the intermediate data and will treat them by some other classification method, for example, by LDA. However, this cannot be done directly because matrix $\mathbf{Y}_{\mathrm{c}}^{\mathrm{hat}}$ has rank $K-1$, and the corresponding covariance matrix (see Eq. 11.3) is singular. Therefore, it is necessary to subject data $\mathbf{Y}_{\mathrm{c}}^{\mathrm{hat}}$ to PCA before applying LDA, just as we did in Section 11.3.1. The pertaining calculations are shown in sheet PLSDA-PCA-LDA. The results are shown in Fig. 11.23. In this way, we obtain the results, which have only one case of misclassification in the training set. Namely, one sample in the second class (*Versicolor*) is mistakenly attributed to the third class (*Virginica*). The test set is classified correctly.

Here, we have carried out a chain of operations. We started with PLS2-regression on a dummy response matrix, then we continued with PCA projection of the predicted results, and finished with LDA. This chain can be considered as preprocessing of the original data $\mathbf{X}$, that is, as a filter that discloses new data features directly related to the differences between the classes. It is very important that the PCA-LDA method is applied to the matrix of the predicted dummy responses $\mathbf{Y}^{\mathrm{hat}}$ but not to the matrix of PLS2-scores.

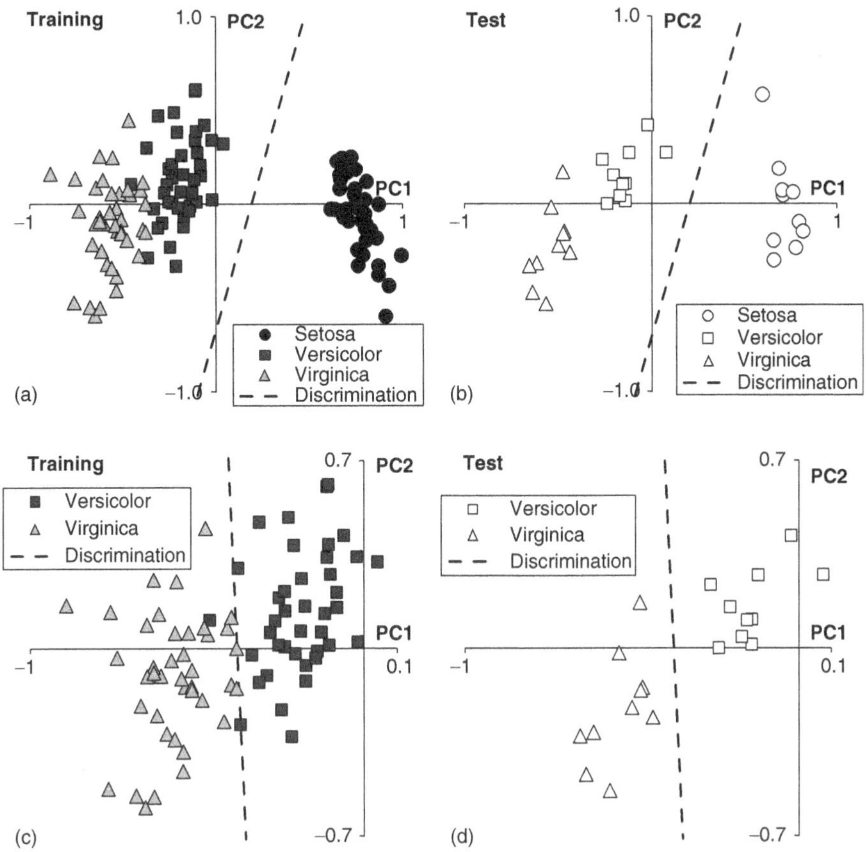

Figure 11.23 Results of PLSDA-PCA-LDA classification.

Disadvantages of PLSDA

1. PLSDA requires preliminary regression analysis.
2. The result depends on the number of PCs in the PLS2 regression.
3. PLSDA is sensitive to outliers.
4. PLSDA does not work for a small number of samples in the training set.

Advantages of PLSDA

1. It does not use the distribution type assumption.
2. PLSDA is applicable to a large number of variables and resistant to the curse of dimensionality.

11.3.4 SIMCA

In this section, we discuss a method with a long name *soft independent modeling of class analogy*, in short, *SIMCA*. This method has several distinctive features.

Firstly, each class is modeled separately, *independent* of the others, which is why SIMCA is a one-class method. Secondly, SIMCA classification is ambiguous (*soft*), that is, each sample can be simultaneously assigned to several classes. Thirdly, SIMCA provides a unique opportunity to set the type I error value and develop an error-driven classification (Section 11.1.1). In the methods discussed above, the type I and II errors are out of control; they cannot be changed. For example, in the LDA method, the type I (α) and the type II (β) errors can be calculated (with difficulty) but they cannot be varied.

As SIMCA is a one-class method, we explain it using only one class represented by matrix $\mathbf{X}$ of dimension I (samples) by J (variables). Our goal is to develop a classifier (a rule), which decides on a new sample $\mathbf{x}$. The sample is either accepted (belonging to class $\mathbf{X}$) or rejected (not belonging to class $\mathbf{X}$). To solve this problem, we present matrix $\mathbf{X}$ as a cloud of I points in J dimensional variable space. The space origin is in the center of gravity of the cloud. This cloud often has a specific shape. Caused by strong correlations between the variables, the points are located close to a hyperplane (subspace) of dimension $A < J$. To determine the dimension, A, and to construct this hyperplane the PCA is applied.

Each data element can be presented (Section 9.2.13) as a sum of two vectors. One of them is located on the hyperplane (projection) and the other one is perpendicular to the hyperplane (residual). The lengths of these vectors are important indicators of whether an object belongs to the class (see Fig. 11.24). They are called the *leverage*, h, and the *deviation*, v. As soon as a new object is considered a candidate for the class membership, the object is projected onto the existing subspace, and the object's leverage and deviation are calculated. Comparing these values with the critical (cutoff) levels, which have been determined from the training set, it is possible to decide on the class membership of the new object.

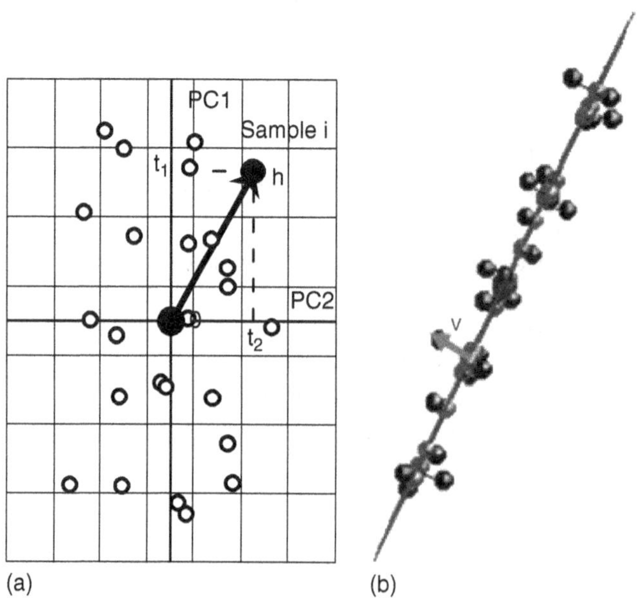

Figure 11.24 PCA leverage and deviation.

Let us consider a PCA decomposition of the training matrix **X**

$$\mathbf{X} = \mathbf{T}\mathbf{P}^{\mathrm{t}} + \mathbf{E}$$

Each sample **x** (training, test, or new) can be projected onto the PC space, that is, its score vector can be determined as

$$\mathbf{t} = \mathbf{x}\mathbf{P} = \left(t_1, \dots, t_A\right)^{\mathrm{t}}$$

The leverage h is calculated as

$$h = \mathbf{t}^{\mathrm{t}}\left(\mathbf{T}^{\mathrm{t}}\mathbf{T}\right)^{-1}\mathbf{t} = \sum_{a=1}^{A}\left(t_a/s_a\right)^2 \tag{11.9}$$

where s_a is a singular value, that is, $s_a = \sqrt{\lambda_a}$ (section 5.2.11).

The difference between the original vector **x** and its projection

$$\mathbf{e} = \mathbf{x}\left(\mathbf{I}-\mathbf{P}\mathbf{P}^{\mathrm{t}}\right) \tag{11.10}$$

is the residual. Deviation v is the mean squared residual

$$v = (1/J)\sum_{j=1}^{J} e_j^2. \tag{11.11}$$

There is reason to believe[3] that both quantities h and v follow the chi-squared distribution, that is,

$$N_h\left(h/h_0\right) \sim \chi^2\left(N_h\right), \quad N_v\left(v/v_0\right) \sim \chi^2\left(N_v\right) \tag{11.12}$$

where h_0 and v_0 are the mean values, and N_h and N_v are the numbers of degrees of freedom of h and v, respectively (see Section 6.2.4).

Employing the training set $\mathbf{X}_{\mathrm{c}} = (x_1, \dots, x_I)^{\mathrm{t}}$, we can calculate the leverages $h_1, \dots, h_I$, and deviations $v_1, \dots, v_I$, which are used to estimate the corresponding means

$$h_0 = (1/I)\sum_{i=1}^{I} h_i \equiv A/I, \quad v_0 = (1/I)\sum_{i=1}^{I} v_i \tag{11.13}$$

as well as the variances

$$S_h = (1/(I-1))\sum_{i=1}^{I}\left(h_i - h_0\right)^2, \quad S_v = (1/(I-1))\sum_{i=1}^{I}\left(v_i - v_0\right)^2. \tag{11.14}$$

The numbers of degrees of freedom N_h and N_v are found by the method of moments (Section 6.3.9)

$$\widehat{N}_h = 2h_0^2/S_h, \quad \widehat{N}_v = 2v_0^2/S_v \tag{11.15}$$

[3] A.L. Pomerantsev Acceptance areas for multivariate classification derived by projection methods (2008). Journal of Chemometrics 22: 601–609.

The decision statistics (compare with Section 6.2.2) f, which is used for classification, is calculated by the equation

$$f = N_h \left(h/h_0\right) + N_v \left(v/v_0\right) \tag{11.16}$$

From Eq. 11.12, it follows that

$$f \sim \chi^2 \left(N_h + N_v\right)$$

Let α be a given value of the type I error (false rejection), then

$$f_{\text{crit}} = \chi^{-2} \left(\alpha, N_h + N_v\right) \tag{11.17}$$

where $\chi^{-2}(\alpha, n)$ is the α-th quantile of the chi-squared distribution with n degrees of freedom (Section 6.2.4). The classification rule in the SIMCA method reads as follows. A sample is attributed to the class, if

$$f < f_{\text{crit}}$$

This rule ensures that $(1-\alpha)100\%$ of the training and test samples will be correctly classified.

Let us look at SIMCA method application to the Iris dataset. We start with the first class (*Setosa*) and develop a SIMCA classifier for this class. Calculations are presented in worksheet SIMCA_1.

In the first step, we apply PCA using matrix X1c (the part of matrix $\mathbf{X}_c$ related to Class 1) as the training set and the entire matrix Xt as the test set. Similarly to the previous methods, we are using two PCA components (see Fig. 11.25).

The areas that contain the scores (training and testing) and loadings are given local names Tc, Tt, and Pc, respectively.

Then we calculate singular values by summing up the squares of the scores values for each PC and then taking the square root of the sums. Next, we compute the leverage values h for the training and the test sets using Eq. 11.9 as shown in Fig. 11.26.

The calculation of deviations v is more complicated. Here, we combine Eqs 11.10 and 11.11 in the manner described in Section 13.1.8. The test set deviations are calculated by a similar formula replacing X1c by Xt and Tc by Tt (see Fig. 11.27).

After values h and v are obtained, we can find their means (Eq. 11.13) and variances (Eq. 11.14) and estimate the number of degrees of freedom using the formula given in

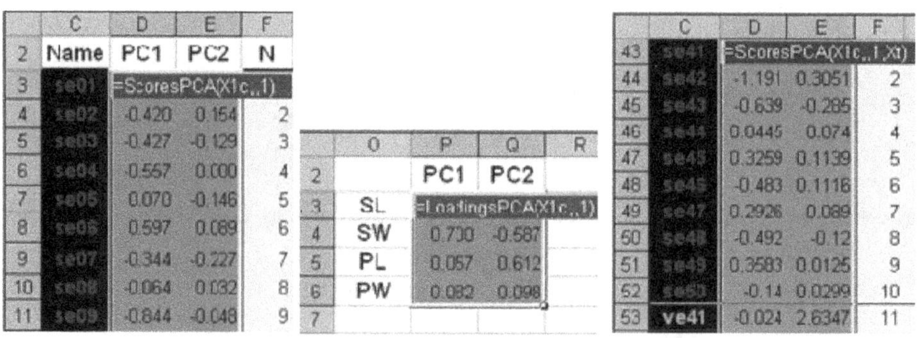

Figure 11.25 PCA scores and loadings calculations.

	C	D	E	F	G	H	I	J	K	L	M	N	O	P	Q
2	Name	PC1	PC2	N	h	v	h/h0	v/v0	f	fcrit	Discr			PC1	PC2
3	se01	0.071	-0.035	1	0.001	0.001	0.03	0.16	0.21	11.34	1		SL	0.707	0.521
4	se02	-0.420	0.154	2	0.035	0.007	0.70	0.94	2.34	11.34	1		SW	0.700	-0.587
5	se03	-0.427	-0.129	3	=SUMSQ(D5:E5/P$9:Q$9)				1.42	11.34	1		PL	0.057	0.612
6	se04	-0.557	0.000	4	0.035	0.002	0.69	0.28	1.67	11.34	1		PW	0.082	0.098
7	se05	0.070	-0.146	5	0.014	0.000	0.29	0.05	0.63	11.34	1				
8	se06	0.597	0.089	6	0.045	0.013	0.90	1.75	3.54	11.34	1			PC1	PC2
9	se07	-0.344	-0.227	7	0.047	0.008	0.93	1.06	2.92	11.34	1		SingV	2.9949	1.2449
10	se08	-0.064	0.032	8	0.001	0.000	0.02	0.06	0.11	11.34	1				

Figure 11.26 Calculation of leverages.

	C	D	E	F	G	H	I	J	K	L	M	N	O	P	Q
2	Name	PC1	PC2	N	h	v	h/h0	v/v0	f	fcrit	Discr			PC1	PC2
3	se01	0.071	-0.035	1	0.001	0.001	0.03	0.16	0.21	11.34	1		SL	0.707	0.521
4	se02	-0.420	0.154	2	0.035	0.007	0.70	0.94	2.34	11.34	1		SW	0.700	-0.587
5	se03	-0.427	-0.129		0.031	0.001	0.62	0.17	1.42	11.34	1		PL	0.057	0.612
6	se04	-0.557	0.000	4	0.035	0.002	0.69	0.28	1.67	11.34	1		PW	0.082	0.098
7	se05	0.070	-0.146	5	0.014	=SUMSQ(INDEX(X1c-mean1-MMULT(Tc,TRANSPOSE(Pc)),F7,))/J									
8	se06	0.597	0.089	6	0.045	0.013	0.90	1.75	3.54	11.34	1			PC1	PC2
9	se07	-0.344	-0.227	7	0.047	0.008	0.93	1.06	2.92	11.34	1		SingV	2.9949	1.2449
10	se08	-0.064	0.032	8	0.001	0.000	0.02	0.06	0.11	11.34	1				

Figure 11.27 Calculation of deviations.

	O	P	Q	R
11		h	v	
12	mean	0.050	0.008	
13	var	0.0028	0.0001	
14	DoF	=ROUND(2*P12*P12/P13,0)		
15				

	O	P	Q	R
11		h	v	
12	mean	0.050	0.008	
13	var	0.0028	0.0001	
14	DoF	2	1	
15				
16				
17	alpha	0.01		
18	fcrit	=CHIINV(P17, DoF_h + DoF_v)		
19				

Figure 11.28 Calculation of the number of degrees of freedom (left), calculation of the critical level (right).

Eq. 11.15. This is shown in the left panel of Fig. 11.28. Using the obtained results, we calculate statistics f (Eq. 11.16) for any α value, for example, $\alpha = 0.01$. Later on we can vary α and observe the changes in the classification results. The critical level f_{crit} is given by Eq. 11.17. In order to calculate it, we use the standard Excel function **CHIINV** described in Section 6.2.4. See the right panel of Fig. 11.28.

Figure 11.29 shows the classification results. The test set plot (Fig. 11.29b) is modified in order to present all samples. For this purpose, the axes are modified by the power transformation $x^{1/p}, p = 3$. All training set samples are classified correctly. One sample from the first class (*Setosa*) in the test set is not recognized.

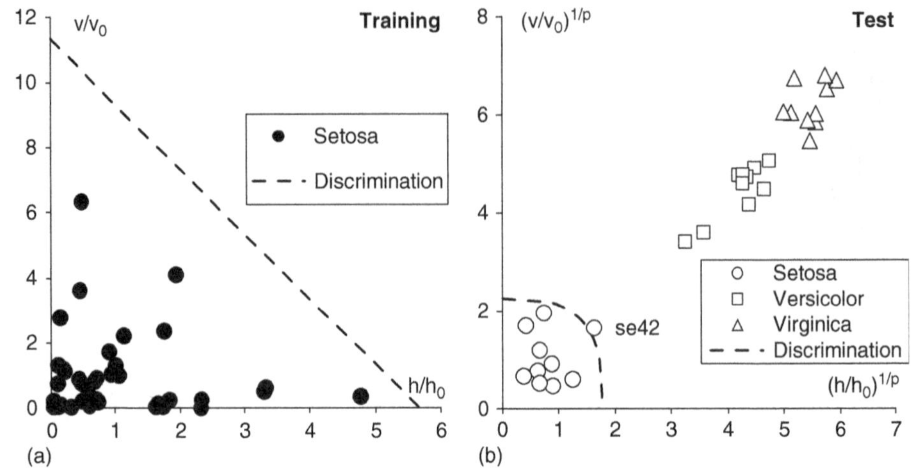

Figure 11.29 The results of SIMCA classification for Class 1.

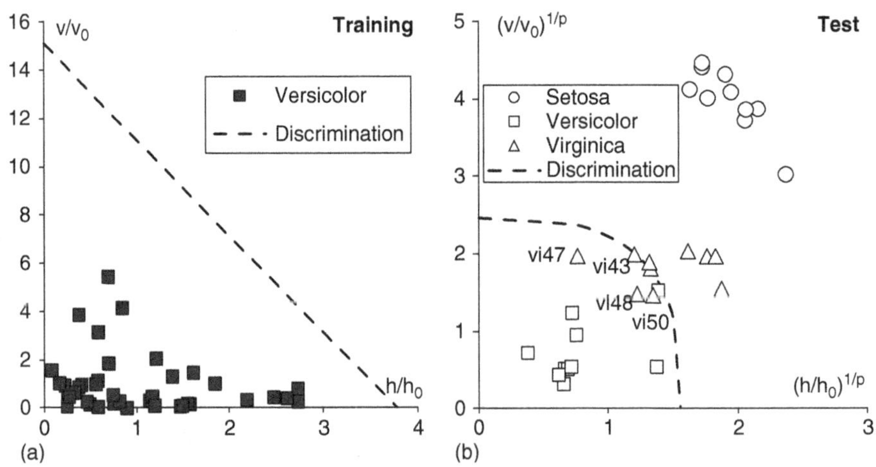

Figure 11.30 The results of SIMCA classification for Class 2.

SIMCA classification of the other classes is done in a similar way. For Class 2 (worksheet SIMCA_2) matrix, X2c is used as the training set and for Class 3 (worksheet SIMCA_3), we use matrix X3c. The average values mean2 and mean3, which are used for deviations calculations, are changed correspondingly. The final results are shown in Figs 11.30 and 11.31.

The results are as follows. One sample from the third training set was not accepted (the type I error). In the test sets, one sample from the first class is not classified (the type I error), two samples from the second class were assigned to the third class (the type II error), four samples from the third class were classified to the second class (the type II error). The actual type I error is $2/150 \approx 0.013$ that is close to the preset $\alpha = 0.01$ value.

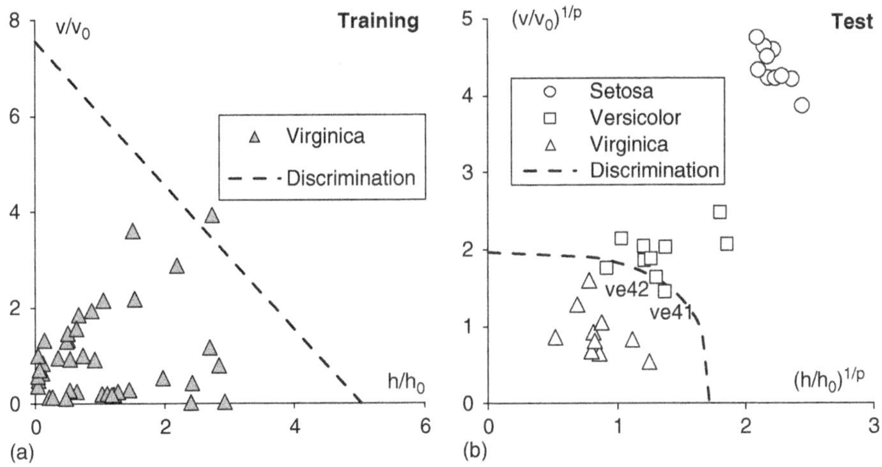

Figure 11.31 The results of SIMCA classification for Class 3.

Disadvantages of SIMCA

1. It requires prior analysis by PCA.
2. The result depends on the choice of PCs. However, this is facilitated by the fact that we can take the minimum number for which the training set is correctly classified.
3. SIMCA is sensitive to outliers, but they can be recognized by the method.
4. SIMCA does not work for a small number of samples in the training set.

Advantages of SIMCA

1. SIMCA does not use a type of distribution assumption.
2. SIMCA is a one-class method.
3. The type I error can be adjusted.
4. SIMCA applicable to a large number of variables and resistant to the curse of dimensionality.

11.3.5 *k*-Nearest Neighbors (kNN)

kNN is one of the simplest methods of classification. As previously, let $\mathbf{X}_c$ be the training set divided into class subsets, and let $\mathbf{x}$ be a new unknown object, which should be classified. At first we calculate the distances (usually Euclidean) from $\mathbf{x}$ to all samples of the training set ($\mathbf{x}_1, \ldots, \mathbf{x}_I$). Then we select kNN, which are located at minimal distances. A new object $\mathbf{x}$ belongs to a class, which encompasses the majority of the k neighbors. The parameter k is chosen empirically. An increase in k reduces the impact of errors, whereas its decrease worsens classification.

The application of the kNN method to the Iris dataset is presented in worksheet kNN. We start with the calculation of distances between each of the test set samples and all samples in the training set. The results are collected in a table shown in Fig. 11.32. The rows of this table correspond to the training set and the columns represent the test set.

	A	B	C	D	E	F	G	H	I	J	K	L	M
2			name	se41		name	se42		name	se43		name	se43
3	1		se01	=SUMSQ(INDEX(Xc,MATCH(C3,nC,0),)-INDEX(Xt,MATCH(D$2,nT,0),))									
4	2		se02	0.28		se02	0.67		se02	0.3		se02	0.3
5	3		se03	0.19		se03	0.86		se03	0.09		se03	0.09
6	4		se04	0.37		se04	0.7		se04	0.09		se04	0.09
7	5		se05	0.03		se05	1.96		se05	0.53		se05	0.53
8	6		se06	0.49		se06	3.54		se06	1.69		se06	1.69
9	7		se07	0.18		se07	1.23		se07	0.1		se07	0.1
10	8		se08	0.06		se08	1.51		se08	0.44		se08	0.44
11	9		se09	0.74		se09	0.39		se09	0.1		se09	0.1
12	10		se10	0.25		se10	0.88		se10	0.31		se10	0.31
13	11		se11	0.25		se11	2.82		se11	1.29		se11	1.29
14	12		se12	0.15		se12	1.4		se12	0.29		se12	0.29
15	13		se13	0.34		se13	0.63		se13	0.22		se13	0.22
16	14		se14	0.82		se14	0.61		se14	0.1		se14	0.1
17	15		se15	0.91		se15	4.6		se15	2.61		se15	2.61

Figure 11.32 Sample distances calculation.

To calculate the distances, we use two global names nC and nT (sheet Data), which refer to the names of the sample in the training and the test set, correspondingly.

As soon as all distances are calculated, each column is sorted in the ascending order, which leaves the smallest distance at the top. The rest is simple. One should just go through the columns and determine which class a test sample belongs to. In the end, all test samples are correctly classified (Fig. 11.33).

Disadvantages of kNN

1. kNN does not work for a large number of features as the curse of dimensionality eliminates the differences between samples.
2. The result depends on the choice of metric and the number of neighbors.

Advantages of kNN

1. kNN does not use the type of distribution assumption.
2. kNN can be used for a small number of samples in the training set.

	BQ	BR	BS	BT	BU	BV	BW	BX	BY	BZ	CA	CB	CC	CD
2	name	vi41		name	vi42		name	vi43		name	vi44		name	vi45
3	vi21	0.07		vi40	0.13		vi02	0		vi21	0.05		vi21	0.09
4	vi13	0.12		vi13	0.22		vi14	0.07		vi25	0.1		vi25	0.16
5	vi05	0.13		vi11	0.26		vi22	0.1		vi05	0.15		vi37	0.19
6	vi25	0.14		vi16	0.3		ve34	0.13		vi03	0.17		vi05	0.23
7	vi40	0.17		vi21	0.37		vi28	0.23		vi13	0.24		vi01	0.25
8	vi16	0.2		ve28	0.42		vi39	0.23		vi01	0.31		vi13	0.3
9	vi33	0.22		vi25	0.48		vi15	0.26		vi40	0.31		vi16	0.3
123														

Figure 11.33 Seven neighbors for test samples.

11.4 UNSUPERVISED CLASSIFICATION

11.4.1 PCA Again (Revisited)

The PCs method is the simplest and the most popular method of unsupervised classification. In this section, we are using only the training set of the Iris data, leaving out the test set. The calculations are presented in worksheet PCA-Explore.

Here, we do not know to which class a sample belongs (i.e., the class membership). Moreover, even the number of classes is unknown.

However, studying the PCA scores plot built for the training set (Fig. 11.34), we can easily select a group of samples (marked by an ellipse), which is clearly separated from other objects. It is natural to assume that these samples belong to a single class.

Let us remove all these objects from the training set and apply PCA to the remaining samples. In the PC1–PC2 scores plot shown in Fig. 11.35, a person with some imagination can distinguish two clusters shown by the two ellipses. However, it is difficult to find anything that looks like a class in the PC1–PC3 scores plot.

Hence, the PC explorative data analysis can sometimes reveal hidden classes and sometimes cannot. In any case, further testing of the hypotheses using other unsupervised classification methods is necessary.

11.4.2 Clustering by *K*-Means

There is a large group of methods that perform *clustering*. Clustering is an operation that divides a sample set into subsets (called *clusters*) in a way that all samples in the same cluster are somehow similar. Similarity of samples $\mathbf{x}_1$ and $\mathbf{x}_2$ is usually associated with the distance $d(\mathbf{x}_1, \mathbf{x}_2)$ between the samples. As a measure of distance, the Euclidean metric can be used.

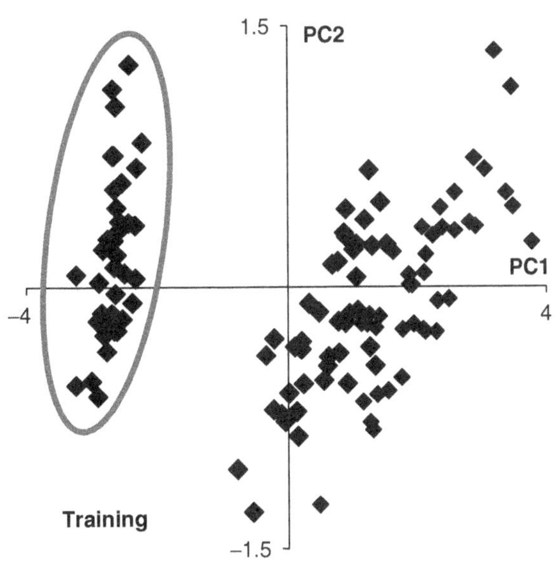

Figure 11.34 PCA of the training set.

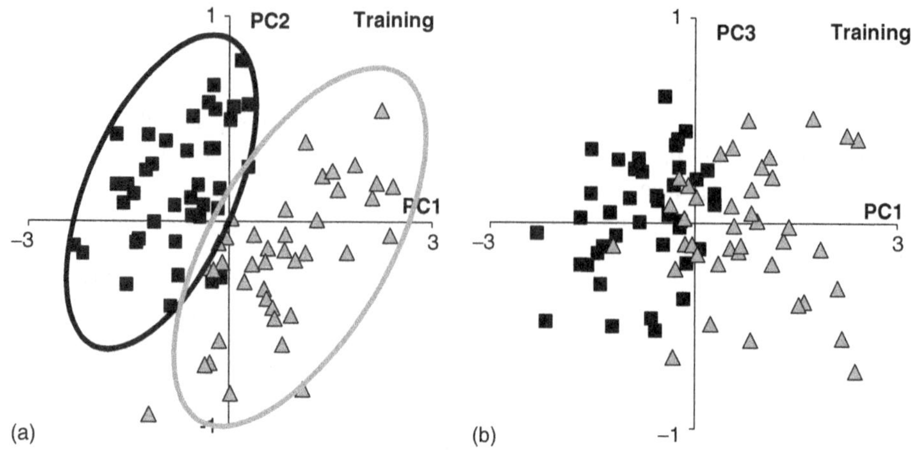

Figure 11.35 PCA of the shortened training set.

A *K-means* method is the simplest and, therefore, the most popular method. It splits the original set of samples into K clusters, where number K is known. Here, each sample $\mathbf{x}_i$ should be attributed to one of the clusters S_k, $k = 1, \ldots, K$. Each cluster is characterized by its centroid $\mathbf{m}_k$, which is the center of mass of all samples in the cluster. The K-means method uses an iterative algorithm, which performs the following operations.

1. Distances $d(\mathbf{x}_j, \mathbf{m}_k)$, $\quad j = 1, \ldots, J; k = 1, \ldots, K$ between the centroids and the samples are calculated.
2. Each sample is reattributed to the nearest cluster in accordance with the minimal distance.
3. New centroids $\mathbf{m}_k$ are calculated for each cluster in line with the new partition

$$\mathbf{m}_k = \left(1/J_k\right) \sum_{\mathbf{x}_j \in S_k} \mathbf{x}_j,$$

where J_k is the number of samples in cluster S_k.

Steps 1–3 are repeated until convergence.

To initialize the algorithm, one must set all initial centroids $\mathbf{m}_k$. This can be done arbitrarily, for example, by selecting the first K samples, that is,

$$\mathbf{m}_1 = \mathbf{x}_1, \quad \mathbf{m}_2 = \mathbf{x}_2, \quad \ldots.., \quad \mathbf{m}_K = \mathbf{x}_K.$$

Let us see how the K-means method works in the Iris example. The full dataset is too large for our purposes, and the first class (*Setosa*) can be easily separated from the rest. Therefore, we will analyze only a shortened training set that consists of Class 2 (*Versicolor*) and Class 3 (*Virginica*) represented by the first two PCA components, as shown in Fig. 11.35b. The score values are calculated in worksheet PCA-Explore (local name T23c) and then copied to worksheet kMeans. Naturally, we conduct clustering for two classes ($K = 2$). The centroids are presented in ranges (see Fig. 11.36) with local names kMean1 and kMean2. Their initial values correspond to the first two data objects.

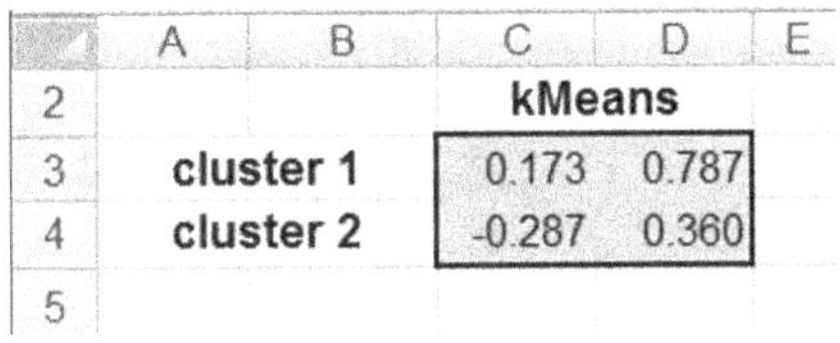

	A	B	C	D	E
2			kMeans		
3	cluster 1		0.173	0.787	
4	cluster 2		-0.287	0.360	
5					

Figure 11.36 Initial values of centroids.

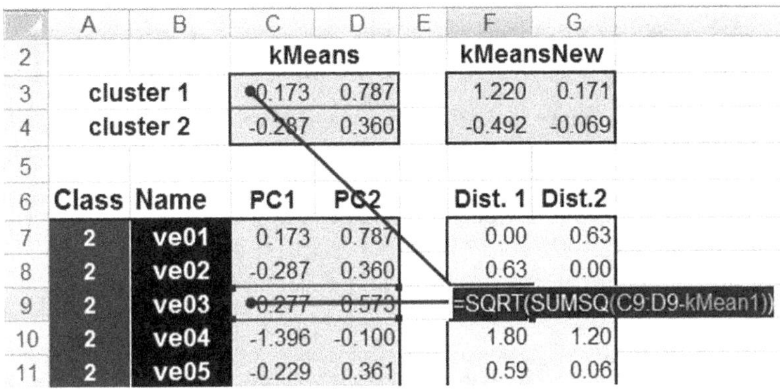

	A	B	C	D	E	F	G
2			kMeans			kMeansNew	
3	cluster 1		0.173	0.787		1.220	0.171
4	cluster 2		-0.287	0.360		-0.492	-0.069
5							
6	Class	Name	PC1	PC2		Dist. 1	Dist.2
7	2	ve01	0.173	0.787		0.00	0.63
8	2	ve02	-0.287	0.360		0.63	0.00
9	2	ve03	0.277	0.573		=SQRT(SUMSQ(C9:D9-kMean1))	
10	2	ve04	-1.396	-0.100		1.80	1.20
11	2	ve05	-0.229	0.361		0.59	0.06

Figure 11.37 Calculation of the distances to centroids.

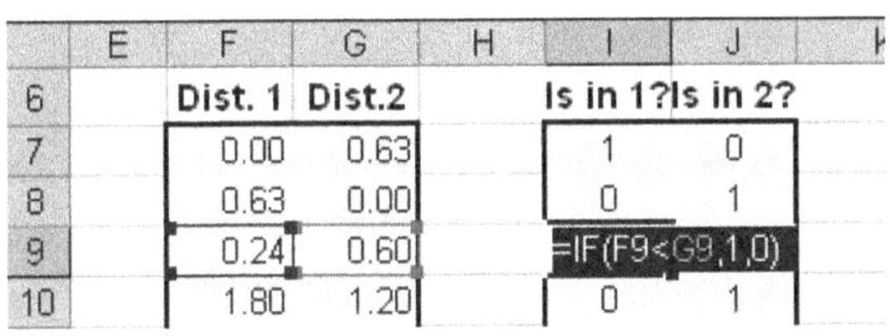

	E	F	G	H	I	J
6		Dist. 1	Dist.2		Is in 1?	Is in 2?
7		0.00	0.63		1	0
8		0.63	0.00		0	1
9		0.24	0.60		=IF(F9<G9,1,0)	
10		1.80	1.20		0	1

Figure 11.38 Determination of the cluster belonging.

Distances from all samples to these centroids are calculated by the Euclidean metric (Step 1, Fig. 11.37).

Then, the cluster membership of each sample is determined (Step 2, Fig. 11.38).

Next, the new centroids are calculated in a range with local name kMeansNew (Step 3, Fig. 11.39).

To finalize the iterative sequence, it is necessary to copy values from KMeansNew range and paste them (as values!) into kMeans range. This should be repeated as many times as necessary, until all values kMeans – kMeansNew come close to zero. Worksheet kMeans has a button Calculate. It calls a simple Visual Basic for Applications (VBA) macro, which copies the content of kMeansNew and inserts it into the range kMeans. This operation is repeated as many times as specified in cell P2. The iterative procedure always converges, but the result may be different depending on the choice of the initial centroids.

If we choose the first two points **ve01** and **ve02** as the initial values, we get the result shown in Fig. 11.40. Figure 11.40a shows samples distribution at the beginning of the algorithm. Figure 11.40b presents the final result.

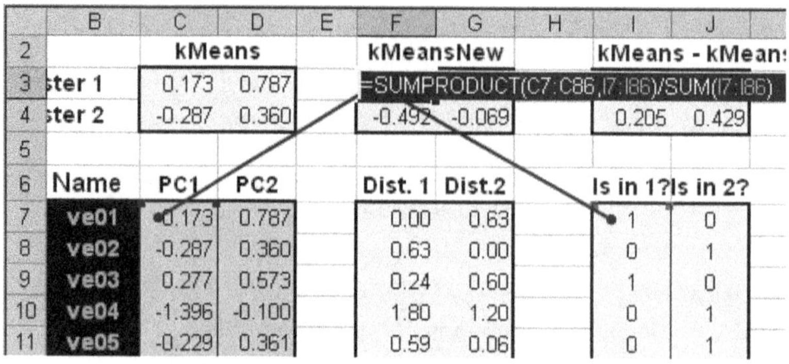

Figure 11.39 Calculation of new centroids.

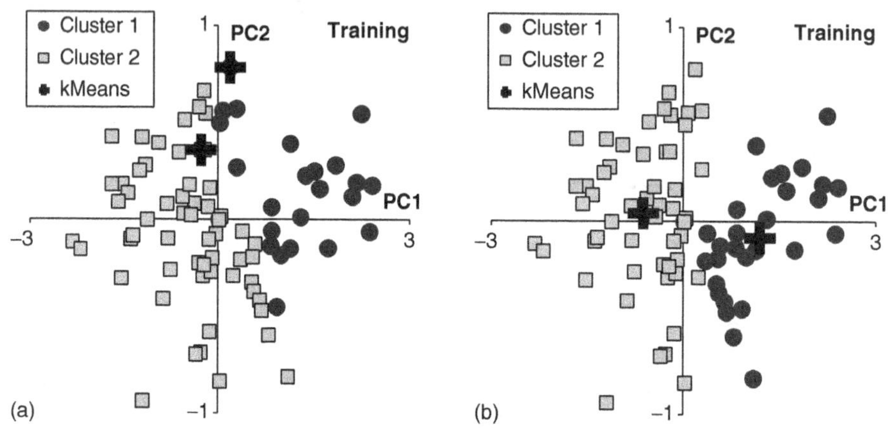

Figure 11.40 *K*-means clustering. The beginning (a) and end (b) of the algorithm. The first two samples serve as a starting point.

Figure 11.41 shows the result of clustering, which is obtained when the initial values are the last two samples **vi39** and **vi40**. It can be seen that the clusters are swapped and a few points changed their membership. In order to understand what solution is better, we use an objective function

$$S = \sum_{k=1}^{K} \sum_{\mathbf{x}_j \in S_k} (\mathbf{x}_j - \mathbf{m}_k)^2$$

to be minimal. In the first case, $S = 52.830$, while in the second case $S = 52.797$. Thus, the second solution is better.

It is natural to link the first cluster to Class 2 (*Versicolor*) and the second cluster to Class 3 (*Virginica*). Then the results can be interpreted as follows. Two objects from Class 2 are incorrectly attributed and 11 samples from Class 3 are wrongly assigned to Class 2.

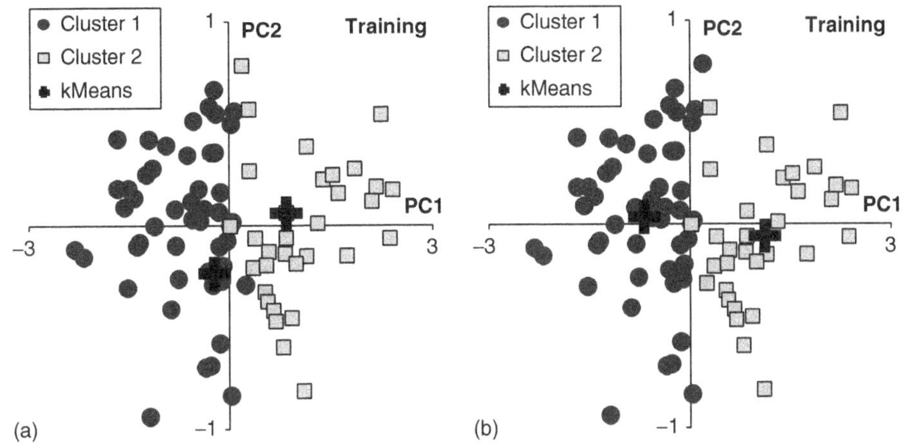

Figure 11.41 *K*-means clustering. The beginning (a) and end (b) of the algorithm. The last two samples serve as a starting point.

The *K*-Means Method has Several Drawbacks

1. The number of clusters *K* is unknown and it is not clear how to determine it. We can only change the value of *K* and examine the results.
2. The result depends on the initial selection of centroids. It is necessary to go through different options.
3. The result depends on the choice of metric.

CONCLUSION

We have explored some of the methods used for classification. This area of chemometrics, like no other, offers a plethora of approaches. Inevitably, many interesting techniques, such as unequal class modelling (UNEQ), classification and regression tree (CART), and others, have been left out. One can study them independently using this chapter as a guide.

12

MULTIVARIATE CURVE RESOLUTION

This chapter describes the most popular methods used in solving the multivariate curve resolution (MCR) problem.

Preceding chapters	5, 7–10
Dependent chapters	14
Matrix skills	Advanced
Statistical skills	Basic
Excel skills	Advanced
Chemometric skills	Basic
Chemometrics Add-In	Yes
Accompanying workbook	MCR.xls

12.1 THE BASICS

12.1.1 Problem Statement

Consider a system consisting of several chemical substances (*components*) A, B, ..., the *concentrations* of which, $c_A, c_B, \ldots$, regularly evolve with time t as shown in Fig. 12.1. One possible reason for the change can be a chemical reaction then $c(t)$ is kinetics. Alternatively, the change can happen during a chromatographic experiment then $c(t)$ represents an elution profile.

We presume that each component can be characterized by its *spectrum* (understood in a generalized sense) $s_A(\lambda), s_B(\lambda), \ldots$, where λ is a conventional wave number. See Fig. 12.2 for an example.

Chemometrics in Excel, First Edition. Alexey L. Pomerantsev.

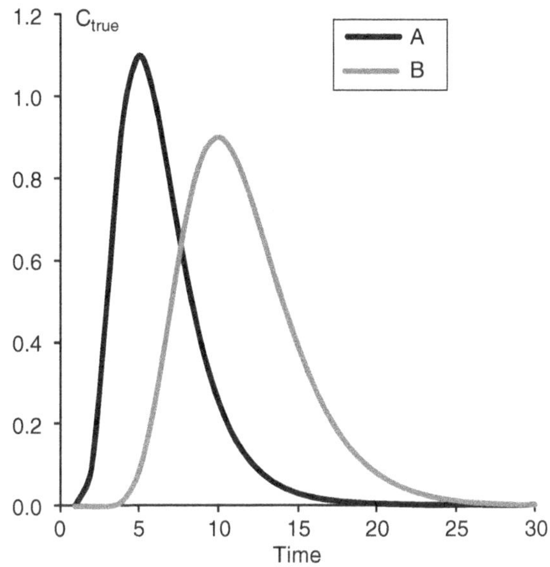

Figure 12.1 Concentration profiles.

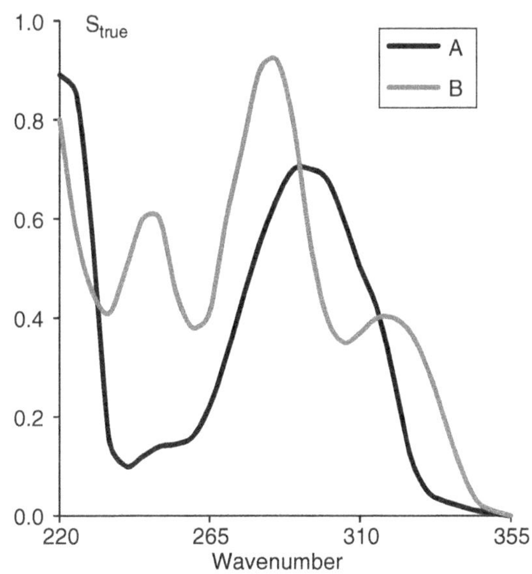

Figure 12.2 Spectra of pure components.

We assume that during the experiment, at time t_i, and at wave number λ_j, we can measure a value

$$x\left(t_i, \lambda_j\right) = c_{\mathrm{A}}\left(t_i\right) s_{\mathrm{A}}\left(\lambda_j\right) + c_{\mathrm{B}}\left(t_i\right) s_{\mathrm{B}}\left(\lambda_j\right) + \cdots$$

which is a linear combination of concentration and spectrum products, in line with the principle of linearity presented in Section 10.2.1. See Fig. 12.3 for an illustration.

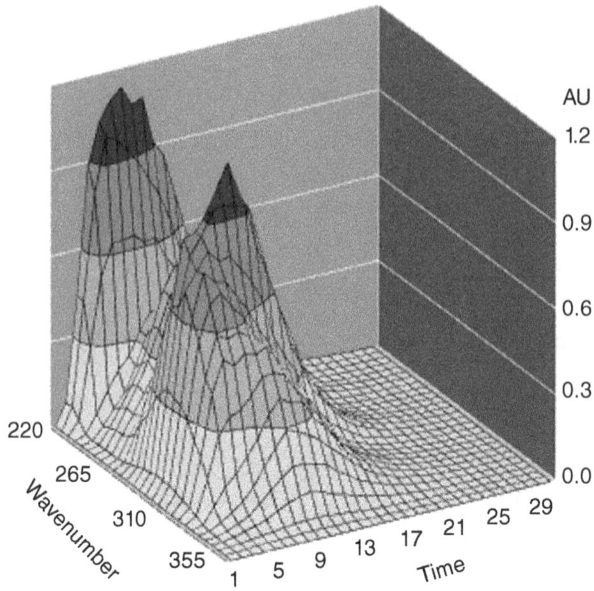

Figure 12.3 Mixed data.

Let I be the number of time observations $t_1, \ldots, t_I$, and J be the number of wave numbers $\lambda_1, \ldots, \lambda_J$. Then the data obtained in the experiment can be presented in a matrix form

$$\mathbf{X} = \mathbf{C}\mathbf{S}^{\mathrm{t}} + \mathbf{E}. \tag{12.1}$$

Here, the matrices of data **X** and errors **E** have the same dimension of $I \times J$. If the system has A chemical components, the concentration matrix **C** has I rows and A columns. Each column of matrix **C** stands for the concentration profile of a component. The pure spectra matrix $\mathbf{S}^{\mathrm{t}}$ (it is more convenient to present this matrix in a transposed form) has A rows and J columns. Each row of matrix $\mathbf{S}^{\mathrm{t}}$ is a pure spectrum of the corresponding component. Figure 12.4 provides an illustration.

The *MCR* problem is as follows. Given the data matrix **X**, determine the number of chemical components (A), find the concentration matrix **C**, and the pure spectra matrix **S**. From this it can be assumed that we do not have *significant* prior knowledge of matrices **C** and **S**, besides the most common, natural constraints, such as nonnegativity, continuity (both in t and λ), unimodality, etc. Of course, in this formulation, a solution may not exist or be nonunique. This is discussed below.

12.1.2 Solution Ambiguity

Let us assume that the solution to the problem given by Eq. (12.1) exists, that is, there is a pair of matrices **C** and **S**, whose product approximates the original matrix **X**

$$\mathbf{X} = \mathbf{C}\mathbf{S}^{\mathrm{t}} + \mathbf{E}.$$

It is known that this pair is not necessarily unique. Instead of matrices **C** and **S**, we can use other matrices $\widetilde{\mathbf{C}}$ and $\widetilde{\mathbf{S}}$, which provide a similar decomposition

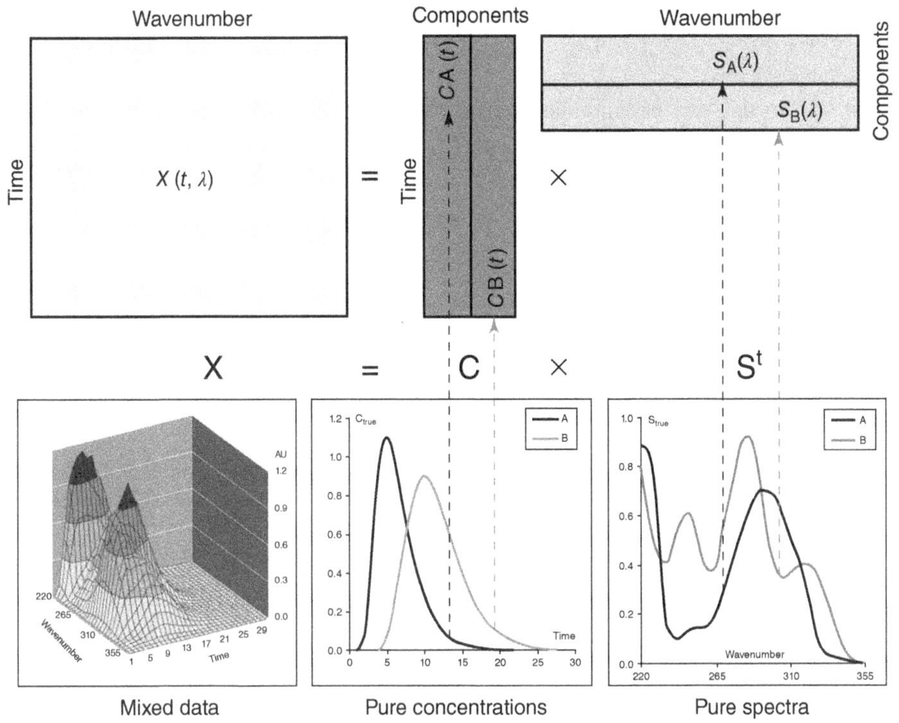

Figure 12.4 Pattern of the "chromatographic data".

$$\mathbf{X} = \widetilde{\mathbf{C}}\widetilde{\mathbf{S}}^{\mathrm{t}} + \mathbf{E}$$

with the same errors matrix **E**. Two types of ambiguity are of most importance.

Let **R** be a rotation matrix of dimension $A \times A$, that is, such a matrix that $\mathbf{R}^{\mathrm{t}} = \mathbf{R}^{-1}$. Then,

$$\mathbf{C}\mathbf{S}^{\mathrm{t}} = \mathbf{C}\mathbf{R}\mathbf{R}^{\mathrm{t}}\mathbf{S}^{\mathrm{t}} = (\mathbf{C}\mathbf{R})(\mathbf{S}\mathbf{R})^{\mathrm{t}} = \widetilde{\mathbf{C}}\widetilde{\mathbf{S}}^{\mathrm{t}} \tag{12.2}$$

This case is called the *rotational ambiguity*. This means that, in principle, there is an infinite number of MCR solutions, and some of them have a physical meaning (e.g., nonnegativity) but others do not. Implementation of the rotational ambiguity principle can be very helpful. It allows finding a reasonable MCR solution. An example is presented below.

In addition to the rotational ambiguity, there is *scalar ambiguity*. In this case, matrix **R** is a diagonal one, so

$$\mathbf{C}\mathbf{S}^{\mathrm{t}} = \mathbf{C}\mathbf{R}\mathbf{R}^{\mathrm{t}}\mathbf{S}^{\mathrm{t}} = (\mathbf{C}\mathbf{R})(\mathbf{S}\mathbf{R})^{\mathrm{t}} = \widetilde{\mathbf{C}}\widetilde{\mathbf{S}}^{\mathrm{t}} \tag{12.3}$$

It is important to note that there is no way to get rid of the scalar ambiguity. Any MCR solution is defined up to a multiplier, for example,

$$x_{ij} = \sum_{a=1}^{A} c_{ia}s_{aj} = \sum_{a=1}^{A} \left(c_{ia}/r_a\right)\left(s_{aj}r_a\right)$$

In other words, each concentration profile $\mathbf{c}_a$ can be r_a times smaller if the corresponding pure spectrum $\mathbf{s}_a$ is increased by r_a times. Certainly, the profile and the spectrum shapes remain unchanged.

It is known that any nonsingular square matrix can be uniquely represented by polar decomposition (Section 5.2.5). Therefore, any ambiguity is a combination of scaling and rotation.

12.1.3 Solvability Conditions

The MCR problem may have no solution. To explain the conditions of solvability, we introduce a concept of the *concentration* (*spectral*) *window* for a chemical component *a*. This window is a continuous range of variables (t or λ) where the corresponding pure signal (concentration profile or spectrum) is greater than a given small positive ε, provided that another range exists where the signal is less than ε (see Fig. 12.5a).

The MCR problem solvability is defined by the overlapping of the windows related to the different chemical components. R. Manne has established the necessary *conditions* in the following theorems[1].

1. The concentration profile of a chemical component can be restored if all other components that appear inside the concentration window also have regions outside this window.
2. The spectral profile of a chemical component can be restored if its concentration window is not completely located inside the concentration window of all other components.

Figure 12.5b illustrates these theorems. It shows that the concentration window of component B is situated completely within the window of A. Therefore, we can restore the

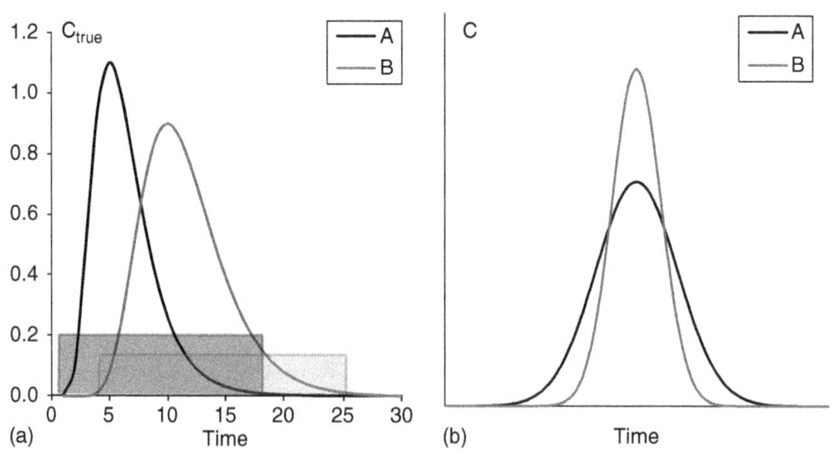

Figure 12.5 Concentration windows.

[1]R. Manne, On the resolution problem in hyphenated chromatography, Chemom. Intell. Lab. Syst. **27** (1995) 89–94

concentration profile of B, but we cannot find its pure spectrum. On the other hand, for component A, we can find a pure spectrum, but we cannot restore its concentration profile.

Quite similar theorems hold, when the word "concentration" is replaced with the word "spectral," and vice versa. This is a general property of the MCR problem – to be invariant under permutation of **C** and **S**.

12.1.4 Two Types of Data

The MCR methods can be applied to data of different types, including multiway data. However, there are two classes of data where MCR is generally applied. They are the "chromatographic" and the "kinetic" patterns.

The data of the *chromatographic* pattern has concentration windows, and it has no spectral windows. An example of such data is shown in Fig. 12.4. On the contrary, the data of the *kinetic pattern* usually have no concentration windows, but it has spectral windows.

An example of the "kinetic" data is shown in Fig. 12.6. We only consider the data of the "chromatographic" pattern. However, all the presented methods can be used for the "kinetic" pattern as well. For this, it is enough to transpose the original MCR task given in Eq. (12.1) and obtain an equation

$$\mathbf{X}^{t} = \mathbf{SC}^{t} + \mathbf{E}^{t}.$$

The "spectra" become "concentrations," and vice versa. After that, all of the methods developed for the "chromatographic" pattern can be applied to the "kinetic" pattern. Note

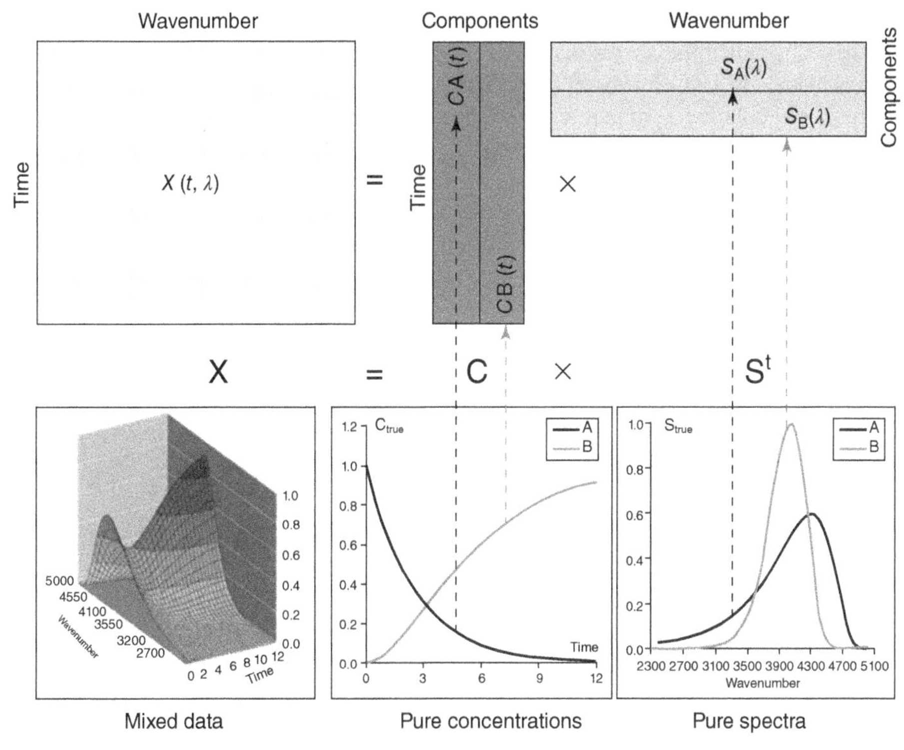

Figure 12.6 Pattern of the "kinetic data".

that R. Manne's theorems are formulated in the same invariant mode. The "kinetic" pattern data are considered in Chapter 14.

Of course, from a chemical point of view, the original data can be of different nature. For example, if spectra are obtained in the titration experiment under the constant conditions, the concentration profiles depend not on time but on some other parameter, for instance, on the pH level. This does not change the essence that the parameter is evolving continuously and monotonically.

12.1.5 Known Spectrum or Profile

Let us suppose that in the MCR task given by Eq. (12.1), one of the matrices **C** or **S** is known. Then another matrix can be easily restored by the least squares method. If matrix **S** is known, then

$$\mathbf{C} = \mathbf{XS}\left(\mathbf{S}^{t}\mathbf{S}\right)^{-1}, \tag{12.4}$$

if matrix **C** is known, then

$$\mathbf{S}^{t} = \left(\mathbf{C}^{t}\mathbf{C}\right)^{-1}\mathbf{C}^{t}\mathbf{X}. \tag{12.5}$$

12.1.6 Principal Component Analysis (PCA)

In the *principal component analysis* (PCA), the $(I \times J)$ data matrix **X** is also decomposed into a product of two matrices

$$\mathbf{X} = \mathbf{TP}^{t} + \mathbf{E} \tag{12.6}$$

Matrix **T** is called the *scores* matrix. It is of dimension $(I \times A)$. Matrix **P** is called the *loadings* matrix. It is of dimension $(J \times A)$. **E** is the *residuals* matrix of dimension $(I \times J)$. The number of columns $\mathbf{t}_a$ in matrix **T** and the number of columns $\mathbf{p}_a$ in matrix **P** are equal to A, which is called the *number of principal components* (PCs). Figure 12.7 provides an illustration.

A detailed description of PCA is presented in Section 9.2.

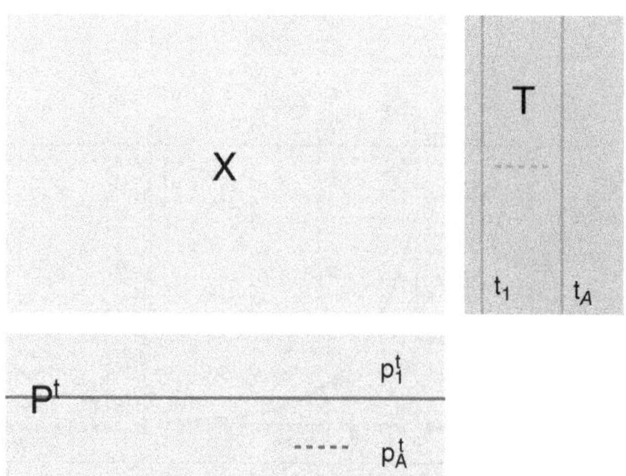

Figure 12.7 Graphical representation of PCA.

It is necessary to distinguish between the *chemical* component (A, B, ... ,) that exists in a mixture and the *principal* component (PC1, PC2, ...) defined by the PCA method. The first one is a real substance having its spectrum, whereas the second one is an abstract variable characterized by its loading. However, those components have much in common. This will be discussed below.

The scores matrix **T** has the following property (see Section 5.2.7):

$$\mathbf{T}^{\mathrm{t}}\mathbf{T} = \Lambda = \operatorname{diag}\left\{\lambda_1, \ldots, \lambda_A\right\},$$

where scalars $\lambda_1 > \ldots > \lambda_A > 0$ are called the *eigenvalues*.

$$\lambda_a = \sum_{i=1}^{I} t_{ia}^2, \quad a = 1, \ldots, A \tag{12.7}$$

The zero-order eigenvalue λ_0 is defined as the sum of all eigenvalues, that is,

$$\lambda_0 = \sum_{a=1}^{A} \lambda_a = \operatorname{sp}\left(\mathbf{X}^{\mathrm{t}}\mathbf{X}\right) = \sum_{i=1}^{I}\sum_{j=1}^{J} x_{ij}^2. \tag{12.8}$$

The square root of the corresponding eigenvalue

$$\sigma_a = \sqrt{\lambda_a} \tag{12.9}$$

is called the *singular value* (see Section 5.2.11).

Number A is the key characteristic that defines the data complexity. Determination of A can be done by various methods, but the most effective methods are based on the analysis of the system behavior when A is increased. In relation to the MCR task, the analysis of two plots – eigenvalues and loadings – is relevant.

12.1.7 PCA and MCR

The PCA and MCR decompositions do not only have external similarities, but they also have deeper ones. Above all, PCA is able to evaluate the most important parameter, namely, the number of chemical components in MCR, which, of course, is equal to the number of PCs in PCA. In addition, PCA provides a basis for a class of the curve resolution methods called *factor analysis*. The word "basis" is used here in its exact sense because the scores matrix **T** and the concentration matrix **C** span the same linear subspace. The same can be claimed for the loadings matrix **P** and the spectral matrix **S**. This means that the matrix of pure spectra **S** can be evaluated as a linear combination of the vectors from the loadings matrix **P**, and, concurrently, the concentration matrix **C** can be composed of the vectors from the scores matrix **T**. In this regard, it is said that the PCA scores and loadings are *abstract* concentrations and spectra.

12.2 SIMULATED DATA

12.2.1 Example

To illustrate and to compare the presented MCR methods, we use a simulated dataset that is placed in workbook MCR.xls. This book includes the following sheets:

Intro: brief introduction
Layout: scheme explaining the range naming
Profiles: true concentration profiles **C**
Spectra: true pure spectra **S**
Data: dataset **X** used in the example
Scores: PCA scores
Loadings: PCA loadings
Number PCs: determining the number of principal components
Procrustes: Procrustes analysis
EFA: Evolving factor analysis
WFA: Windows factor analysis
ITTFA: Iterative target-transformed factor analysis
ALS: Alternating least squares

12.2.2 Data

A prototype for our example is a problem of peaks separation in the high-performance liquid chromatography with the diode array detection (HPLC-DAD). It is known that in such an experiment the linearity principle (the Lambert–Beer law) holds well. Our model system contains two substances A and B having the elution profiles $c_A(t)$ and $c_B(t)$. Then, the spectrum of the mixture is given by the formula

$$x = c_A \cdot s_A + c_B \cdot s_B,$$

where s_A and s_B are the spectra of the pure components A and B.

To simulate the elution profiles of substances A and B, we use overlapping Gaussian peaks calculated for the retention times $t = 0, 1, 2, \ldots, 30$. The profiles are presented in sheet Profiles and shown in Fig. 12.1. The pure spectra of substances A and B are presented in sheet Spectra and shown in Fig. 12.2. They are calculated for 28 wave numbers using a complex superposition of the Gaussian peaks. Mixed data **X** is calculated in sheet Data by Eq. (12.1), where matrix **E** is generated for a relative error of 3%. The error value can be specified in cell E36 having the local name RMSE. We provide an ability to vary the error value and generate a new dataset. Button **Add New Error** launches a Visual Basic for Applications (VBA) macro **MakeErrors** attached to this worksheet. See Section 7.3.1 for the VBA programming explanations.

12.2.3 PCA

PCA of the data should be carried out without centering and normalization (see Section 8.1.4). This is a general rule for the MCR problems. Calculations are performed in worksheets Scores and Loadings. The plot of the first two PCA scores $\mathbf{t}_1$ and $\mathbf{t}_2$ versus time is presented in Fig. 12.8a.

The true concentration profiles of substances A and B are also shown in the plot for reasons of comparison. A plot of $\mathbf{p}_1$ and $\mathbf{p}_2$ loadings (along with the corresponding pure spectra) is shown in Fig. 12.8b. One can notice similarities in the behavior of the PCs and

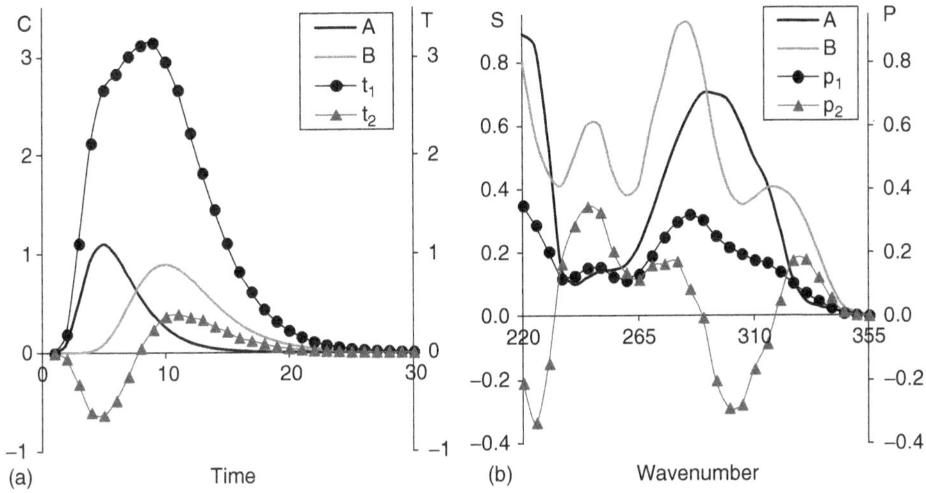

Figure 12.8 PCA results.

the corresponding concentrations. Certainly, the PCA components do not resolve the MCR problems in the least because of their nonpositiveness.

The PC number, of course, equals 2. This is due to the fact that data matrix **X** has been simulated as a mixture of two chemical components A and B. However, it is useful to make sure that PCA can correctly assess that number. Figure 12.9a shows the combined plot of the singular values σ_a (the left y-axis, labeled SnV) and the explained residual variance (the right y-axis, labeled ERV, see Section 9.2.9), in dependence on the possible PC number. It can be seen that PC = 2 is a breaking point for both curves. Figure 12.9b shows the PC3 and PC4 loadings plotted versus the wave numbers. Comparing this plot with the plot of the first two loadings (Fig. 12.8), we can note that the higher-order PCs look rather noisy lacking any trend.

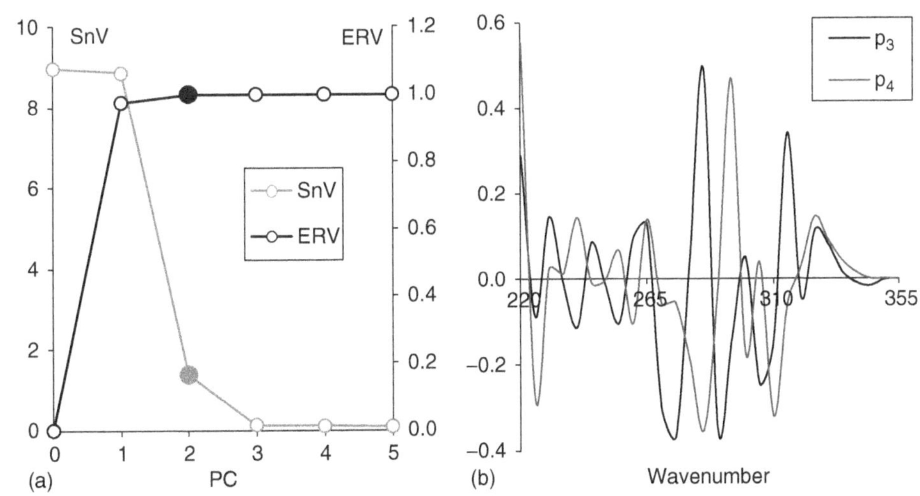

Figure 12.9 Determination of the number of PC.

The analysis of residuals calculated by

$$\mathbf{E} = \mathbf{X} - \mathbf{T}\mathbf{P}^{\mathrm{t}}$$

shows that the relative mean squared error of the PCA modeling

$$\delta = \sqrt{(1/I/J)\sum_{i=i}^{I}\sum_{j=1}^{J} e_{ij}^{2} \Big/ (1/I/J)\sum_{i=i}^{I}\sum_{j=1}^{J} x_{ij}} \qquad (12.10)$$

is equal to 4%, which corresponds to the error (3%) modeled in the data construction.

To calculate the eigenvalues EGV, we use a formula that is presented in worksheet Number PCs and shown in Fig. 12.10.

We explain the meaning of the formula as well as the benefits it delivers. The internal function ScoresPCA(Xdata) (see Section 8.3.1) calculates the (30 × 28) scores matrix **T**. The result is subjected to the standard function INDEX(T,,B6), which extracts one column $\mathbf{t}_a$ from **T**. The column number a is specified by the value given in cell B6 being the current PC number. Then, the vector $\mathbf{t}_a$ is subjected to the standard function SUMSQ, which calculates the sum of the squares of the vector elements, that is, the eigenvalue as it is defined in Eq. (12.7).

Thus, we are able to combine several successive operations in a single formula. If all intermediate results are egested on a worksheet, they take a lot of space, increase the file size, and slow down the computations. The details of such a technique are described in Section 13.1.6

12.2.4 The HELP Plot

It is useful to depict the relationship between the first scores as it is shown in Fig. 12.11a. Each point is labeled by the corresponding time value.

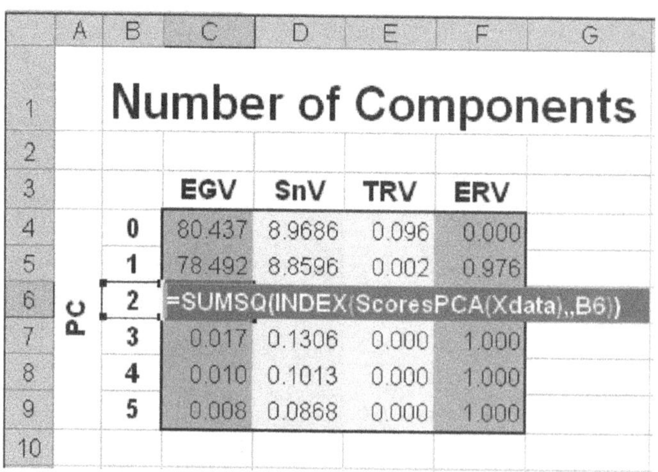

Number of Components
EGV SnV TRV ERV
PC
0 80.437 8.9686 0.096 0.000
1 78.492 8.8596 0.002 0.976
2 =SUMSQ(INDEX(ScoresPCA(Xdata),,B6))
3 0.017 0.1306 0.000 1.000
4 0.010 0.1013 0.000 1.000
5 0.008 0.0868 0.000 1.000

Figure 12.10 Calculation of eigenvalues and singular values.

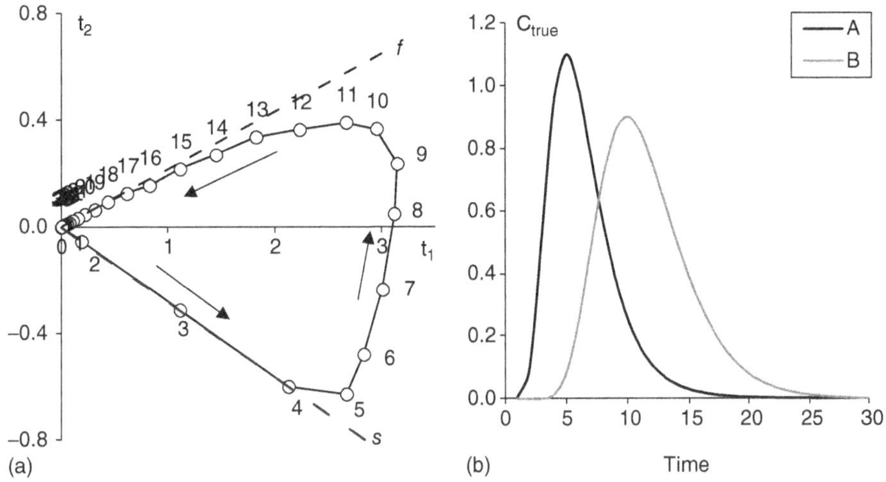

Figure 12.11 The HELP plot.

It shows that the process begins at the coordinate origin and returns to this point at the end. The first four points are located on a dotted line (labeled by s) that starts at the coordinate origin (0, 0). These points correspond to the initial phase when only substance A is eluting and component B is not yet visible. The last points (18–30) are located on another line labeled by f. They correspond to the final stage when only substance B is eluting, while substance A is already gone.

Such diagrams are called the *HELP* (*Heuristic Evolving Latent Projections*) plots. They can assist in solving the MCR tasks for two or three components. Analyzing the HELP plot, one should be guided by the following simple rules:

1. linear sections = pure components;
2. curved sections = coelution;
3. closer to the origin = less intensity;
4. the number of turns = number of pure components.

12.3 FACTOR ANALYSIS

12.3.1 Procrustes Analysis

As noted in Section 12.1.2, the MCR solution is not unique, and it can be altered by rotating and scaling transformations. The *Procrustes analysis* employs this option to convert the PCA decomposition to the MCR solution. The calculations related to this technique are presented in worksheet Procrustes.

Firstly, the scaling is applied. To do this, we multiply the scores matrix **T** by the diagonal matrix $\mathbf{R}_1 = \mathrm{diag}\{r_{11}, r_{12}\}$, that is,

$$\mathbf{T}_1 = \mathbf{T}\mathbf{R}_1$$

Matrix $\mathbf{R}_1$ should be selected in such a way that the transformed asymptotic lines s and f form a right angle in a new HELP plot (Fig. 12.12). At this selection, we encounter

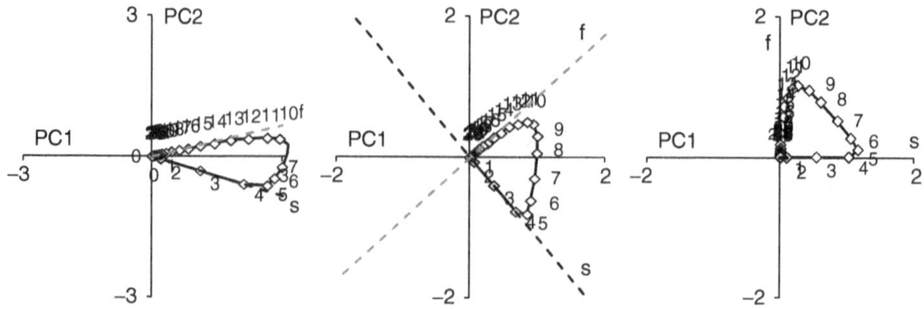

Figure 12.12 Procrustes analysis.

the above-mentioned scalar ambiguity. Actually, the transformation can be done by fitting only one element of matrix $\mathbf{R}_1$, whereas another element can be fixed on any level. In our example, we use a constrain $r_{11} = 1/\max(\mathbf{T})$. The selection of a free r_{12} value can be done using the **GoalSeek** option in Excel.

The second step is rotation. For this purpose, the transformed matrix $\mathbf{T}_1$ is multiplied by a rotation matrix $\mathbf{R}_2$, that is,

$$\mathbf{T}_2 = \mathbf{T}_1\mathbf{R}_2$$

where

$$\mathbf{R}_2 = \begin{vmatrix} \cos(\alpha) & \sin(\alpha) \\ -\sin(\alpha) & \cos(\alpha) \end{vmatrix}$$

The rotation angle α should be chosen to match the PCs axes and the asymptotic lines s and f. By this complex transformation, we convert the PCA PCs to the pure chemical components. The final result is shown in Fig. 12.13.

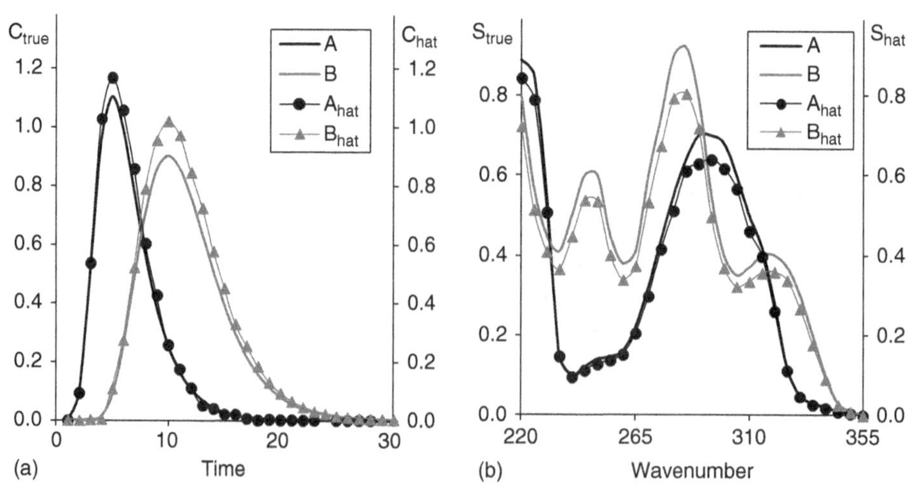

Figure 12.13 MCR by Procrustes analysis.

The estimate of the concentration matrix $\mathbf{C}_{hat}$ is obtained in the result of the scores matrix $\mathbf{T}$ transformation

$$\mathbf{C}_{hat} = \mathbf{TR}_1\mathbf{R}_2 = \mathbf{TR}, \quad \text{where } \mathbf{R} = \mathbf{R}_1\mathbf{R}_2$$

The estimate of the pure spectra $\mathbf{S}_{hat}$ is then calculated from the loadings matrix $\mathbf{P}$ by the formula

$$\mathbf{S}_{hat} = \mathbf{P}\left(\mathbf{R}^t\right)^{-1}.$$

The plots in Fig. 12.13 show that the MCR task is solved with an acceptable accuracy. The analysis of the residuals, which are given by

$$\mathbf{E} = \mathbf{X} - \mathbf{C}_{hat}\left(\mathbf{S}_{hat}\right)^t,$$

confirms that the relative mean squared error defined by Eq. (12.10) is equal to 4%, that is, it coincides with the accuracy of the PCA decomposition. This indicates that the obtained MCR solution is effective, that is, the accuracy cannot be improved anymore.

The obtained concentration profiles $\mathbf{C}_{hat}$ and spectra $\mathbf{S}_{hat}$ do not exactly match with the original $\mathbf{C}_{true}$ and $\mathbf{S}_{true}$ curves. The relative fitting error reaches 19%. This is explained by the scalar ambiguity, which manifested itself in the first step of the Procrustean transformation, where the fixed value of r_{11} can be selected arbitrary. It is easy to see that variation of r_{11} does not change the total error δ, whereas this uncertainty influences the profiles and spectra matching.

The scalar ambiguity influence can be estimated by a component-wise comparing of the magnitudes of the concentration profiles $\mathbf{C}_{true}$ and $\mathbf{C}_{hat}$ as well as the pure spectra $\mathbf{S}_{true}$ and $\mathbf{S}_{hat}$.

Figure 12.14 shows that the concentrations are systematically overestimated by the factor of 1.07 (A) and 1.12 (B), whereas the spectra are underestimated by the factor of 0.93 and 0.90. Nevertheless, the product $c(t)s(\lambda)$ factors are very close to 1; namely it is 0.995 (A) and 1.008 (B).

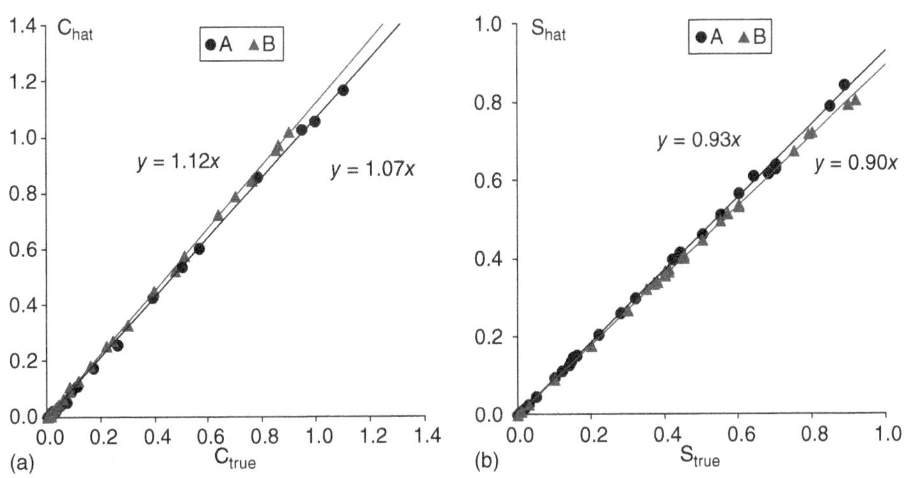

Figure 12.14 Component-wise comparison in Procrustes analysis.

12.3.2 Evolving Factor Analysis (EFA)

In the abstract factor analysis carried out by PCA, the order of rows in matrix **X** does not matter. The rows can be permutated, but the PCA results, that is, matrices **T** and **P** do not change. On the contrary, in MCR, the sequence of data acquisition is very important, and this principle can be used for resolution. The main idea of the *evolving factor analysis* (EFA) is to apply PCA to a sequence of the **X** submatrices.

The classical EFA method uses two sequences of the submatrices. The expanding (forward) sequence includes the following matrices. The first submatrix consists of a single row that is the first (in the order of time) row of matrix **X**, the second submatrix consists of the first two rows, the third submatrix includes three rows, and so on, until the last submatrix, which is equal to the full matrix **X**. The second sequence includes the reducing submatrices (backward). The first submatrix is full **X**, in the second submatrix the first row of X is excluded, in the third submatrix the first two rows of **X** are left out, and so on, up to the submatrix consisting of only the last row of **X**. Each matrix from each sequence is subjected to PCA, and the corresponding singular values, $\sigma_1, \ldots, \sigma_A$, are calculated (see Fig. 12.15).

The most important part of the EFA method is the analysis of the singular values evolution when matrix **X** expands and contracts. Figure 12.16a shows the first and second singular values (the left axis) in the forward EFA applied to our data (see sheet EFA). This plot also presents the A and B concentration profiles (right axis). It shows that singular value is increased at the moment when the corresponding substance appears in the system. A similar plot for the backward EFA (Fig. 12.16b) shows that singular value is decreased at the moment when the corresponding substance leaves the system. Often, the logarithms of singular values are plotted in order to reduce the scale difference. By analyzing two EFA plots together, we can construct the concentration windows $W_a(t)$ for all components. This is done by the following simple algorithm.

Let ε be a small value that corresponds to the zero level of a singular value. For each PC, $a = 1, \ldots, A$, consider a pair of singular values $(\sigma_a^{\mathrm{F}},\ \sigma_{A-a+1}^{\mathrm{B}})$. For the first PC ($a = 1$), this pair consists of the first singular value in the forward EFA and the last singular value in the backward EFA. The next pair ($a = 2$) combines the second singular value in the forward EFA and the last by one singular value in the backward EFA, etc. Then,

$$W_a(t) = \begin{cases} 0, & \min\left(\sigma_a^{\mathrm{F}},\ \sigma_{A-a+1}^{\mathrm{B}}\right) < \varepsilon \\ H_a, & \min\left(\sigma_a^{\mathrm{F}},\ \sigma_{A-a+1}^{\mathrm{B}}\right) \geq \varepsilon \end{cases}$$

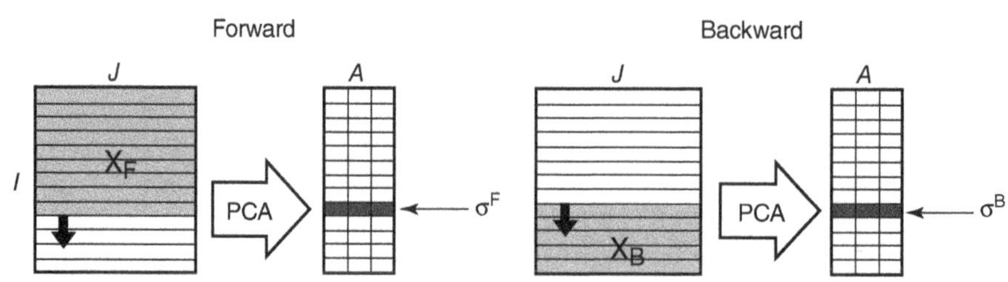

Figure 12.15 EFA method.

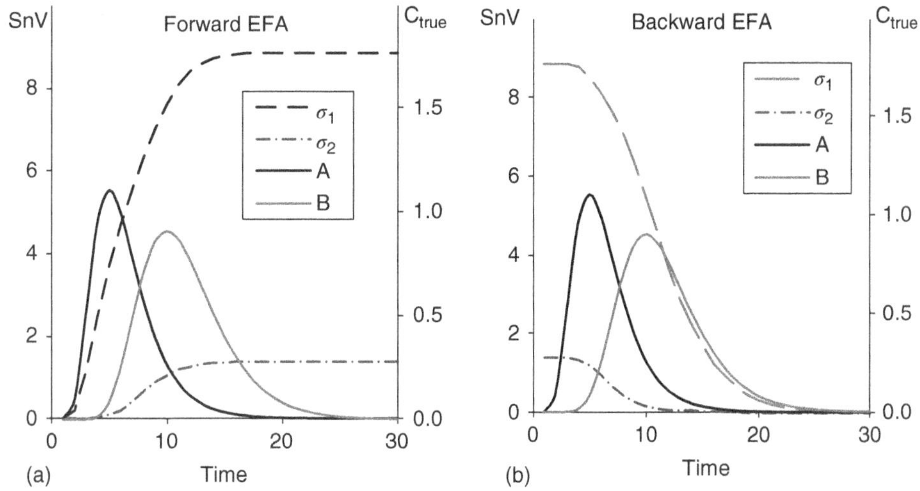

Figure 12.16 Singular values evolution in EFA.

This algorithm assumes that the chemical component, that appears first, is the first one out. The critical level ε is chosen accounting the criterion used in the determination of number A in the full PCA. Figure 12.9 shows that $\varepsilon < 1$. We chose $\varepsilon = 0.05$. The window height H_a is an arbitrary value due to the scalar ambiguity. In particular, it can be chosen as $H_a = A - a + 1$ or $H_a = \min(\sigma_a^{\mathrm{F}}, \sigma_{A-a+1}^{\mathrm{B}})$.

The concentration windows W_a can be considered a rough approximation of the corresponding concentration profiles, as it is shown in Fig. 12.16a. The estimates of pure spectra $\mathbf{S}_{\mathrm{hat}}$ can be found using Eq. (12.5). They are shown in Fig. 12.17b.

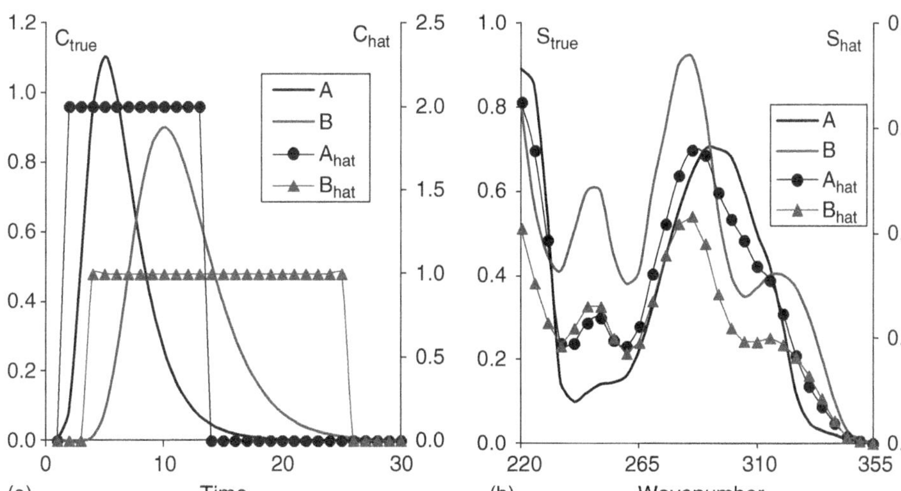

Figure 12.17 Resolution by the EFA method.

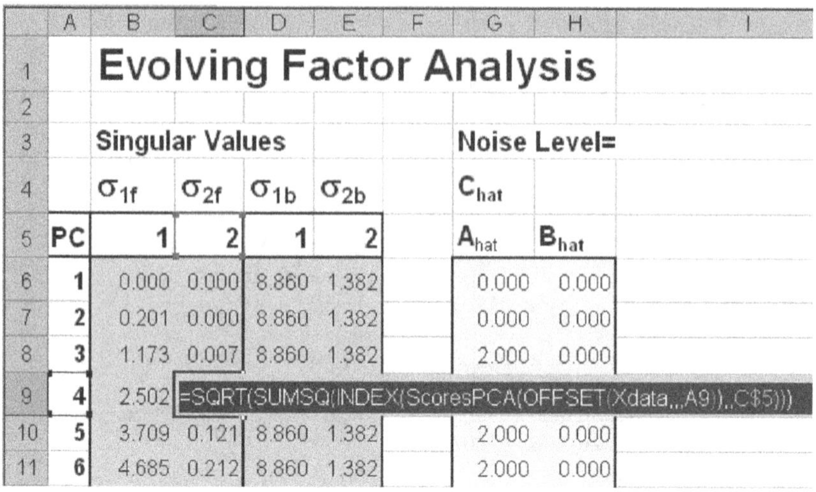

A B C D E F G H I
Evolving Factor Analysis
Singular Values
Noise Level=
σ1f σ2f σ1b σ2b
Chat
PC 1 2 1 2
Ahat Bhat
1 0.000 0.000 8.860 1.382 0.000 0.000
2 0.201 0.000 8.860 1.382 0.000 0.000
3 1.173 0.007 8.860 1.382 2.000 0.000
4 2.502
=SQRT(SUMSQ(INDEX(ScoresPCA(OFFSET(Xdata,,,A9)),,C$5)))
5 3.709 0.121 8.860 1.382 2.000 0.000
6 4.685 0.212 8.860 1.382 2.000 0.000

Figure 12.18 Singular values calculation in EFA.

We see that fitting is not very good. This is confirmed by the error $\delta = 62\%$, which is much larger than that in the Procrustes analysis. However, the EFA method is mostly interesting as a foundation for other more accurate methods. From this point of view, the main result of EFA is the concentration windows it can provide.

The EFA worksheet contains another complex formula used to calculate the singular values (Fig. 12.18). It is similar to the formula presented in Fig. 12.10, but it has its own features that are worthy of explanations.

The inner part of the formula is OFFSET(Xdata,,,A9). The standard function **OFFSET** (see Section 7.2.4) selects from matrix **X** as many first rows, as indicated in cell A9. The value in cell A9 stands for the number of the time point t_i. The formula returns a truncated data matrix, which is used in the special function **ScoresPCA** (see Section 8.3.1). The latter calculates the scores matrix **T**, used in the standard function **INDEX** (see Section 7.2.4) that selects a single column of **T**, whose number is specified in cell C$5, etc. In backward EFA, the inner formula is OFFSET(Xdata,A9-1,,I-A9+1). This formula selects I-A9+1 rows from matrix **X** starting with row A9-1.

This example demonstrates the benefits of employment such as complex but compact formulas. If all intermediate results were egested, just for the calculation of the singular values, we would need worksheet space for 30 submatrices $\mathbf{X}_F$ as well as space for 30 submatrices $\mathbf{X}_B$ and space for 60 scores matrices.

12.3.3 Windows Factor Analysis (WFA)

The *windows factor analysis* (WFA) is based on the following principle. Using the concentration window given by EFA, we can determine the time domain where the concentration of chemical component a is greater than zero. The corresponding rows are deleted from matrix **X**, and thus a reduced matrix $\mathbf{X}_a$ is obtained. This matrix is decomposed by PCA to find the loadings matrix $\mathbf{P}_a$. Because one chemical component has been removed from the system, the number of the PCA components is reduced by one and becomes equal to $A-1$. The loadings matrix $\mathbf{P}_a$ contains spectral information about all chemical components other

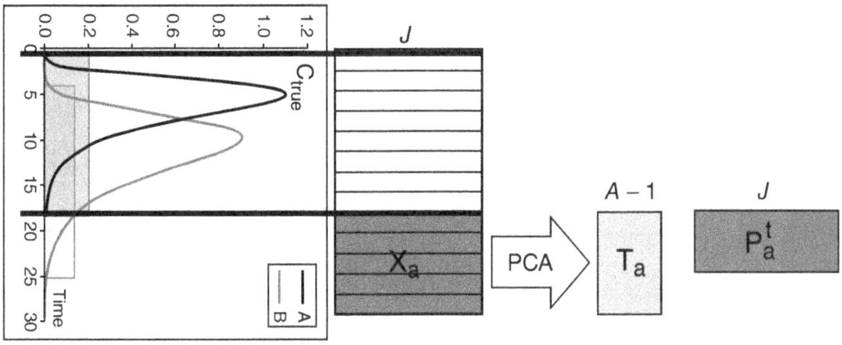

Figure 12.19 Windows factor analysis.

than the excluded one. Therefore, in order to determine the spectrum of component a, it is necessary to find a vector orthogonal to $\mathbf{P}_a$. This is illustrated by Fig. 12.19.

The WFA algorithm consists of the following steps that are repeated for each chemical component, $a = 1, \ldots, A$:

1. Matrix $\mathbf{X}_a$ is formed.
2. Matrix $\mathbf{X}_a$ is decomposed by PCA: $\mathbf{X}_a = \mathbf{T}_a(\mathbf{P}_a)^{\mathrm{t}} + \mathbf{E}_a$ and matrix $\mathbf{P}_a$ is determined.
3. Matrix $\mathbf{Z}_a = \mathbf{X} - \mathbf{X}\mathbf{P}_a(\mathbf{P}_a)^{\mathrm{t}}$ is calculated.
4. Matrix $\mathbf{Z}_a$ is decomposed by PCA: $\mathbf{Z}_a = \mathbf{t}_{\mathrm{Z}}(\mathbf{p}_{\mathrm{Z}})^{\mathrm{t}} + \mathbf{E}_{\mathrm{Z}}$ with one PC and the scores vector $\mathbf{t}_{\mathrm{Z}}$ is obtained.
5. All negative elements in $\mathbf{t}_{\mathrm{Z}}$ are replaced by zeros; this is an estimate of the concentration profile $\mathbf{c}_a$ for a chemical component a.

Step 4 of the algorithm deserves a special comment. The theoretical rank of matrix $\mathbf{Z}_a$ is $A-(A-1) = 1$, hence every column of $\mathbf{Z}_a$ has to be proportional to the sought concentration profile $\mathbf{c}_a$. However, the unavoidable calculation errors can distort this rule. Step 4 is performed to eliminate the error influence. In principle, Step 4 can be replaced by selecting the column with the greatest norm from matrix $\mathbf{Z}_a$.

After the algorithm has been performed for each component, and the matrix of concentration profiles $\mathbf{C}_{\mathrm{hat}}$ has been estimated, the pure spectra $\mathbf{S}_{\mathrm{hat}}$ can be evaluated by Eq. (12.5).

In our example, we have two components, A and B; therefore, WFA should be carried out twice. These calculations are presented in worksheet WFA. To determine substance A, we should remove several rows from the original data **X**, for example, a block starting from the second row till the 13th row. For substance B, we remove a block starting with the 4th row till row 25. This selection follows from Fig. 12.17. Figure 12.20 presents the MCR solution found by WFA. The approximation accuracy is $\delta = 4\%$, which is the same as in the Procrustes rotation.

To calculate the loadings matrix $\mathbf{P}_a$ (in our case, it is a vector), we applied a technique (see Fig. 12.21), which is similar to the complex formula approach used earlier.

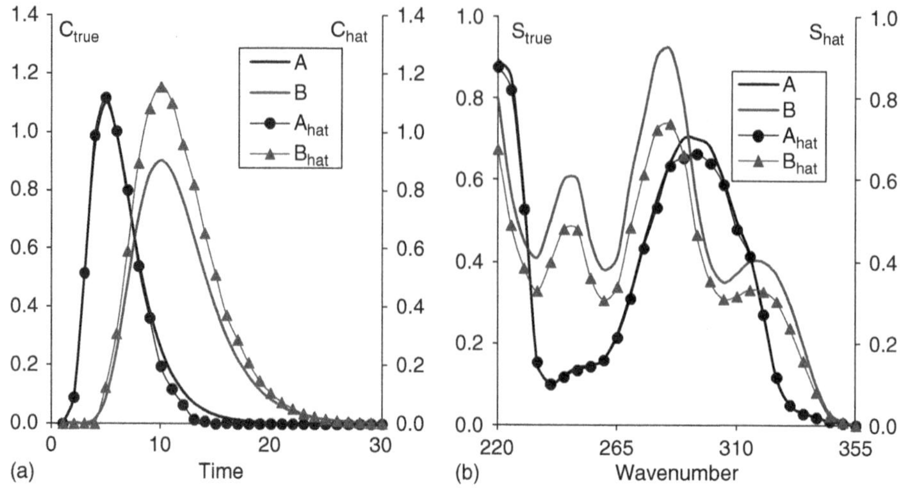

Figure 12.20 MCR solution by WFA.

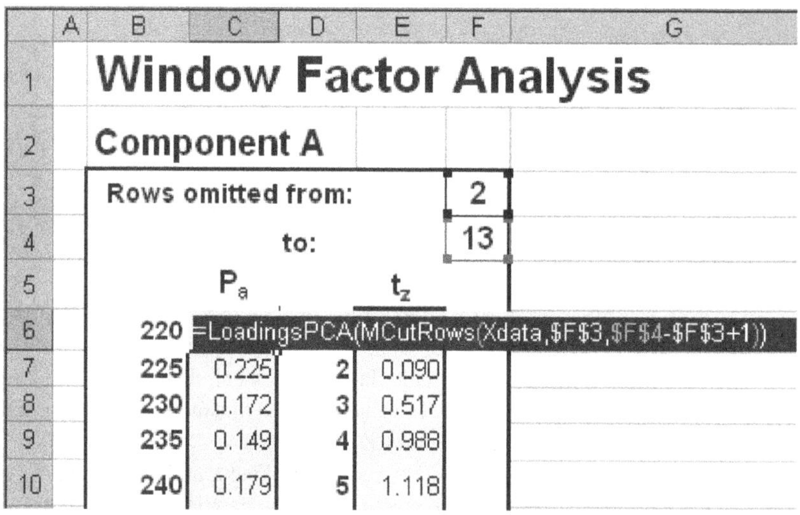

	A	B	C	D	E	F	G
1		Window Factor Analysis					
2		Component A					
3		Rows omitted from:				2	
4				to:		13	
5			P_a		t_z		
6		220	=LoadingsPCA(MCutRows(Xdata,$F$3,$F$4-$F$3+1))				
7		225	0.225	2	0.090		
8		230	0.172	3	0.517		
9		235	0.149	4	0.988		
10		240	0.179	5	1.118		

Figure 12.21 Calculation of loadings in WFA.

A special function (see section 8.6.3)

MCutRows(Xdata, $F$3, $F$4-$F$3 + 1)

deletes a block from **X**, starting from the row specified in cell $F$3 up to the row given in cell $F$4; $F$4-$F$3+1 rows in total. Then, a special function **LoadingsPCA** (see Section 8.3.2) is applied to the truncated dataset in order to calculate the PCA loadings.

Calculation of the scores vector $\mathbf{t}_Z$ is shown in Fig. 12.22.

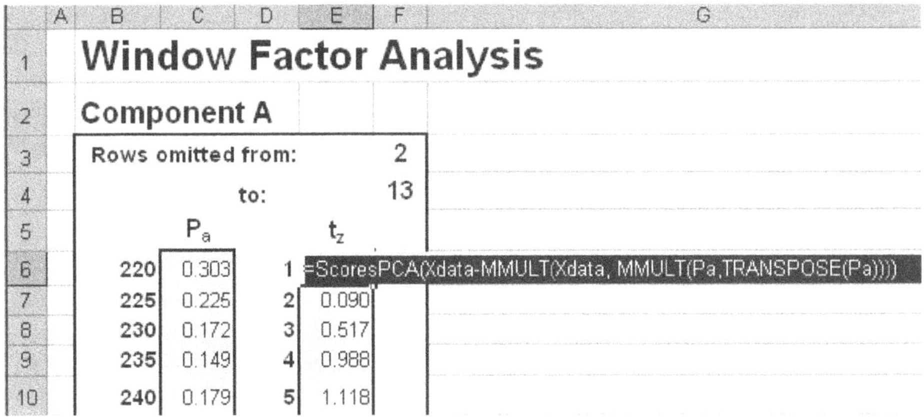

Figure 12.22 Calculation of the score vector in WFA.

12.4 ITERATIVE METHODS

12.4.1 Iterative Target Transform Factor Analysis (ITTFA)

Iterative target transform factor analysis (ITTFA) solves the MCR problem by a consequential optimization of the concentration profiles matrix estimate $\mathbf{C}_{\text{hat}}$. The intuition of the method is as follows. The initial estimate of matrix $\mathbf{C}_{\text{in}}$ is projected onto the scores subspace spanned by the vectors of the PCA scores matrix **T**. For this projection we use a matrix of the normalized scores **U**, which coincides with the matrix of the left eigenvectors in the singular values decomposition (SVD, see Section 5.2.11). To find **U**, each vector of **T** is divided by the corresponding singular value

$$\mathbf{u}_a = \mathbf{t}_a/\sigma_a.$$

The projected matrix $\mathbf{C}_{\text{out}}$ is then adjusted accounting for certain constrains, such as non-negativity, unimodality, etc. The intermediate concentration profile $\mathbf{C}_{\text{hat}}$ is projected on the scores space again, etc. The ITTFA algorithm consists of the following steps:

1. The original matrix **X** is decomposed by PCA: $\mathbf{X} = \mathbf{TP}^{\text{t}} + \mathbf{E}$; then the normalized scores matrix **U** is calculated from matrix **T**.
2. The initial approximation of the concentration matrix $\mathbf{C}_{\text{in}}$ is selected.
3. Matrix $\mathbf{C}_{\text{in}}$ is projected onto the scores subspace spanned by **T**, resulting in matrix $\mathbf{C}_{\text{out}}$
$$\mathbf{C}_{\text{out}} = \mathbf{UU}^{\text{t}}\mathbf{C}_{\text{in}}.$$
4. Matrix $\mathbf{C}_{\text{out}}$ is adjusted accounting for constrains, for example, if an element is negative, it is replaced by zero. The result is matrix $\mathbf{C}_{\text{hat}}$.
5. Matrix $\mathbf{C}_{\text{in}}$ is replaced by matrix $\mathbf{C}_{\text{hat}}$.
6. Steps 3–5 are repeated until convergence.

This algorithm contains a critical Step 2, during which the initial approximation of the concentration matrix is selected. When the choice is poor, the algorithm diverges. At the third step, the algorithm finds a mathematical solution of MCR, that is, an abstract profile matrix $\mathbf{C}_{out}$, which, however, can have a conflict with the physical meaning. The subsequent Step 4 makes the abstract solution agree with these constrains, but at the same time pushes matrix $\mathbf{C}_{hat}$ out of the PCA scores subspace. Therefore, the algorithm starts again, until a solution will satisfy both mathematical and physical requirements.

Once the matrix of concentration profiles $\mathbf{C}_{hat}$ is estimated, the pure spectra matrix $\mathbf{S}_{hat}$ can be evaluated using formula given in Eq. (12.5).

In worksheet ITTFA, we demonstrate how the ITTFA method is applied to our example. Figure 12.23 shows a part of the sheet. The button Calculate launches a simple VBA macro (Section 7.3.3), which copies the content of range $\mathbf{C}_{hat}$ and pastes the values into range $\mathbf{C}_{in}$. This operation is repeated as many times as specified in cell F2. Thus, a desired number of iterations is conducted.

The initial approximation is the needle peak profiles shown in Fig. 12.24.

After 20 iterations we obtain solutions shown in Fig. 12.25.

The solution accuracy is $\delta = 4\%$, that is, the ITTFA method provides the same accuracy as WFA, and Procrustes. It is easy to check that the ITTFA algorithm will diverge, if you select a poor initial approximation of $\mathbf{C}_{in}$.

12.4.2 Alternating Least Squares (ALS)

Alternating least squares (ALS) is a simple, but powerful method for the MCR task solving. Unlike most other MCR methods, ALS is not based on PCA. Instead, it successively uses the principle of least squares, applying Eqs (12.4) and (12.5) at each step. Just as ITTFA, the ALS method is iterative, but its convergence is better.

The ALS algorithm is as follows.

1. The number of components is determined by PCA, and the concentration windows are found by EFA.

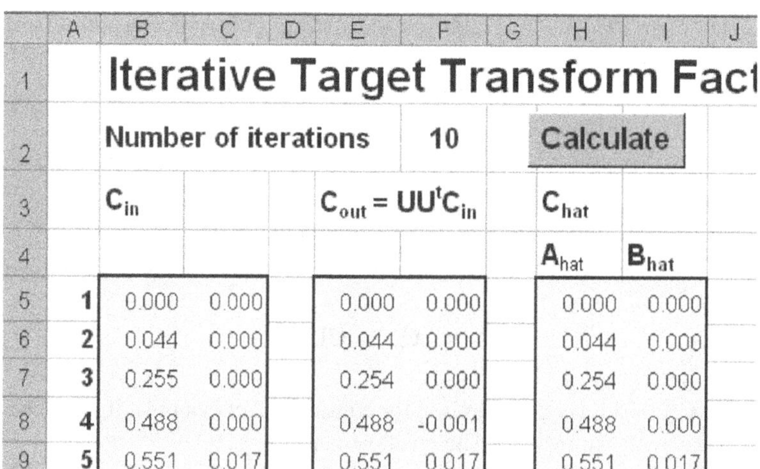

	A	B	C	D	E	F	G	H	I	J
1		Iterative Target Transform Fact								
2		Number of iterations				10		Calculate		
3		C_{in}			$C_{out} = UU^tC_{in}$			C_{hat}		
4								A_{hat}	B_{hat}	
5	1	0.000	0.000		0.000	0.000		0.000	0.000	
6	2	0.044	0.000		0.044	0.000		0.044	0.000	
7	3	0.255	0.000		0.254	0.000		0.254	0.000	
8	4	0.488	0.000		0.488	-0.001		0.488	0.000	
9	5	0.551	0.017		0.551	0.017		0.551	0.017	

Figure 12.23 ITTFA realization.

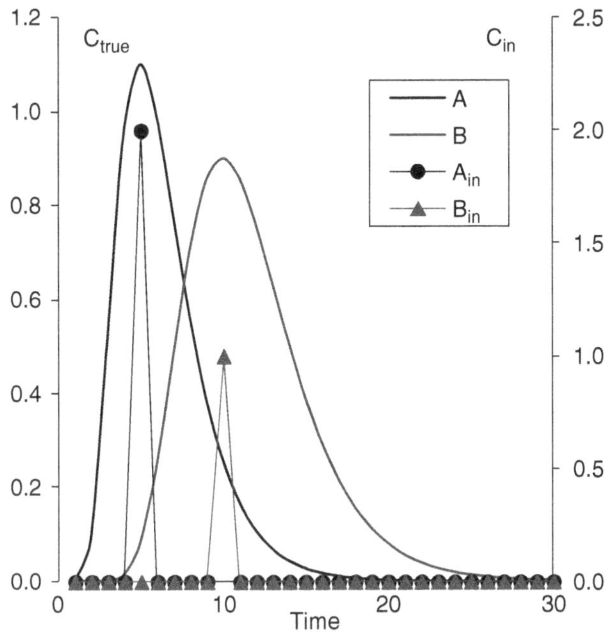

Figure 12.24 Initial matrix $\mathbf{C}_{in}$ defined by the needle peak profiles.

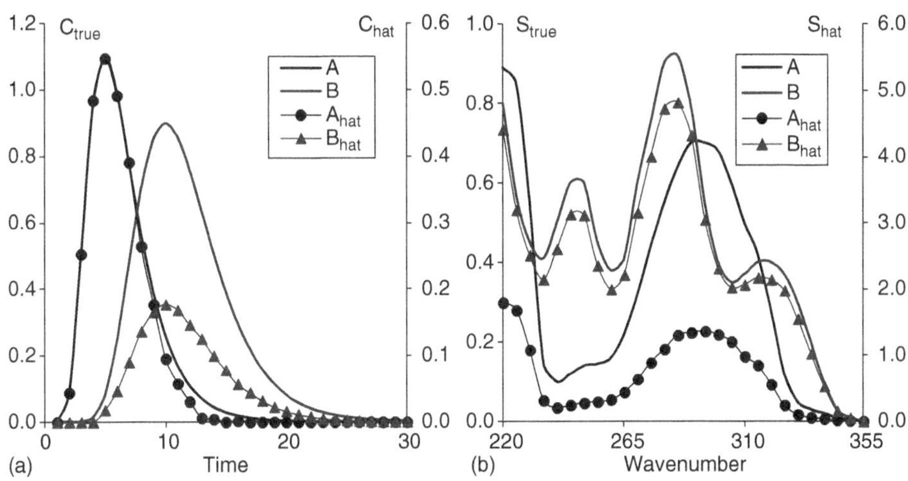

Figure 12.25 Solution by ITTFA.

2. Initial approximation of the concentration profiles matrix $\mathbf{C}_{in}$ is selected.
3. Matrix $\mathbf{C}_{in}$ is adjusted accounting for constrains, for example, negative elements are replaced by zeros. In addition, the matrix is normalized in such a way that its maximal element equals 1. As a result, we obtain matrix $\mathbf{C}_{hat}$.
4. Using Eq. (12.5), the pure spectra matrix $\mathbf{S}_{in}$ is estimated by formula

$$\mathbf{S}_{\text{in}} = \mathbf{X}^{\text{t}}\mathbf{C}_{\text{hat}}\left(\mathbf{C}_{\text{hat}}^{\text{t}}\mathbf{C}_{\text{hat}}\right)^{-1}$$

5. Matrix $\mathbf{S}_{\text{in}}$ is adjusted accounting for constrains, for example, negative elements are replaced by zeros. As a result, we have matrix $\mathbf{S}_{\text{hat}}$.
6. Using Eq. (12.4), the concentration profile matrix $\mathbf{C}_{\text{out}}$ is evaluated by formula

$$\mathbf{C}_{\text{out}} = \mathbf{X}\mathbf{S}_{\text{hat}}\left(\mathbf{S}_{\text{hat}}^{\text{t}}\mathbf{S}_{\text{hat}}\right)^{-1}.$$

7. Matrix $\mathbf{C}_{\text{in}}$ is replaced by matrix $\mathbf{C}_{\text{out}}$.
8. Steps 3–7 are repeated until convergence.

It is easy to see that the main difference between ALS and ITTFA is the active employment of the concentration windows. Practice shows that the choice of the initial matrix ($\mathbf{C}_{\text{in}}$) is not as critical as in the ALS method. In particular, one can use the rough solutions provided by EFA.

Application of ALS to our example is presented in worksheet ALS and in Fig. 12.26. Here, we also use the button **Calculate** to run a VBA macro that copies the contents of range $\mathbf{C}_{\text{out}}$ and pastes the values into range $\mathbf{C}_{\text{in}}$. The number of iterations is given in cell F2.

The concentration window profiles $W_a(t)$ were determined by EFA for the noise level of $\varepsilon = 0.05$. They are also used as the initial guess for matrix $\mathbf{C}_{\text{in}}$. The algorithm converges in 10 iterations to a result, which is fully ($\delta = 4\%$) consistent with other methods.

CONCLUSION

We have observed just a few methods among a plethora of other methods used in solving the MCR tasks. Inevitably, many interesting techniques, such as simple-to-use interactive

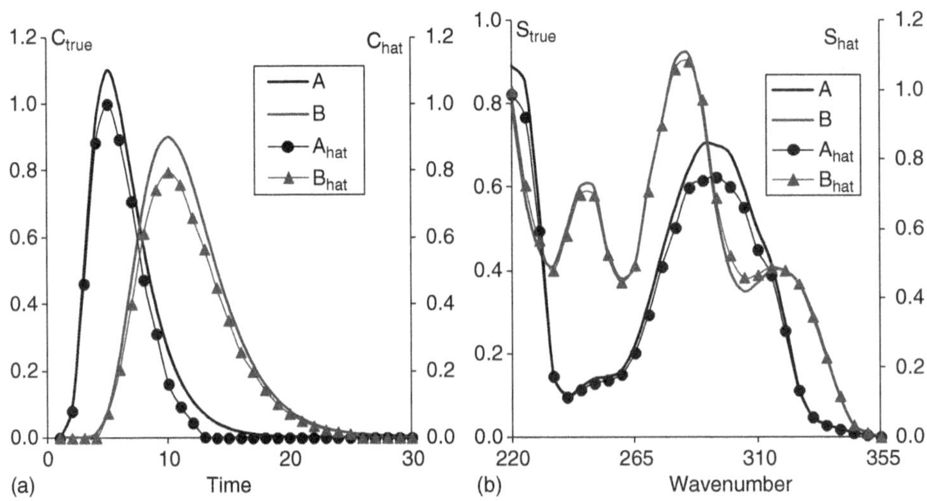

Figure 12.26 Solution by ALS.

self-modeling mixture analysis (SIMPLISMA) and orthogonal projection approach (OPA), have been left out. You can study them separately using this chapter as a guide.

A special MCR approach that closely relates to the ALS method is worthy of a special study. This method combines a soft chemometric approach (black modeling) with a substantial kinetic description (white modeling). It is presented in Chapter 14.

PART IV

SUPPLEMENTS

13

EXTENSION OF CHEMOMETRICS ADD-IN

This chapter describes the calculation of auxiliary values used in multivariate data analysis. Applying Chemometrics Add-In, which is described in Chapter 8, it is possible to obtain the main characteristics such as scores and loadings. However, other derivative values, for example, the residuals, eigenvalues, leverages, are also important. The text builds on the ideas of virtual arrays and Visual Basic for Applications (VBA) programming; therefore, it is a natural continuation of Chapter 7.

Preceding chapters	7–11
Dependent chapters	
Matrix skills	Basic
Statistical skills	Basic
Excel skills	Basic
Chemometric skills	Basic
Chemometrics Add-In	Used
Accompanying workbook	Tricks.xls

13.1 USING VIRTUAL ARRAYS

13.1.1 Simulated Data

To illustrate the methods and techniques described below, we use simulated data presented in worksheet Data and in Fig. 13.1. The data consist of two parts.

The first set is the training (calibration) set that is located in region named X. This is a global name available throughout the whole workbook. Matrix X consists of 15 rows

Chemometrics in Excel, First Edition. Alexey L. Pomerantsev.

	A	B	C	D	E	F	G	H	I
3		V01	V02	V03	V04	V05	V06	V07	V08
4	S01	4.046	0.813	0.003	0.181	0.153	0.154	3.373	1.341
5	S02	5.472	0.225	0.003	-0.859	-0.104	0.797	4.634	2.922
6	S03	4.901	0.456	0.003	0.185	0.115	0.125	3.376	1.284
7	S04	5.359	0.314	0.003	0.370	0.197	0.320	3.413	0.985
8	S05	5.651	0.116	0.002	0.264	0.113	0.184	3.465	1.511
9	S06	3.778	-0.358	0.002	0.456	0.207	0.097	3.269	0.609
10	S07	5.366	0.260	0.003	0.399	0.172	0.063	3.305	1.036
11	S08	6.034	0.005	0.002	0.556	0.184	-0.033	3.203	0.492
12	S09	5.642	0.110	0.002	0.976	0.262	-0.258	2.621	0.382
13	S10	5.588	0.201	0.002	0.034	0.098	0.063	3.171	1.167
14	S11	5.467	0.214	0.003	0.399	0.161	0.263	3.542	1.486
15	S12	4.883	0.446	0.003	0.131	0.139	0.163	3.299	1.096
16	S13	3.975	0.136	0.002	0.402	0.166	0.111	3.314	0.955
17	S14	4.978	0.529	0.003	0.301	0.157	0.176	3.373	1.378
18	S15	5.406	0.180	0.002	-0.616	-0.050	0.777	4.571	2.255
19									
20	Mean	5.103	0.243	0.002	0.212	0.131	0.200	3.462	1.260
21	SDev	0.678	=IF(PP>1,STDEV(C4:C18),1)				0.272	0.507	0.651
22									
23	S16	5.352	0.289	0.002	0.256	0.159	0.221	3.331	0.983
24	S17	4.998	0.342	0.003	0.291	0.157	0.348	3.741	1.463
25	S18	5.292	0.201	0.002	0.162	0.156	0.121	3.217	1.048
26	S19	6.765	0.879	0.003	0.253	0.160	0.040	3.226	1.109
27	S20	6.376	-0.164	0.001	0.350	0.182	0.178	3.285	0.763
28									
29	Number of samples					I=	15		
30									
31	Number of new samples					Inew=	5		
32									
33	Number of variables					J=	25		
34									
35	Preprocessing			☑ Centring		PP=	3		
36									
37				☑ Scaling					
38									

Figure 13.1 Simulated data.

(samples) and 25 columns (variables). These dimensions are calculated in cell G29 (named I) and cell G31 (named J). The I value is calculated by function `ROWS`(X) and the J value by function `COLUMNS`(X).

The second part is a new (or test) set named Xnew. This name is also available throughout the whole workbook. It consists of five rows (samples). The number of rows is calculated in cell G31 and has a global name Inew.

Before applying projection methods, the data are usually subjected to preprocessing. They are centered and/or scaled. Functions incorporated into Chemometrics Add-In include an argument **`CentWeight`** responsible for automatic autoscaling. The argument can occupy the following values (see Section 8.2.2).

0 – no centering and no scaling (default value)
1 – only centering, that is, subtraction of column-wise mean values
2 – only scaling by column-wise standard deviations
3 – centering and scaling, that is, autoscaling

Worksheet Data uses cell G35 with a global name PP, the value of which is calculated on the basis of the selected option from those listed next to it.

In general, there is no need for explicit preprocessing of data in a worksheet for principal component analysis (PCA) decomposition. At the same time, sample mean values and standard deviations may be useful for the calculation of various model characteristics. These values are calculated in worksheet Data in the region B20:Z20 (named Mean) and in the region B21:Z21 (named SDev). Formulas employed for the calculations use PP values.

Formula

$$\mathsf{AVERAGE}(\mathsf{C4} : \mathsf{C}18) * \mathsf{MOD}(\mathsf{PP}, 2)$$

is used to calculate mean values. Function MOD(PP,2) returns 1 if PP=1 or PP=3, and returns 0 if PP=0 or PP=2.

Formula

$$\mathsf{IF}(\mathsf{PP} > 1, \mathsf{STDEV}(\mathsf{C4} : \mathsf{C}18), 1)$$

is used to calculate standard deviations.

13.1.2 Virtual Array

In the course of data analysis there often appears a problem of storing the intermediate results, which are not important themselves but have to be calculated to get to the ultimate result. For example, residuals in the PCA decomposition are rarely analyzed *per se* but used for the calculation of explained variance, orthogonal distances, etc. At the same time, such intermediate arrays may be very large and must be calculated for various numbers of principal components. They cause flooding of worksheets by unnecessary, intermediate information. This situation can be avoided by applying virtual arrays. Section 7.2.9 demonstrates this with a simple example.

Below, we consider several cases that can be considered exemplary for similar problems. All these examples are performed for the PCA (Chapter 9), but they can also be applied to calibration (Chapter 10) or classification (Chapter 11) tasks.

13.1.3 Data Preprocessing

The following simple array formula (see also Section 7.2.2) can be used to preprocess data in accordance with the selected options

$$\mathsf{(X\text{-}Mean)/SDev}$$

Accounting to the way the values Mean and SDev were defined in worksheet Data, the preprocessing result is correct for any combination of centering and scaling options. The examples of preprocessing of the training set (X) and the test set (Xnew) areas are given in sheet Scaling.

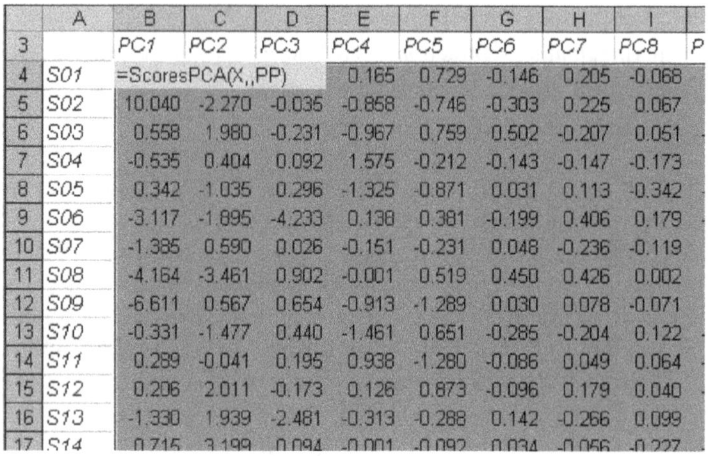

	A	B	C	D	E	F	G	H	I
3		PC1	PC2	PC3	PC4	PC5	PC6	PC7	PC8
4	S01	=ScoresPCA(X,,PP)			0.165	0.729	-0.146	0.205	-0.068
5	S02	10.040	-2.270	-0.035	-0.858	-0.746	-0.303	0.225	0.067
6	S03	0.558	1.980	-0.231	-0.967	0.759	0.502	-0.207	0.051
7	S04	-0.535	0.404	0.092	1.575	-0.212	-0.143	-0.147	-0.173
8	S05	0.342	-1.035	0.296	-1.325	-0.871	0.031	0.113	-0.342
9	S06	-3.117	-1.895	-4.233	0.138	0.381	-0.199	0.406	0.179
10	S07	-1.385	0.590	0.026	-0.151	-0.231	0.048	-0.236	-0.119
11	S08	-4.164	-3.461	0.902	-0.001	0.519	0.450	0.426	0.002
12	S09	-6.611	0.567	0.654	-0.913	-1.289	0.030	0.078	-0.071
13	S10	-0.331	-1.477	0.440	-1.461	0.651	-0.285	-0.204	0.122
14	S11	0.289	-0.041	0.195	0.938	-1.280	-0.086	0.049	0.064
15	S12	0.206	2.011	-0.173	0.126	0.873	-0.096	0.179	0.040
16	S13	-1.330	1.939	-2.481	-0.313	-0.288	0.142	-0.266	0.099
17	S14	0.715	3.199	0.094	-0.001	-0.092	0.034	-0.056	-0.227

Figure 13.2 Scores calculation.

13.1.4 Decomposition

Worksheet Scores presents the PCA score values calculated by function **ScoresPCA** (see Section 8.3.1 and Fig. 13.2).

The first argument X stands for the name of data matrix, the second argument (the number of principal components) is omitted; that is, the function calculates PCA scores for all possible PCs. The last argument PP is the value responsible for data scaling. Function **ScoresPCA** is an array formula and it must be completed by **CTRL+SHIFT+ENTER**.

The test set scores are calculated by a similar formula

ScoresPCA (X, , PP, Xnew) ,

which, however, contains the fourth argument Xnew that indicates that the scores are calculated for this new dataset. Data arrays with PCA scores have global names T and Tnew, respectively.

Using the similar function **LoadingsPCA**, one can calculate the PCA loadings. We do not show these results as they might be very large. When it happens, the loadings matrix occupies a large area in a worksheet. If the loadings matrix is not used explicitly, it is better to not display it but rather apply a technique of virtual arrays.

13.1.5 Residuals Calculation

Worksheet Residuals and Fig. 13.3 illustrate how a virtual array technique is employed for the calculation of the residual values for a given number of PCs. The following array formula is used:

(X-Mean) /SDev-

MMULT (ScoresPCA (X, PC, PP) , TRANSPOSE (LoadingsPCA (X, PC, PP))) .

The first part of the formula, that is, (X-Mean)/SDev performs the data preprocessing with respect to PP value. The last part of the formula, that is,

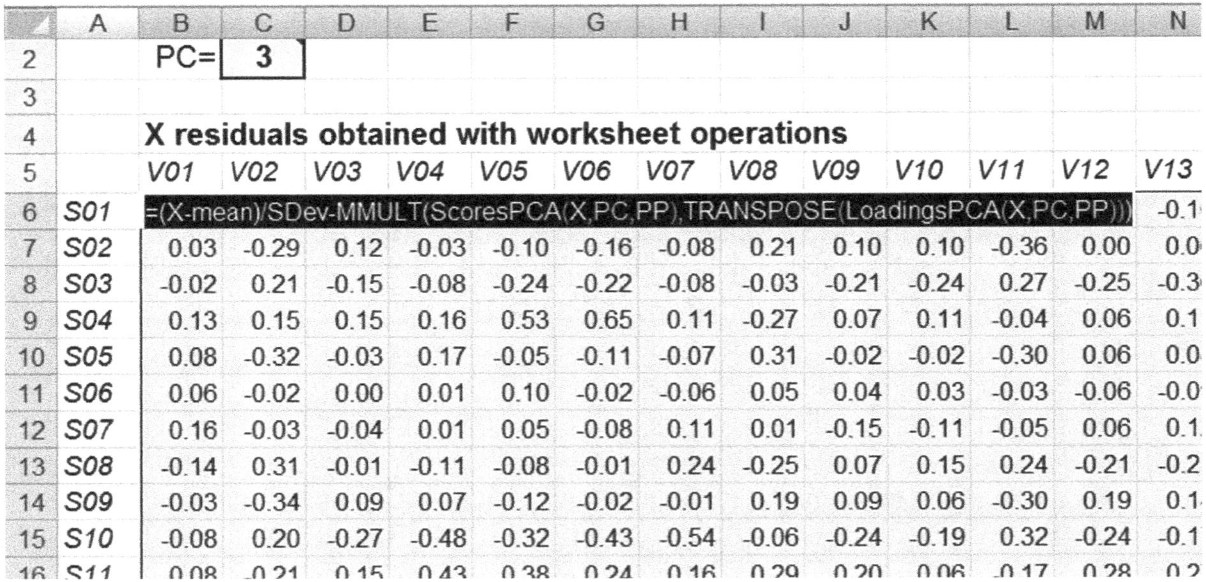

	A	B	C	D	E	F	G	H	I	J	K	L	M	N
2		PC=	3											
3														
4		X residuals obtained with worksheet operations												
5		V01	V02	V03	V04	V05	V06	V07	V08	V09	V10	V11	V12	V13
6	S01	=(X-mean)/SDev-MMULT(ScoresPCA(X,PC,PP),TRANSPOSE(LoadingsPCA(X,PC,PP)))												-0.1
7	S02	0.03	-0.29	0.12	-0.03	-0.10	-0.16	-0.08	0.21	0.10	0.10	-0.36	0.00	0.0
8	S03	-0.02	0.21	-0.15	-0.08	-0.24	-0.22	-0.08	-0.03	-0.21	-0.24	0.27	-0.25	-0.3
9	S04	0.13	0.15	0.15	0.16	0.53	0.65	0.11	-0.27	0.07	0.11	-0.04	0.06	0.1
10	S05	0.08	-0.32	-0.03	0.17	-0.05	-0.11	-0.07	0.31	-0.02	-0.02	-0.30	0.06	0.0
11	S06	0.06	-0.02	0.00	0.01	0.10	-0.02	-0.06	0.05	0.04	0.03	-0.03	-0.06	-0.0
12	S07	0.16	-0.03	-0.04	0.01	0.05	-0.08	0.11	0.01	-0.15	-0.11	-0.05	0.06	0.1
13	S08	-0.14	0.31	-0.01	-0.11	-0.08	-0.01	0.24	-0.25	0.07	0.15	0.24	-0.21	-0.2
14	S09	-0.03	-0.34	0.09	0.07	-0.12	-0.02	-0.01	0.19	0.09	0.06	-0.30	0.19	0.1
15	S10	-0.08	0.20	-0.27	-0.48	-0.32	-0.43	-0.54	-0.06	-0.24	-0.19	0.32	-0.24	-0.1
16	S11	0.08	-0.21	0.15	0.43	0.38	0.24	0.16	0.29	0.20	0.06	-0.17	0.28	0.2

Figure 13.3 Residuals calculation.

MMULT (ScoresPCA (X, PC, PP), TRANSPOSE (LoadingsPCA (X, PC, PP)))

corresponds to ordinary PCA equation (see Section 9.2.1)

$$\mathbf{TP}^{t},$$

where **T** and **P** are the scores and loadings matrices, respectively. The PC value is taken from cell C2 and used when calculating the residuals. By changing this value, one can get the residual values for various numbers of PCs.

The presented formula is an array formula and must be completed by **CTRL+SHIFT+ ENTER**.

13.1.6 Eigenvalues Calculation

Worksheet Eigenvalues presents calculation of eigenvalues (Section 9.2.4) and other relative characteristics. The following array formula is used (see Fig. 13.4).

SUMSQ(INDEX(ScoresPCA(X, , PP), , F3))

This formula is constructed as follows. Firstly, the virtual array of all possible scores is calculated by formula **T** = ScoresPCA(X,,PP). Then a virtual column **t**, corresponding to the number of PC indicated in cell F4 is selected. For this purpose, a formula INDEX(**T**,,F3) with the omitted second argument (Section 7.2.4) is applied. The ultimate result is returned after the application of function SUMSQ(**t**), which calculates the sum of squares of the vector elements (Section 7.1.7). In this way, the formula for the calculation of the a-th eigenvalue

$$\lambda_a = \sum_{i=1}^{I} t_{ia}^2$$

is executed.

Each eigenvalue is calculated individually in accordance with the corresponding PC number indicated in the cells of the fourth row. To insert formulas, we should enter the formula for the first PC first and then "drag" the expression to the right (Section 7.1.9) for all PCs. Each formula is of an array type and must be completed by pressing **CTRL+SHIFT+ENTER**.

Once the eigenvalues are obtained, they can be used for the calculation of other related characteristics such as singular values, as well as total and explained residual variances (Sections 9.2.7 and 9.2.9).

The region of calculated singular values has a global name SgV.

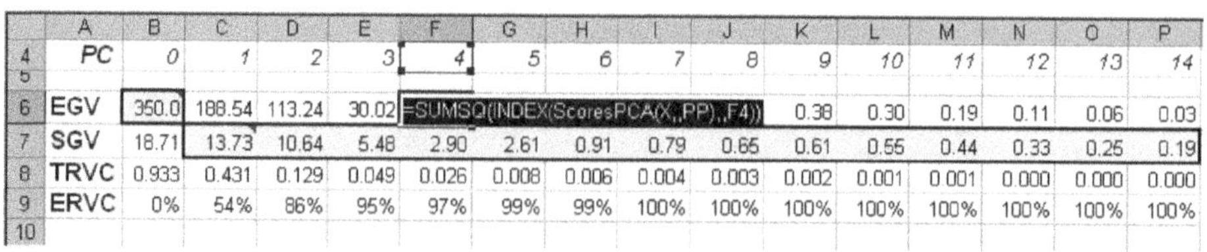

	A	B	C	D	E	F	G	H	I	J	K	L	M	N	O	P
4	PC	0	1	2	3	4	5	6	7	8	9	10	11	12	13	14
5																
6	EGV	350.0	188.54	113.24	30.02	=SUMSQ(INDEX(ScoresPCA(X,,PP),,F4))					0.38	0.30	0.19	0.11	0.06	0.03
7	SGV	18.71	13.73	10.64	5.48	2.90	2.61	0.91	0.79	0.65	0.61	0.55	0.44	0.33	0.25	0.19
8	TRVC	0.933	0.431	0.129	0.049	0.026	0.008	0.006	0.004	0.003	0.002	0.001	0.001	0.000	0.000	0.000
9	ERVC	0%	54%	86%	95%	97%	99%	99%	100%	100%	100%	100%	100%	100%	100%	100%
10																

Figure 13.4 Eigenvalues calculation.

13.1.7 Orthogonal Distances Calculation

Orthogonal PCA distances (aka sample variances) are defined in Section 9.2.13 by the formula

$$\mathrm{OD}_i = (1/J)\sum_{j=1}^{J} e_{ij}^2; \quad i = 1, \ldots, I$$

The most straightforward way to calculate them is to apply the residual matrix as illustrated in column AB in worksheet Residuals. However, in this case, we would have to keep all residual matrices for all PCs. This could be a space-consuming method. The application of virtual arrays is more convenient.

Worksheet OD and Fig. 13.5 demonstrate the respective formula. Its inner part coincides with the formula given in Section 13.1.5. This part calculates a virtual array of residuals **E** for each PC number defined in cell C3

$$\mathbf{E} = \text{(X-Mean)/SDev-}$$
$$\text{MMULT(ScoresPCA(X, PC, PP), TRANSPOSE(LoadingsPCA(X, PC, PP))).}$$

Then a virtual row $\mathbf{e}_i$ is extracted from the virtual array **E**. The row index i is determined in cell A11, hence $\mathbf{e}_i$=INDEX(**E**,A11,). Please note the absence of the last argument in function **INDEX** (see Section 7.2.4). The ultimate result is calculated by function OD_i=SUMSQ($\mathbf{e}_i$)/J, which returns the sum of squares for all residuals for object i.

Distances for the test data Xnew are calculated by a similar formula

SUMSQ(INDEX((Xnew-Mean)/SDev-
MMULT(ScoresPCA(X, PC, PP, Xnew),
TRANSPOSE(LoadingsPCA(X, PC, PP))), A23,))/J,

Figure 13.5 Calculation of orthogonal distances.

which, however, contains the test set name Xnew twice once in the scaling part as well as in the scores calculation.

Each value OD_i is calculated separately for a sample, the number of which is indicated in column A. To insert formulas, we enter a formula for the first object first and afterward "drag" the formula down (Section 7.1.9) for all objects. The formula is of an array type and must be completed by pressing **CTRL+SHIFT+ENTER**.

13.1.8 Leverages Calculation

The technique of virtual arrays can be also applied for the calculation of leverages (or the Mahalanobis distances in the scores space, see Section 9.2.13).

$$SD_i = \mathbf{t}_i^t(\mathbf{T}^t\mathbf{T})^{-1}\mathbf{t}_i = \sum_{a=1}^{A} t_{ia}^2/\lambda_a, \quad i = 1, \ldots, I$$

First of all, the scores matrix and a row of singular values should be obtained. The scores matrix (global name T) is calculated in worksheet Scores and the singular values are calculated in worksheet Eigenvalues (global name SgV).

The SD_i is calculated by the following formula (see also Fig. 13.6):

SUMSQ(INDEX(OFFSET(T,,,,PC)/OFFSET(SgV,,,,PC),A11,)).

The inner functions OFFSET(T,,,,PC) and OFFSET(SgV,,,,PC) extract the first PC columns from arrays T and SgV. The PC value is determined in cell C3. A virtual array of

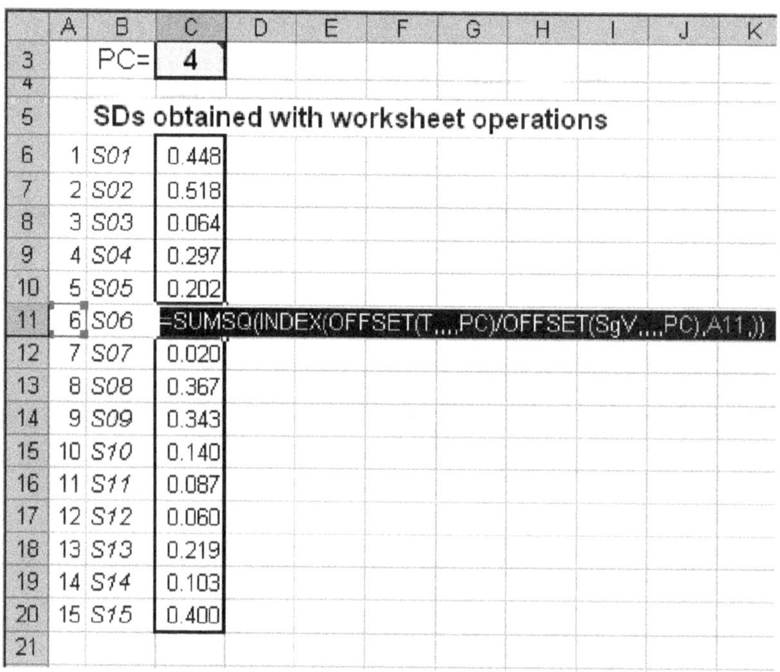

Figure 13.6 Leverage calculation.

normalized scores **U** is obtained as a result of element-wise division of the truncated matrix of scores by the truncated vector of singular values. Afterward function INDEX(**U**,A11,) selects one row $\mathbf{u}_i$ from matrix **U**. Index i is determined in cell A11. The ultimate result is derived by using the function

$$\text{SD}_i = \text{SUMSQ}(\mathbf{u}_i),$$

which returns the sum of squares for object i.

The formula for the test set is similar

$$\text{SUMSQ(INDEX(OFFSET(Tnew,,,,PC)/OFFSET(SgV,,,,PC),A23,))}.$$

Each SD_i is calculated individually, and this calculation is linked to the corresponding object number provided in column A. To add formulas to all objects, one should first enter the formula for the first object in cell C6 and "drag" it down afterward, as described in Section (7.1.9).

In this formula, we cannot substitute matrix **T** by formula ScoresPCA(X,,PP) because function `OFFSET` does not support virtual arrays. Each formula here is not of an array type and should be completed by the **ENTER** key only.

13.2 USING VBA PROGRAMMING

13.2.1 VBA Advantages

VBA programming is an alternative method to extend the Chemometrics Add-In possibilities. VBA programming aims at the development of user-defined functions (UDF) used to solve various subtasks of multivariate data analysis. The principles of this technique are presented in Chapter 7. Several examples help to demonstrate the usage of the Chemometrics Add-In functions in VBA programming. Here, the scope is limited to the same PCA tasks considered in the first part of this chapter. Our goal is to demonstrate the alternative, VBA-based solutions. We have no intention of expanding the list of UDFs, leaving it to the reader to perform such exercises and create his/her own library of UDFs.

While designing a new UDF, which employs the projection functions, one should not forget to establish the reference to Chemometrics Add-In. For this purpose, it is necessary to select option **Tools-References** in Visual Basic Editor and mark **Chemometrics** in the list of available add-ins. As a result, a new section **References** with a reference to **Chemometrics** (see Fig. 13.9) will appear in the **Project Explorer** window.

13.2.2 Virtualization of Real Arrays

The first task that we will consider is a preparatory procedure, which is always necessary in VBA programs.

The worksheets contain the real data arrays, which are used in VBA procedures. These data have to be virtualized, that is, converted into virtual arrays located in the computer's memory. This task is solved by a function `ConvertRange` that is presented in Fig. 13.7. The code is also given in workbook Tricks.xls in Utils modulus.

```
Private Function ConvertRange(X As Variant) As Variant
   Dim vMatrix As Variant

   On Error GoTo ErrorEnd
   If IsObject(X) Then
      vMatrix = X.Value
   Else
      vMatrix = X
   End If

   ConvertRange = vMatrix
   Exit Function
ErrorEnd:
   ConvertRange = CVErr(xlErrValue)
End Function
```

Figure 13.7 Function **`ConvertRange`**.

13.2.3 Data Preprocessing

Function **`ScaleData`** can be employed for data preprocessing. The procedure calculates the sample mean and variance values of matrix **X**, transforms matrix $\mathbf{X}_{new}$, and then returns an array containing the transformed matrix $\mathbf{X}_{new}$.

Syntax

`ScaleData` `(`**`X`** `[,CentWeightX] [, Xnew])`

`X` is an array of **X**-values (calibration set);

`CentWeightX` is an optional argument (integer) that indicates (see Section 8.2.2) whether centering and/or scaling is done;

`Xnew` is an optional argument that presents an array of new values $\mathbf{X}_{new}$ (test set) for which the transformed values are calculated.

Remarks

- Arrays `Xnew` and **`X`** must have the same number of columns;
- If argument `Xnew` is omitted, it is assumed to be the same as **`X`**, thus a transformed **`X`** is returned;
- The result is an array (matrix) with the same dimensions as the matrix `Xnew`;
- **`ScoresPCA`** is a similar function in Chemometrics Add-In.

Example (Fig. 13.8)

The **`ScaleData`** function code is presented in workbook Tricks.xls, Utils modulus, and shown in Fig. 13.9.

`ScaleData` is an array function, which must be completed by **CTRL+SHIFT+ ENTER**.

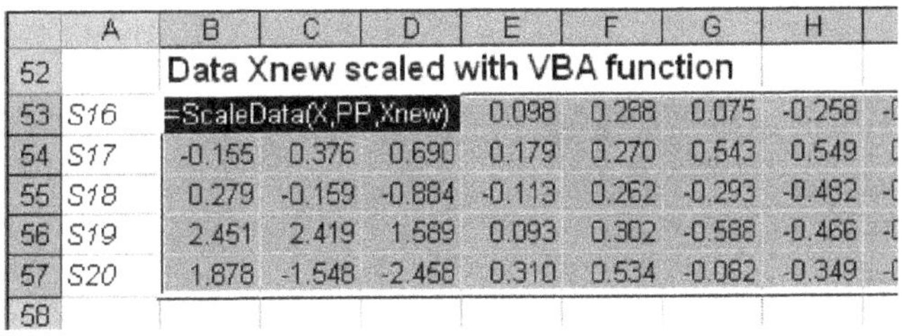

	A	B	C	D	E	F	G	H
52		Data Xnew scaled with VBA function						
53	S16	=ScaleData(X,PP,Xnew)			0.098	0.288	0.075	-0.258
54	S17	-0.155	0.376	0.690	0.179	0.270	0.543	0.549
55	S18	0.279	-0.159	-0.884	-0.113	0.262	-0.293	-0.482
56	S19	2.451	2.419	1.589	0.093	0.302	-0.588	-0.466
57	S20	1.878	-1.548	-2.458	0.310	0.534	-0.082	-0.349
58								

Figure 13.8 Example of function **ScaleData**.

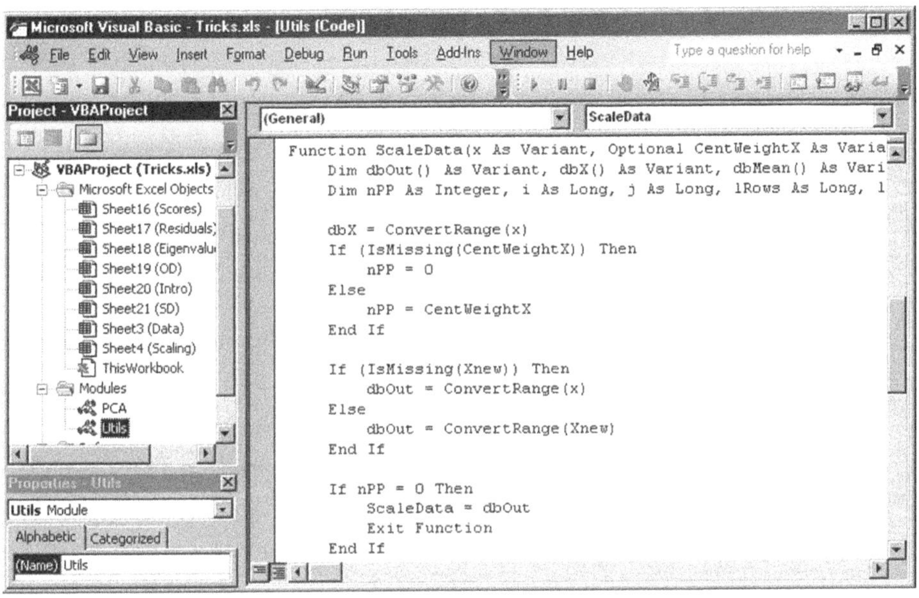

```
Function ScaleData(x As Variant, Optional CentWeightX As Varia
    Dim dbOut() As Variant, dbX() As Variant, dbMean() As Vari
    Dim nPP As Integer, i As Long, j As Long, lRows As Long, l

    dbX = ConvertRange(x)
    If (IsMissing(CentWeightX)) Then
        nPP = 0
    Else
        nPP = CentWeightX
    End If

    If (IsMissing(Xnew)) Then
        dbOut = ConvertRange(x)
    Else
        dbOut = ConvertRange(Xnew)
    End If

    If nPP = 0 Then
        ScaleData = dbOut
        Exit Function
    End If
```

Figure 13.9 Code of function **ScaleData**.

13.2.4 Residuals Calculation

PCA residuals can be obtained by function **ResPCA**. It decomposes matrix **X**, projects matrix $\mathbf{X}_{\text{new}}$ onto the scores space, and calculates the residuals for $\mathbf{X}_{\text{new}}$.

Syntax

`ResPCA ((X [, PC] [,CentWeightX] [, Xnew])`

`X` is an array of X-values (calibration set);

`PC` is an optional argument (integer) that defines the number of principal components (A) used in the PCA decomposition (see Section 8.2.2);

`CentWeightX` is an optional argument (integer) that indicates whether centering and/or scaling is done (see Section 8.2.2);

	A	B	C	D	E	F	G	H	I
48		Xnew residuals obtained with VBA function							
49	S16	=ResPCA(X,PC,PP,Xnew)				0.15	0.29	-0.07	-0.22
50	S17	0.06	-0.20	0.09	0.45	0.49	0.30	0.33	-0.03
51	S18	-0.27	0.30	-0.39	-0.38	0.08	-0.04	-0.26	-0.03
52	S19	-0.09	0.27	-0.45	-0.32	0.14	0.02	0.16	-0.39
53	S20	-0.01	0.35	-0.45	-0.20	0.33	0.36	-0.05	-0.09
54									

Figure 13.10 Example of function **ResPCA**.

`Xnew` is an optional argument that presents an array of new values $\mathbf{X}_{new}$ (test set) for which the scores values are calculated.

Remarks

- Arrays `Xnew` and **X** must have the same number of columns;
- If argument `Xnew` is omitted, it is assumed to be the same as **X**, hence PCA residuals for **X** are returned;
- The result is an array (matrix) with the same dimensions as matrix `Xnew`;
- **ScoresPCA** is a similar function in Chemometrics Add In.

Example (Fig. 13.10)

ResPCA is an array function, which must be completed by **CTRL+SHIFT+ENTER**.

The **ResPCA** function code is presented in workbook Tricks.xls, modulus PCA.

13.2.5 Eigenvalues Calculation

Eigenvalues can be calculated by function **EigenPCA** that returns one eigenvalue for a certain number of PC.

Syntax

EigenPCA ((**X** , **PC** [,CentWeightX])

X is an array of **X**-values (calibration set);

PC is an integer that defines the principal component, for which the eigenvalue is calculated;

`CentWeightX` is an optional argument (integer) that indicates whether centering and/or scaling is done (see Section 8.2.2).

Remarks

- The result is a number.

	A	B	C	D	E	F	G
11		Obtained with VBA function					
12	PC	0	1	2	3	4	5
13	EGV	350	188.54	=EigenPCA(X,D12,PP)			6.80
14	TRVC	0.933	0.431	0.129	0.049	0.026	0.008
15	TRVP	0.877	0.791	0.314	0.079	0.046	0.025
16							

Figure 13.11 Example of function **`EigenPCA`**.

Example (Fig. 13.11)

`EigenPCA` is an array function, which must be completed by **CTRL+SHIFT+ENTER**.

The **`EigenPCA`** function code is presented in workbook Tricks.xls, modulus PCA.

13.2.6 Orthogonal Distances Calculation

Orthogonal distances (aka sample variances) can be calculated by function **`ODisPCA`** that decomposes matrix **X** and then calculates the distances between the samples $\mathbf{X}_{new}$ and their projections in the scores space. The function returns an array (column) of the distances.

Syntax

`ODisPCA` `((`**`X`** `[, PC] [,CentWeightX] [, Xnew])`

`X` is an array of **X**-values (calibration set);

`PC` is an optional argument (integer) that defines the number of principal components (A) used in the PCA decomposition (see Section 8.2.2);

`CentWeightX` is an optional argument (integer) that indicates whether centering and/or scaling is done (see Section 8.2.2);

`Xnew` is an optional argument that presents an array of new values $\mathbf{X}_{new}$ (test set) for which the score values are calculated.

Remarks

- Arrays `Xnew` and **`X`** must have the same number of columns;
- If argument `Xnew` is omitted, it is assumed to be the same as **`X`** and thus ODs for **`X`** are returned;
- The result is an array (vector) with the same number of elements as the number of rows (I_{new}) in matrix `Xnew`;
- **`ScoresPCA`** is a similar function in Chemometrics Add In.

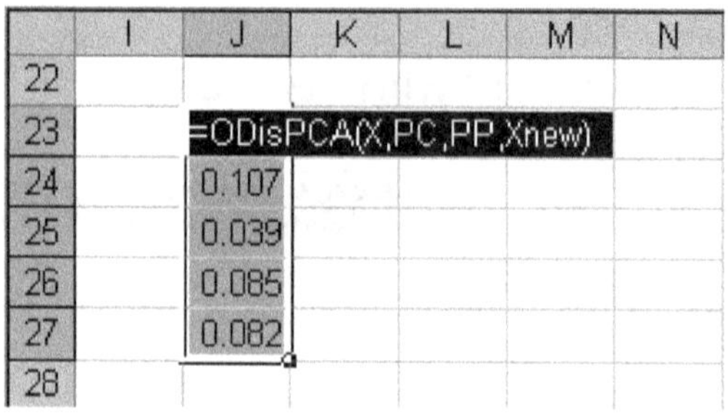

Figure 13.12 Example of function **`ODisPCA`**.

Example (Fig. 13.12)

`ODisPCA` is an array function, which must be completed by **CTRL+SHIFT+ ENTER**.

The **`ODisPCA`** function code is presented in workbook Tricks.xls, modulus PCA.

13.2.7 Leverages Calculation

Sample leverages (aka Mahalanobis distances) can be calculated by function **`SDisPCA`** that decomposes matrix **X**, projects matrix $\mathbf{X}_{new}$ onto the score space, and calculates the leverage values for $\mathbf{X}_{new}$.

Syntax

`SDisPCA` `((`**`X`** `[, PC] [,CentWeightX] [, Xnew])`

`X` is an array of **X**-values (calibration set);

`PC` is an optional argument (integer) that defines the number of principal components (A) used in the PCA decomposition (see Section 8.2.2);

`CentWeightX` is an optional argument (integer) that indicates whether centering and/or scaling is done (see Section 8.2.2)

`Xnew` is an optional argument that presents an array of new values $\mathbf{X}_{new}$ (test set) for which the score values are calculated.

Remarks

- Arrays `Xnew` and **`X`** must have the same number of columns;
- If argument `Xnew` is omitted, it is assumed to be the same as **`X`**, hence SDs for **`X`** are returned;
- The result is an array (vector) with the same number of elements as the number of rows (I_{new}) in matrix `Xnew`;
- **`ScoresPCA`** is a similar function in Chemometrics Add In.

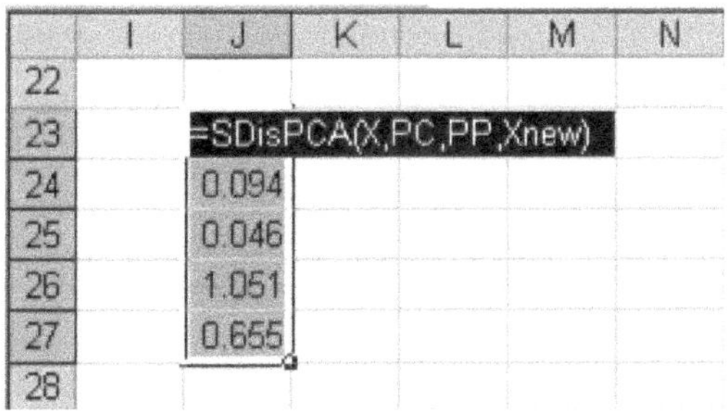

Figure 13.13 Example of function **SDisPCA**.

Example (Fig. 13.13)

SDisPCA is an array function, which must be completed by **CTRL+SHIFT+ENTER**. The **SDisPCA** function code is presented in workbook Tricks.xls, modulus PCA.

CONCLUSION

The virtual arrays technique considered here can look rather sophisticated. However, the presented formulae should be considered as templates, which may be used in a new workbook. A formula can be copied (as text) and pasted into your workbook. A minor revision (in the names of cells and ranges) will make this formula suitable for the analysis of another dataset.

Examples of VBA programming can help a user in working with the Chemometrics Add-In. They can be used as patterns for developing one's own UDF. Once you accumulate a lot of such UDFs, they could be collected in a separate Excel workbook and converted into a new add-in.

14

KINETIC MODELING OF SPECTRAL DATA

In this short chapter, we consider the application of kinetic methods to multivariate curve resolution (MCR) tasks.

Preceding chapters	7, 10, 12
Dependent chapters	
Matrix skills	Basic
Statistical skills	Basic
Excel skills	Advanced, Solver add-in usage
Chemometric skills	Advanced
Chemometrics Add-In	Not used
Accompanying workbook	Grey.xls

14.1 THE "GREY" MODELING METHOD

14.1.1 Problem Statement

The initial problem formulation is similar to that presented in Chapter 12. We consider a system consisting of several chemical substances (*components*) A, B, … , the concentrations of which $c_A(t), c_B(t), \ldots$ evolve with time t in the course of a natural (chemical or physical) process. The novel point is that this evolvement is subjected to a known kinetic mechanism. Each component is characterized by a spectrum, $s_A(\lambda), s_B(\lambda), \ldots$ where λ is a wave number. The setup of the experiment assumes that at time t_i and at wave number λ_j, we measure a value

Chemometrics in Excel, First Edition. Alexey L. Pomerantsev.

$$x(t_i, \lambda_j) = c_A(t_i)s_A(\lambda_j) + c_B(t_i)s_B(\lambda_j) + \cdots$$

This value is a linear superposition of the concentrations and spectra products constructed in line with the principle of linearity presented in Section 10.2.1.

Let I be the number of time observations $t_1, \ldots, t_I$, and J be the number of wave numbers $\lambda_1, \ldots, \lambda_J$.. The experimental data $\mathbf{X}$ can be presented in a matrix form, that is,

$$\mathbf{X} = \mathbf{C}\mathbf{S}^{\mathrm{t}} + \mathbf{E}. \tag{14.1}$$

Here, the data matrix $\mathbf{X}$ and the error matrix $\mathbf{E}$ are of the same dimension $I \times J$. If the system has A chemical components, the concentration matrix $\mathbf{C}$ has I rows and A columns. Each column of matrix $\mathbf{C}$ is a kinetic profile of a component. The pure spectra matrix $\mathbf{S}^{\mathrm{t}}$ (it is more convenient to present this matrix in a transposed form) has A rows and J columns. Each row of matrix $\mathbf{S}^{\mathrm{t}}$ is a pure spectrum of the corresponding component. See Fig. 14.1.

The *MCR problem* is as follows. Given the data matrix $\mathbf{X}$, determine the number of chemical components (A), find the concentration matrix $\mathbf{C}$, and the pure spectra matrix $\mathbf{S}$. To solve this task, we can use general soft methods presented in Chapter 12. However, in the present case, we have additional prior knowledge as opposed to a general MCR problem. We assume that the concentration profiles are described by kinetics equations

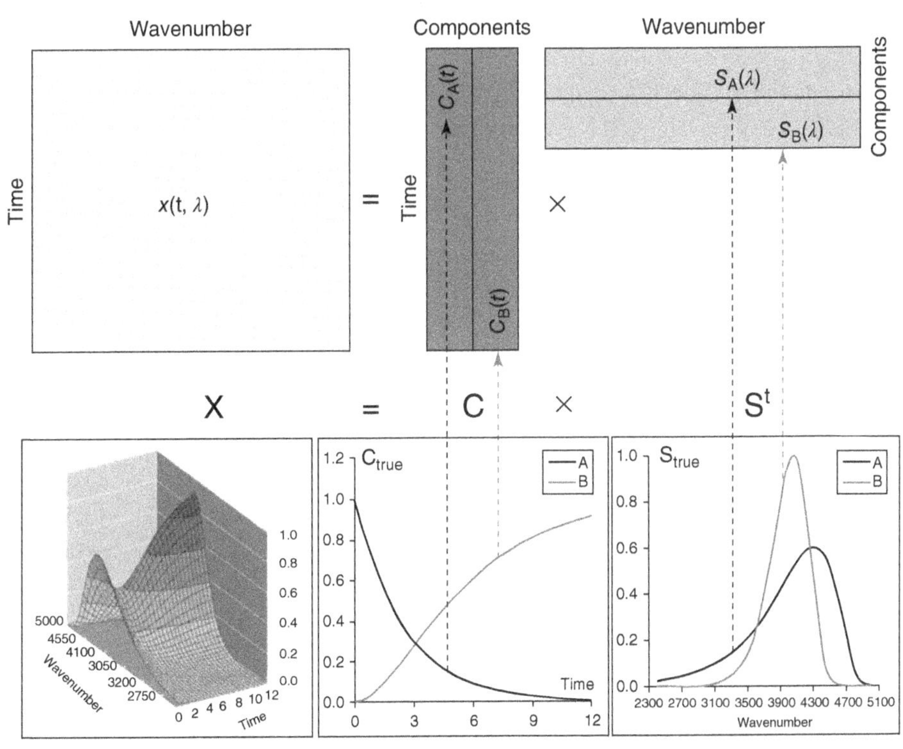

Figure 14.1 Pattern of the "kinetic data".

$$c_{\mathrm{A}}(t) = f_{\mathrm{A}}(t, k_1, \ldots, k_{\mathrm{p}})$$
$$c_{\mathrm{B}}(t) = f_{\mathrm{B}}(t, k_1, \ldots, k_{\mathrm{p}})$$
$$\ldots$$

The vector $\boldsymbol{k} = (k_1, \ldots, k_{\mathrm{p}})$ consists of unknown parameters that are the constant rates of corresponding reactions. All we know about the spectral matrix **S** is that it is subject to common constraints, such as nonnegativity, continuity, etc.

This formulation of the MCR task looks like an inverse problem of chemical kinetics, that is, a task to determine the unknown kinetic parameters of a known kinetic mechanism in the presence of nuisance parameters, being the spectral readings (extinction coefficients). However, a conventional inverse problem is usually based on the data, which are not spectra, but measured concentrations of a certain chemical reaction components.

In this chapter, we demonstrate how to solve the inverse spectral kinetic problem using a "grey" modeling method, which combines soft ("black") algorithms to estimate the spectra matrix **S** and hard ("white") methods to estimate the kinetic profile matrix **C**. Note that in this case, the MCR solution is unique (if it exists).

14.1.2 Example

To illustrate the method we use a simulated dataset placed in workbook Grey.xls. This book includes the following sheets:

Intro: a brief introduction
Layout: a key explaining the range naming
Kinetics: true concentration profiles **C**
Spectra: truc pure spectra **S**
Data: dataset **X** used in the example
Soft: soft method of alternating least squares (Soft-ALS)
Hard: hard method of alternating least squares (Hard-ALS)

14.1.3 Data

The prototype of our example is a problem of estimating the kinetic parameters of a consecutive first-order reaction

$$\mathrm{A} \xrightarrow{k_1} \mathrm{B} \xrightarrow{k_2} \mathrm{C}.$$

This mechanism is described by the following system of differential equations

$$dA/dt = -k_1 A; \qquad A(0) = A_0$$
$$dB/dt = k_1 A - k_2 B; \qquad B(0) = 0$$
$$dC/dt = k_2 B; \qquad C(0) = 0$$

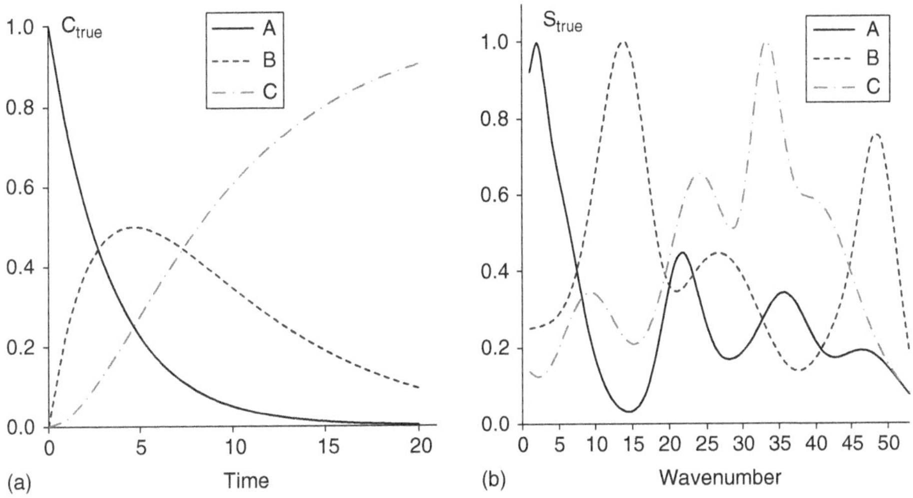

Figure 14.2 Concentration kinetics (a) and pure component spectra (b).

which has an explicit solution

$$
\begin{aligned}
A &= A_0 \exp(-k_1 t) \\
B &= (k_1 A_0 / k_1 - k_2)[\exp(-k_2 t) - \exp(-k_1 t)] \\
C &= A_0 + (A_0 / k_1 - k_2)[k_2 \exp(-k_1 t) - k_1 \exp(-k_2 t)]
\end{aligned}
\tag{14.2}
$$

shown in Fig. 14.2a and calculated for $A_0 = 1$, $k_1 = 0.30$, $k_2 = 0.15$.

These kinetic profiles are calculated by Eq. 14.2 in worksheet Kinetics for the parameters k_1 and k_2 provided in cells H3 (Rate1) and H4 (Rate2), respectively. The pure component spectra are presented in sheet Spectra and shown in Fig. 14.2b. The spectra are calculated for 53 conventional wave numbers using a complex superposition of the Gaussian peaks. Mixed data **X** are calculated in sheet Data by Eq. (14.1), where matrix **E** is generated for a relative error of 5%. This error level is specified in cell E27 and has a local name RMSE. We provide an ability to vary the error value and generate a new dataset. The button **Add New Error** launches a Visual Basic for Applications (VBA) macro **MakeErrors** attached to this sheet. See Section 7.3.1 for VBA programming explanation. The resulting data are shown in Fig. 14.3.

14.1.4 Soft Method of Alternating Least Squares (Soft-ALS)

We can apply a conventional method of alternating least squares (Soft-ALS), described in Section 12.4.2, to solve the MCR task for our simulated dataset. However, it would be better to improve the standard algorithm by using the a priori available information. We should take into account the following.

1. The number of components is known; there are three of them, A, B, and C;
2. The spectra are nonnegative;
3. The concentrations are nonnegative. Also $A(0) = 1$, $B(0) = C(0) = 0$;

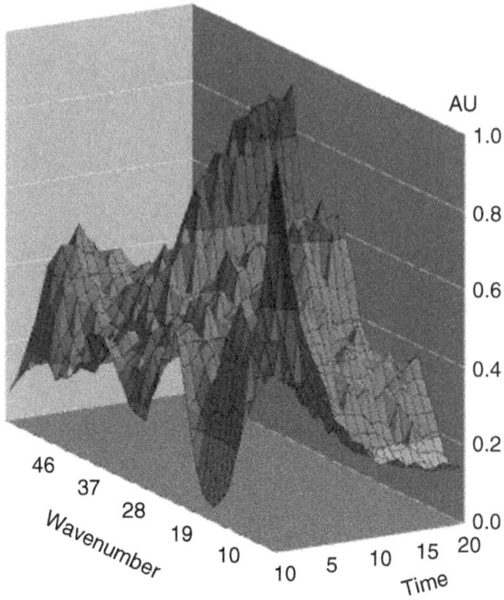

Figure 14.3 Model data.

4. The system is closed, that is, at any time t:

$$A(t) + B(t) + C(t) = 1. \quad (14.3)$$

Given this information, the algorithm is as follows:

1. A concentration template window is specified for each component. Within this window, every component concentration is a constant, except for the points where the concentration is at known zero values. These points should be kept during the iterative process. Here, these are $B(0) = C(0) = 0$. All template windows put together form a concetration template matrix.
2. The concetration template matrix is used as an initial approximation for the concentration matrix $\mathbf{C}_{\rm in}$.
3. Matrix $\mathbf{C}_{\rm in}$ is adjusted in line with the following rules. If its element is negative or the corresponding element of the concetration template matrix is zero, the element is replaced by zero. In addition, the matrix is normalized in such a way that its upper left element $A(0)$ is set equal to 1. The result is matrix $\mathbf{C}_{\rm adj}$.
4. Using matrix $\mathbf{C}_{\rm adj}$, the row-wise sums of its elements are calculated to produce a vector (Closure) that reflects the system's closure in line with Eq. 14.3.
5. Matrix $\mathbf{C}_{\rm adj}$ is adjusted to satisfy the closure constraint. For this purpose, each row of matrix $\mathbf{C}_{\rm adj}$ is divided by the corresponding element of vector Closure. The result is matrix $\mathbf{C}_{\rm hat}$.
6. Using the equation

$$\mathbf{S}_{\rm in} = \mathbf{X}^{\rm t}\mathbf{C}_{\rm hat}(\mathbf{C}_{\rm hat}^{\rm t}\mathbf{C}_{\rm hat})^{-1}$$

a pure spectra matrix $\mathbf{S}_{\rm in}$ is estimated.

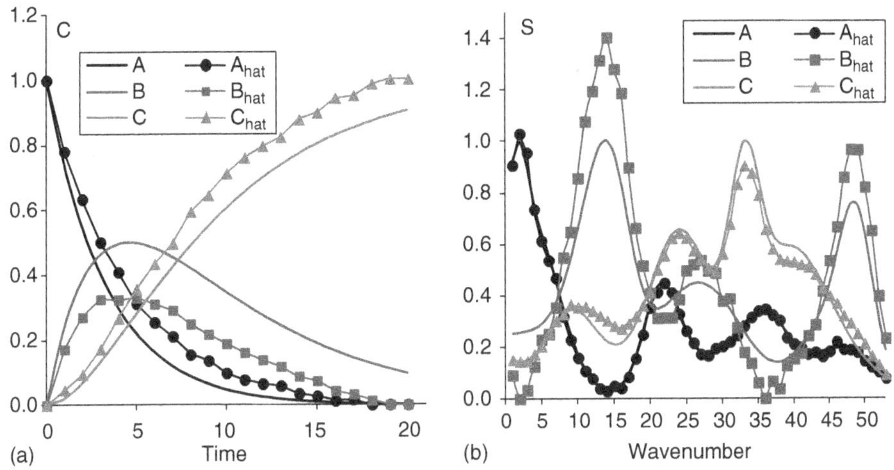

Figure 14.4 Solution by Soft-ALS.

7. Matrix $\mathbf{S}_{\text{in}}$ is adjusted accounting for constraints, that is, its negative elements are replaced by zeros. The result is matrix $\mathbf{S}_{\text{hat}}$.
8. Using formula

$$\mathbf{C}_{\text{out}} = \mathbf{X}\mathbf{S}_{\text{hat}}(\mathbf{S}_{\text{hat}}^{\text{t}}\mathbf{S}_{\text{hat}})^{-1}.$$

   the concentration profile matrix $\mathbf{C}_{\text{out}}$ is evaluated.
9. Matrix $\mathbf{C}_{\text{in}}$ is replaced by matrix $\mathbf{C}_{\text{out}}$.
10. Steps 3–9 are repeated until convergence.

The application of ALS to our dataset is presented in sheet Soft. Here, we use the button **Calculate** to launch a VBA macro that copies the contents of range $\mathbf{C}_{\text{out}}$ and pastes the values into range $\mathbf{C}_{\text{in}}$. The number of repetitions is given in cell K1 (local name iTer). When applied to our data, the algorithm converges in 10 iterations to a result that is exactly ($\delta = 5\%$) equal to the error specified in worksheet Data. Here, δ is the relative mean square error of modeling from Section 12.2.3, Eq. (12.10).

The resolved curves presented in Fig. 14.4 differ from their known true counterparts. Naturally, this is a consequence of the ambiguity (Section 12.1.2), which is unavoidable in soft MCR methods.

14.1.5 Hard Method of Alternating Least Squares (Hard-ALS)

Using the Soft-ALS method, we have not accounted for the known kinetic formulas given in Eq. 14.2. Taking this information into account complicates the algorithm but obviously improves the data analysis. Here is a new algorithm.

1. An initial guess of the concentration profile matrix $\mathbf{C}_{\text{in}}$ is specified. This can be a solution obtained by the Soft-ALS method.
2. Initial values of kinetic parameters $k_1, \ldots, k_{\text{p}}$ are specified.

3. Matrix $\mathbf{C}_{\text{in}}$ is adjusted, that is, all negative elements are replaced by zeros. The result is matrix $\mathbf{C}_{\text{hat}}$.
4. The component concentrations are calculated by Eq. 14.2. They constitute matrix $\mathbf{C}_{\text{fit}}$.
5. New values of the kinetic parameters $k_1, \ldots, k_p$ are found such that they minimize the sum of squares

$$\|\mathbf{C}_{\text{hat}} - \mathbf{C}_{\text{fit}}\|^2 = \sum_{i,j} [c_{\text{hat}}(i,j) - c_{\text{fit}}(i,j)]^2$$

   In the result, the optimized matrix $\mathbf{C}_{\text{fit}}$ is calculated.
6. Using the formula

$$\mathbf{S}_{\text{in}} = \mathbf{X}^{\text{t}}\mathbf{C}_{\text{fit}}(\mathbf{C}^{\text{t}}_{\text{fit}}\mathbf{C}_{\text{fit}})^{-1}$$

   the pure spectra matrix $\mathbf{S}_{\text{in}}$ is estimated.
7. Matrix $\mathbf{S}_{\text{in}}$ is adjusted accounting for constraints, that is, all negative elements are replaced by zeros. The result is matrix $\mathbf{S}_{\text{hat}}$.
8. Using the formula

$$\mathbf{C}_{\text{out}} = \mathbf{X}\mathbf{S}_{\text{hat}}(\mathbf{S}^{\text{t}}_{\text{hat}}\mathbf{S}_{\text{hat}})^{-1}$$

   the concentration profile matrix $\mathbf{C}_{\text{out}}$ is evaluated.
9. Matrix $\mathbf{C}_{\text{in}}$ is replaced by matrix $\mathbf{C}_{\text{out}}$.
10. Steps 3–9 are repeated until convergence.

Figure 14.5 shows a solution found by this algorithm.

This result looks better, that is, the estimates are much closer to their true values. At the same time, the global error δ is slightly worse compared with the Soft-ALS. It is now 0.0200 versus 0.0197 observed previously. The closure and nonnegativity constrains hold

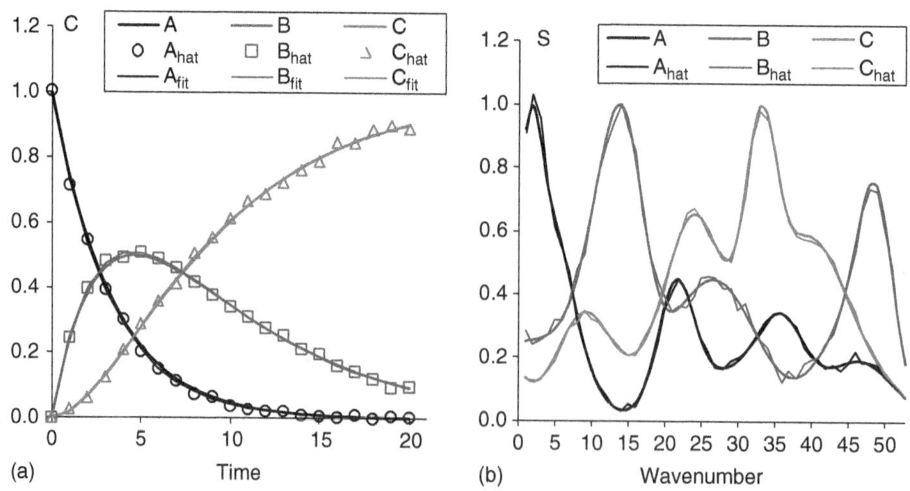

Figure 14.5 Solution by the Hard-ALS.

automatically. The convergence is barely influenced by the choice of the initial matrix $\mathbf{C}_{in}$ (step 1).

14.1.6 Using Solver Add-In

Step 5 in the Hard-ALS algorithm is the most difficult to implement. The problem is to find such values of parameters k_1 and k_2 (given in cells Rates), which minimize the sum of squares

$$\|\mathbf{C}_{hat}-\mathbf{C}_{fit}\|^2$$

calculated in cell named Fit.

To solve this optimization problem, we use a standard Excel add-in named Solver. To install this add-in, you should follow the procedure described in Section 7.3.6, starting with the second step shown in Fig. 14.6.

Solver can be launched from the menu **Tools-Solver** (Excel 2003) or **Data-Solver** (Excel 2007). In the dialog box that appears (Fig. 14.7), you should specify the type of optimization (**Min**), the target cell (Fit), and the changing cells (Rates).

Solver is not an efficient optimization method, but in case it is set up properly (as shown in Fig. 14.8) the search will proceed quickly.

In general, Hard-ALS is a rather slow method. For example, about 300 iterations are required to converge from an initial point of $k_1 = 0.5$, $k_2 = 0.1$. Therefore, it can be implemented in practice only if automated, that is, using the button **Calculate** that performs the specified number of iterations. This automation employs standard Solver functions **SolverOptions**, **SolverOk**, and **SolverSolve**. Their descriptions can be found on

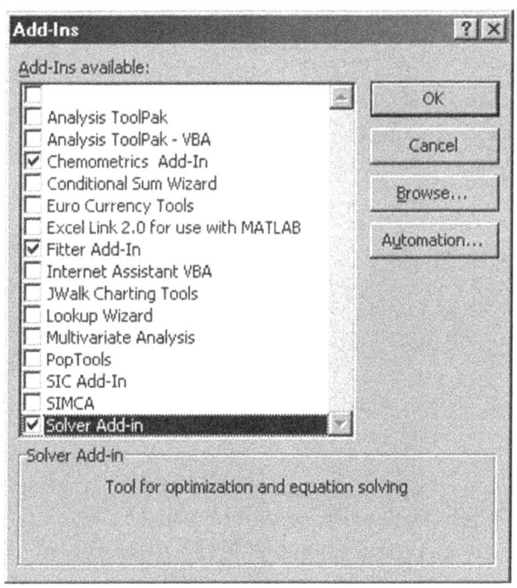

Figure 14.6 Solver add-in installation.

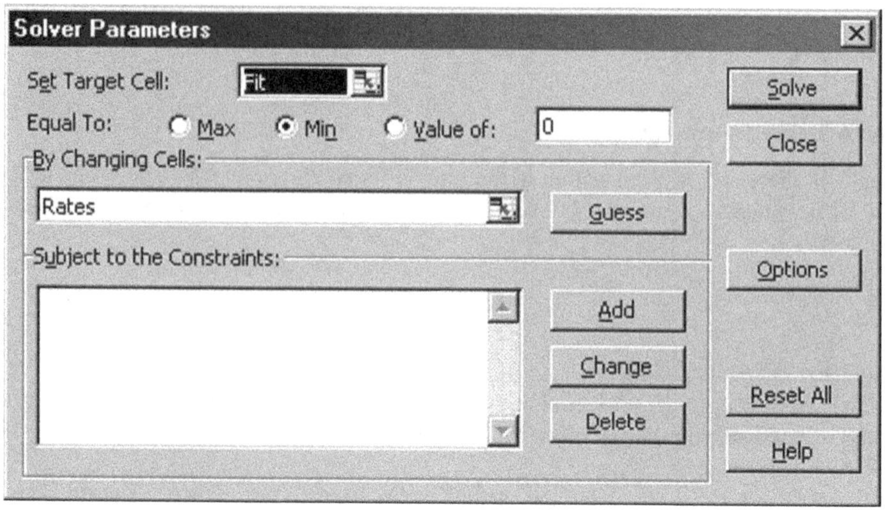

Figure 14.7 Solver parameters.

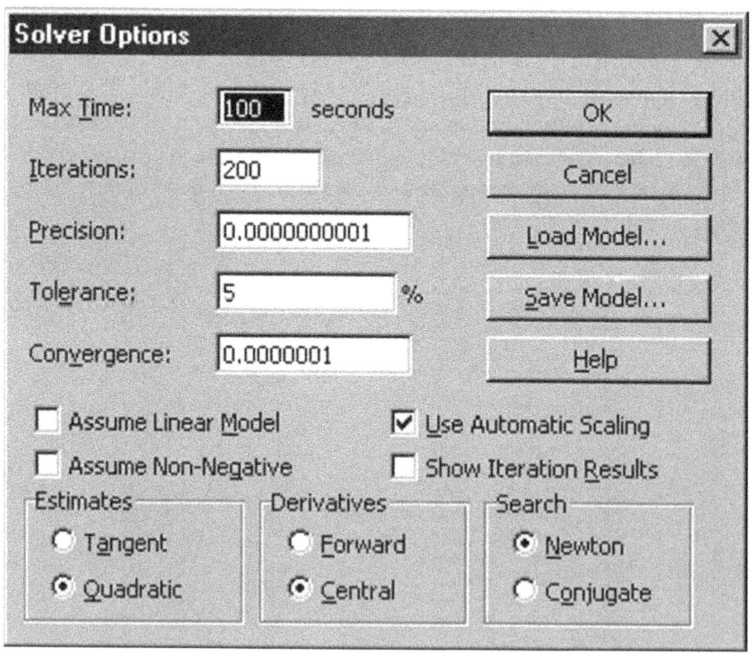

Figure 14.8 Solver options.

the Internet.[1]For VBA to recognize and use these functions, one should set a reference in Solver using the VBA Editor. It is not enough to install Solver add-in alone. Otherwise, an attempt to use the automation (button **Calculate**) results in an error window shown in Fig. 14.9.

[1]http://msdn.microsoft.com/en-us/library/office/ff196600.aspx

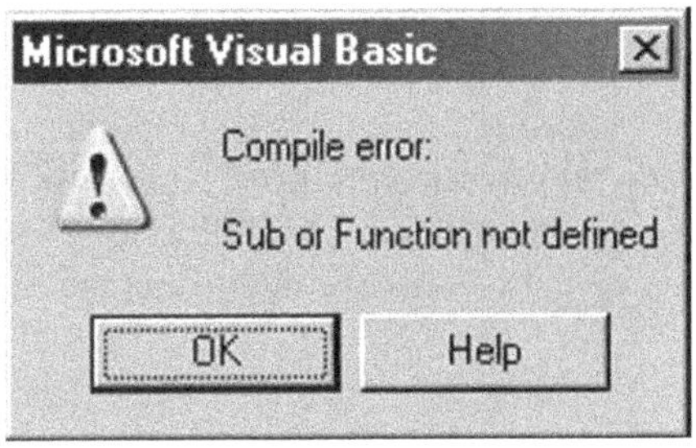

Figure 14.9 Reference error.

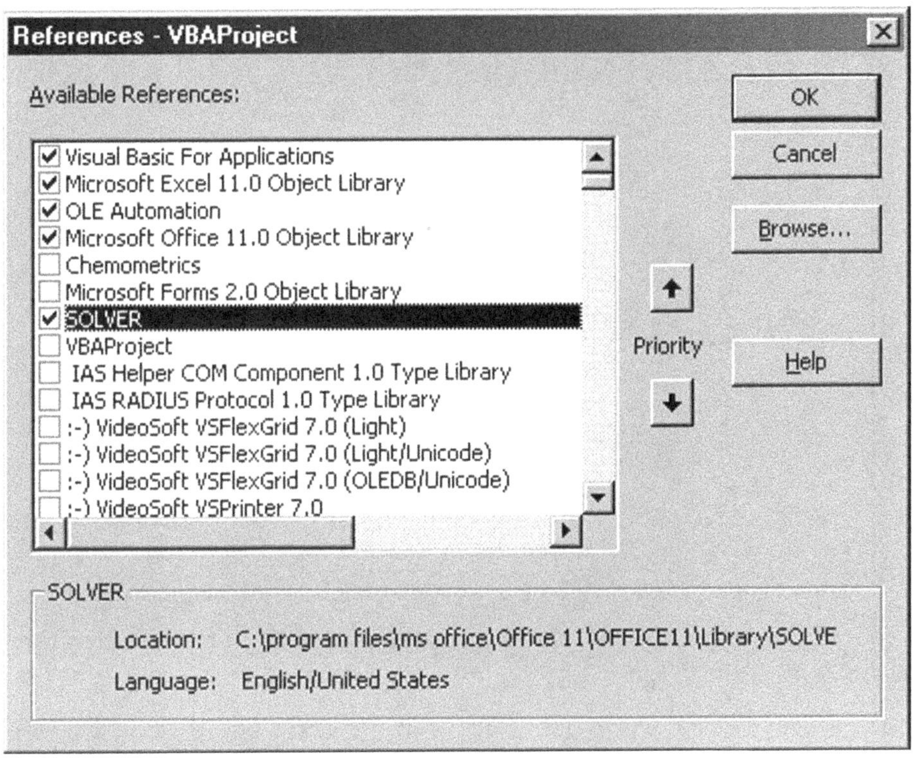

Figure 14.10 Reference to Solver.

To establish this reference, it is necessary to open the VBA editor (see Section 7.3.3) and use menu **Tools-References**. One will see a window shown in Fig. 14.10.

In this window, one should find and mark the SOLVER name.

Accompanying workbook Grey.xls does not have the Solver add-in installed. The reference to Solver is missing too. This is done intentionally, as the add-in's links are computer specific. This means that once correct for our computer, the links become wrong when an XLS file is transferred to another computer. Therefore, you should install Solver yourself and set a reference to it in the VBA editor using the instructions above. This Excel problem is also discussed in Section 7.3.5.

CONCLUSIONS

We have considered "grey" modeling to resolve kinetic profiles presented by the spectral data. This is a new, evolving area of chemometrics, which still has many white spots. A reader is encouraged to explore the method's possibilities by changing the simulated data, for example, the parameters k_1 and k_2 (sheet Kinetics) and the error level (sheet Data).

15

MATLAB®: BEGINNER'S GUIDE*

This chapter describes the use of a MATLAB® package for multivariate data analysis. The text is not a textbook on MATLAB. It serves as an introduction on how to work in an environment and provides the minimal information necessary for basic algorithms implementation.

Preceding chapters	5, 6, 8, 9
Dependent chapters	
Matrix skills	Basic
Statistical skills	Low
Excel skills	Basic
Chemometric skills	Basic
Chemometrics Add-In	Not used
Accompanying workbook	None

15.1 THE BASICS

15.1.1 Workspace

To start the program, double-click the MATLAB icon and you will see a window as shown in Fig. 15.1.

*With contributions from Yevgeny Mikhailov.

Chemometrics in Excel, First Edition. Alexey L. Pomerantsev.

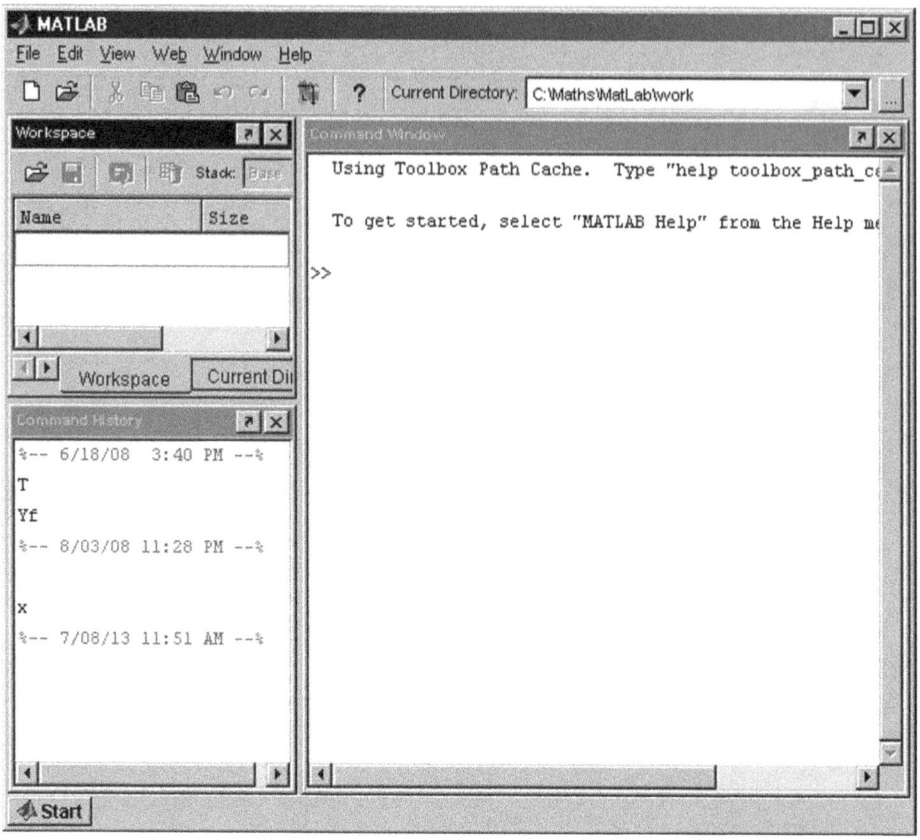

Figure 15.1 MATLAB desktop window.

MATLAB *6.x* desktop menu includes the following items:

- toolbar buttons with drop-down lists;
- window **Workspace** from which one can access various modules of **Toolbox** and one's own workspace;
- **Command History** windows intended for viewing and keeping track of previously executed commands and **Current Directory** tab used to determine the current directory;
- **Command Window** with an input prompt ≫ and a flashing vertical cursor;
- the status bar.

If any windows shown in Fig. 15.1 are missing in your workspace, they can be added via the **View** menu (**Command Window**, **Command History**, **Current Directory**, **Workspace**, **Launch Pad**).

Commands should be typed in the **Command Window**. Symbol ≫ shows the current command. This prompt sign should not be typed. To browse the command window area, it is convenient to use the scroll bar or the cursor keys **Home**, **End**, **PageUp**, **PageDown**.

If for any reason you have moved far away from the command line with the flashing cursor, just press **Enter** to get back. It is important to remember that any command or expression typed in the command window must be followed by pressing **Enter**. This instructs MATLAB to execute the command or to evaluate an expression.

15.1.2 Basic Calculations

Type `1+2` at the prompt and press **Enter**. The command window will display the result shown in Fig. 15.2.

What MATLAB did? Firstly, it calculated the sum of `1+2` and saved it in a special variable `ans`, the value of which `(3)` was displayed in the command window. The command line with a flashing cursor below the result indicates that MATLAB is ready for further calculations. New expressions can be typed for MATLAB to evaluate. If you want to continue working with the previous expression, for example, to calculate `(1+2)/4.5`, the easiest way is to use the existing result, which is stored in variable `ans`. Just type `ans/4.5` (a period is used as a decimal point) and press **Enter**. There follows a calculation as shown in Fig. 15.3.

15.1.3 Echo

Each MATLAB command is accompanied by an echo. In the example above, the echo is `ans=0.6667`. Often the echo makes it difficult to follow a program execution. It can be disabled. To suppress the echo, the command line must be ended with a semicolon. An example is shown in Fig. 15.4.

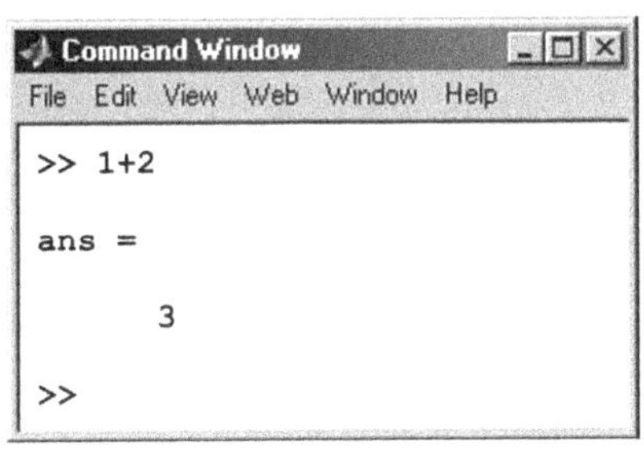
Command Window
File Edit View Web Window Help
>> 1+2
ans =
3
>>

Figure 15.2 Simple calculations.

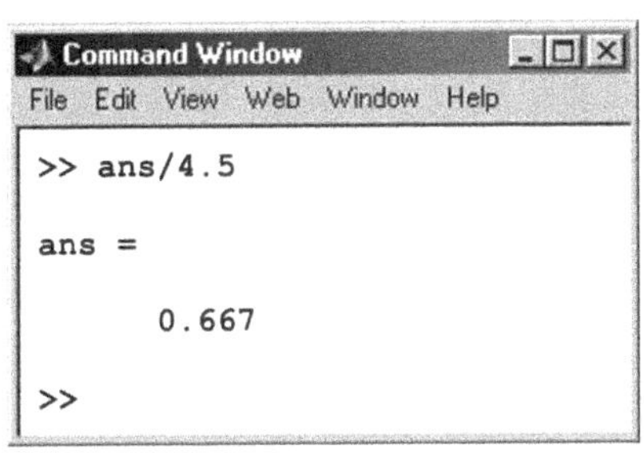
Command Window
File Edit View Web Window Help
>> ans/4.5
ans =
0.667
>>

Figure 15.3 Using `ans`.

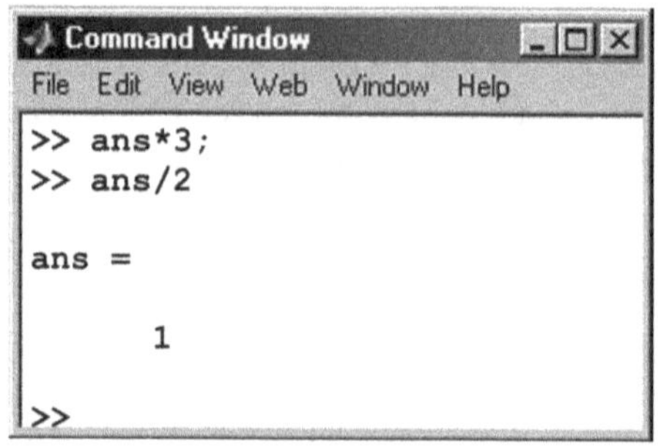
Command Window
File Edit View Web Window Help
>> ans*3;
>> ans/2
ans =
1
>>

Figure 15.4 Echo suppressing.

15.1.4 Workspace Saving: MAT-Files

The easiest way to save a workspace together with its variables is by using the submenu **Save Workspace As** in the **File** menu. The opening dialog box **Save Workspace Variables** will prompt the user to specify the directory and file name. A default location for the files to be stored is a subfolder **work** in the root MATLAB folder. The workspace is saved in a file with extension **mat**. After that MATLAB can be closed. To restore the workspace, the saved MAT-file can be opened via the submenu **Open** in the **File** menu. Now, all the variables used in the previous working session become available and can be used in the commands further on.

15.1.5 Diary

MATLAB provides a possibility to record both the executed commands and the results in a text file (to keep a log of work), which can be read or printed using an ordinary text editor. The command `diary` is used to start logging. The command argument specifies the name of the file where the log is stored. From this moment, all keyboard inputs and outputs will be recorded in the file. For example, a sequence of commands shown in Fig. 15.5. Performs the following actions:

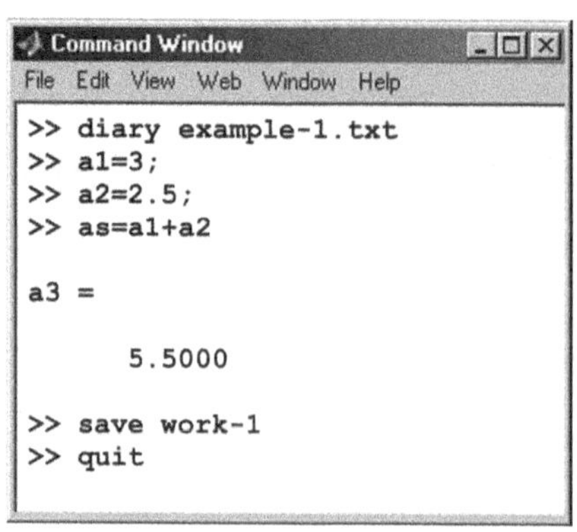
Command Window
File Edit View Web Window Help
>> diary example-1.txt
>> a1=3;
>> a2=2.5;
>> as=a1+a2
a3 =
5.5000
>> save work-1
>> quit

Figure 15.5 Example of `diary`.

1. Open the log in file `exampl-1.txt`;
2. Perform calculations;
3. Save the workspace in file `work-1.mat`;
4. Save the log file `exampl-1.txt` in subfolder **work** of the root MATLAB folder and close MATLAB.

The file `exampl-1.txt` can be viewed with any text editor. The file content will look as follows.

```
a1=3;
a2=2.5;
a3=a1+a2

a3 =

    5.5000

save work-1
quit
```

15.1.6 Help

MATLAB's help system can be reached through the **Help Window** item in the **Help** menu or by clicking the question mark button (?) on the toolbar. The same operation can be performed by typing the **`helpwin`** command. To get help on a specific topic you should type **`helpwin topic`**. The help window provides the same information as the command `help`, but the window-based interface provides a more convenient access to other help topics. Additional information can be viewed on the Math Works web page.[1]

15.2 MATRICES

15.2.1 Scalars, Vectors, and Matrices

Scalars, vectors, and matrices can be used in MATLAB. To enter a scalar, it is enough to assign a value to a variable, as shown in Fig. 15.6.

It is important that MATLAB distinguishes between uppercase and lowercase letters; therefore, `p` and `P` are treated as two different variables. To input an array (a vector or a matrix), its elements should be presented in brackets. For example, a (1×3) row vector can be input using the command shown in Fig. 15.7.

Note the spaces (or commas) that separate the elements. Column vector elements should be separated by a semicolon as shown in Fig. 15.8.

It is convenient to input small-sized matrices from the command line directly. A matrix can be presented as a column vector, the elements of which are row vectors. An example is shown in Fig. 15.9.

Alternatively, a matrix can be presented as a row vector, the elements of which are column vectors, as shown in Fig. 15.10.

[1] http://www.mathworks.com/products/matlab/

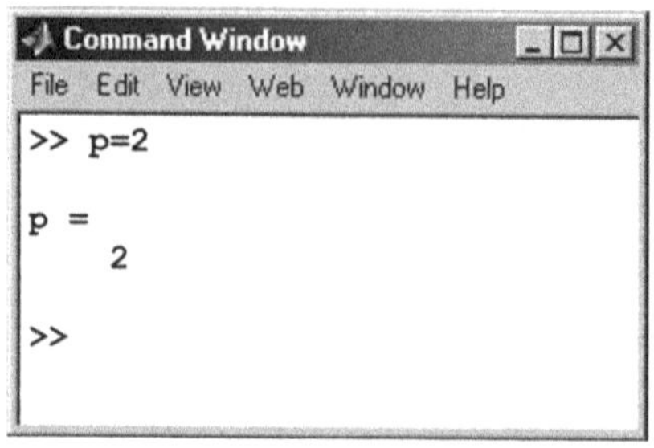
Command Window
File Edit View Web Window Help
>> p=2
p =
2
>>

Figure 15.6 Scalar input.

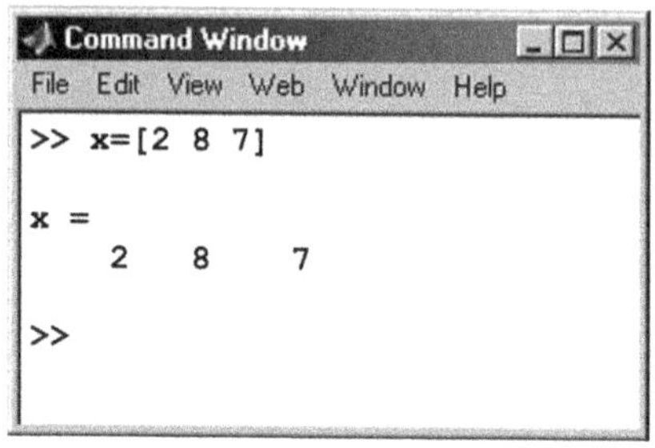
Command Window
File Edit View Web Window Help
>> x=[2 8 7]
x =
2 8 7
>>

Figure 15.7 Row vector input.

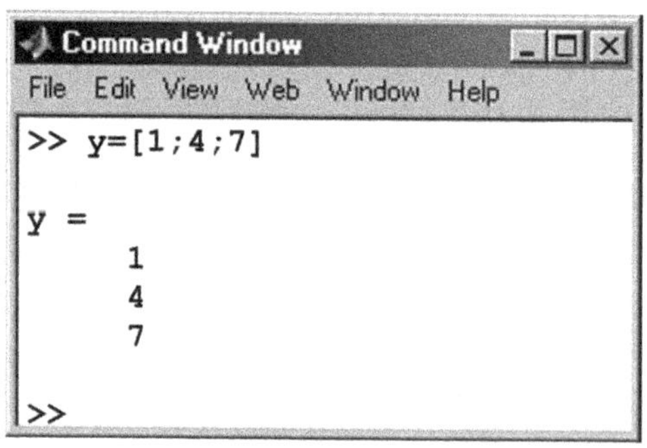
Command Window
File Edit View Web Window Help
>> y=[1;4;7]
y =
1
4
7
>>

Figure 15.8 Column vector input.

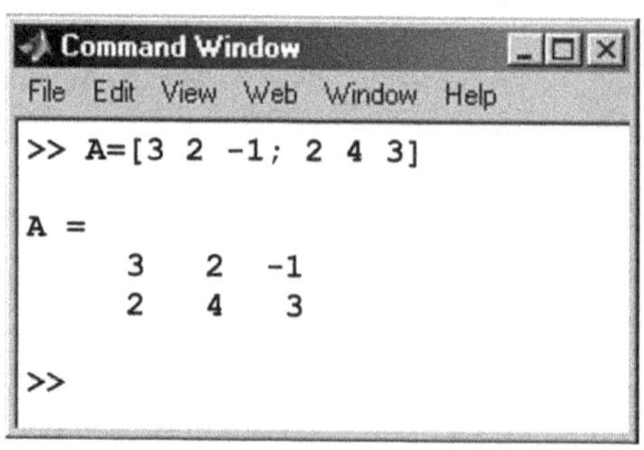
Command Window
File Edit View Web Window Help
>> A=[3 2 -1; 2 4 3]
A =
3 2 -1
2 4 3
>>

Figure 15.9 Matrix input. Method 1.

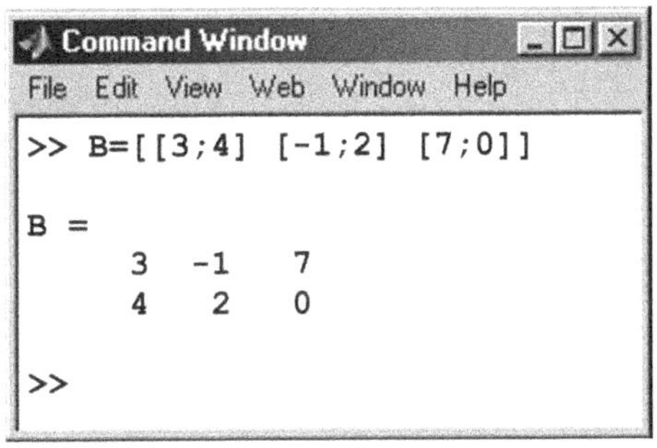
Command Window
File Edit View Web Window Help
>> B=[[3;4] [-1;2] [7;0]]
B =
3 -1 7
4 2 0
>>

Figure 15.10 Matrix input. Method 2.

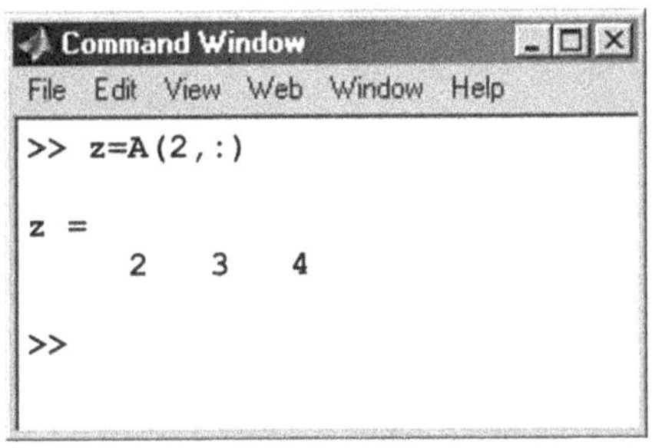
Command Window
File Edit View Web Window Help
>> z=A(2,:)
z =
2 3 4
>>

Figure 15.11 Access to a matrix row.

15.2.2 Accessing Matrix Elements

Matrix elements can be referred to by means of two indices, a row and a column, the numbers of which are enclosed in parentheses. For example, command `B(2,3)` returns the element from the second row and the third column of matrix `B`. To access a matrix column (or a row), one index should indicate the number of a column (or a row), whereas the other one should be replaced by a colon. In the example presented in Fig. 15.11, the command copies the second row of matrix `A` to vector `z`.

Matrix blocks can be accessed in a similar way. For example, let us extract a marked block from matrix `P` (see Fig. 15.12).

The command **`whos`** displays the workspace variables as shown in Fig. 15.13.

Note that the workspace contains one scalar (`p`), four matrices (`A`, `B`, `P`, `P1`), and one row vector (`z`).

15.2.3 Basic Matrix Operations

Matrix operations require working with correct dimensions. Addition or subtraction is only possible for matrices of the same size. If multiplying, the number of columns in the first matrix must be equal to the number of rows in the second matrix. Addition and subtraction of matrices, as well as operations with scalars and vectors, are performed by using the plus and minus signs. Examples are shown in Fig. 15.14.

Multiplication of matrices is done with the asterisk `*` sign as presented in Fig. 15.15.

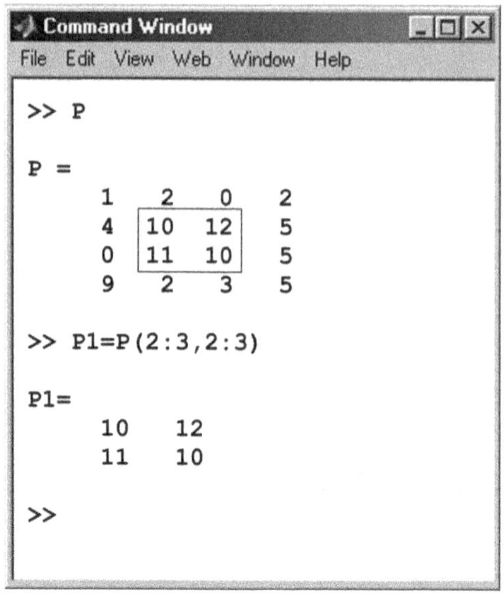

Figure 15.12 Access to a block.

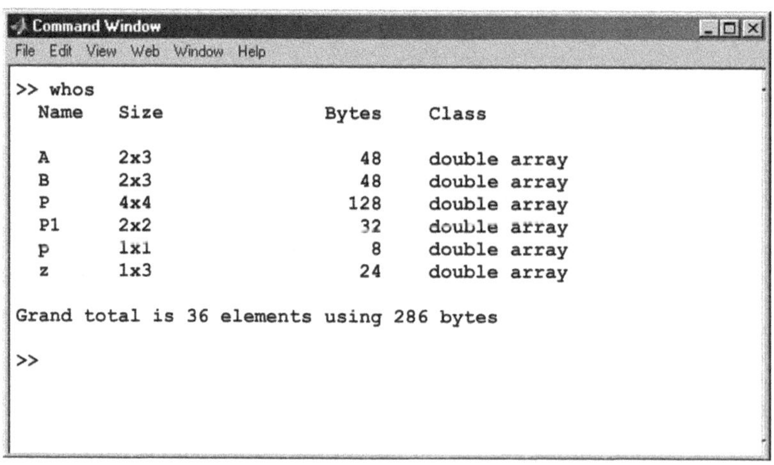

Command Window
File Edit View Web Window Help
>> whos
Name Size Bytes Class
A 2x3 48 double array
B 2x3 48 double array
P 4x4 128 double array
P1 2x2 32 double array
p 1x1 8 double array
z 1x3 24 double array
Grand total is 36 elements using 286 bytes
>>

Figure 15.13 Using `whos`.

Multiplication of a matrix by a number is also performed with an asterisk. Both left and right multiplications are possible.

Bringing a square matrix into a power is performed using the operator `^`. An example is given in Fig. 15.16.

The result can be verified by multiplying matrix `P` by itself.

15.2.4 Special Matrices

Filling a rectangular matrix with zeros is done by the **`zeros`** command as shown in Fig. 15.17.

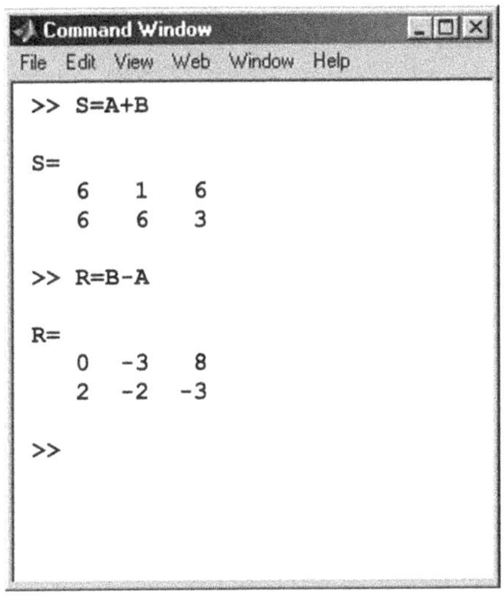
Command Window
File Edit View Web Window Help
>> S=A+B
S=
6 1 6
6 6 3
>> R=B-A
R=
0 -3 8
2 -2 -3
>>

Figure 15.14 Matrices addition.

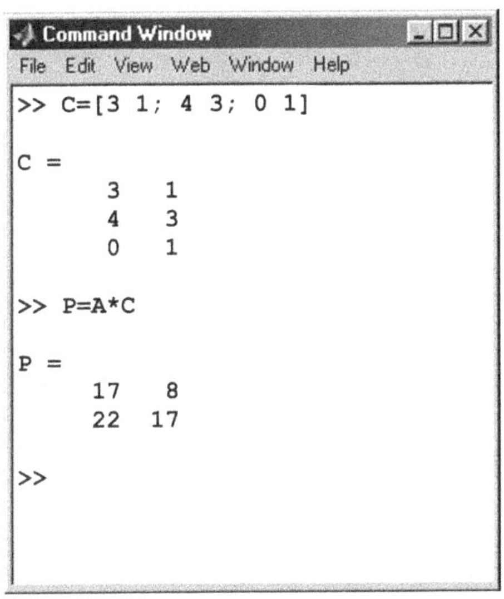
Command Window
File Edit View Web Window Help
>> C=[3 1; 4 3; 0 1]
C =
3 1
4 3
0 1
>> P=A*C
P =
17 8
22 17
>>

Figure 15.15 Matrices multiplication.

The identity matrix is created by using the **eye** command as illustrated in Fig. 15.18.

The matrix consisting of ones is formed by the **ones** command (see Fig. 15.19).

MATLAB provides a possibility to fill a matrix with random numbers. The output of the **rand** command is a matrix, the elements of which are uniformly distributed between zero and one. The output of **randn** is a matrix, the elements of which have the standard normal distribution with zero mean and unit variance.

The **Diag** command produces a diagonal matrix from a vector in such a way that the vector's elements are placed on the diagonal of the matrix.

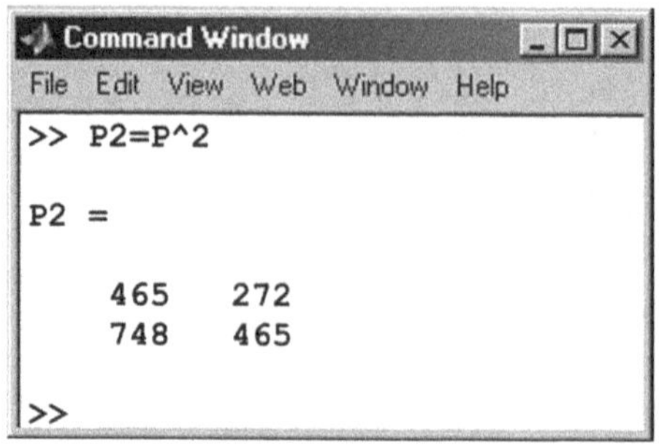

Figure 15.16 Powering of a matrix.

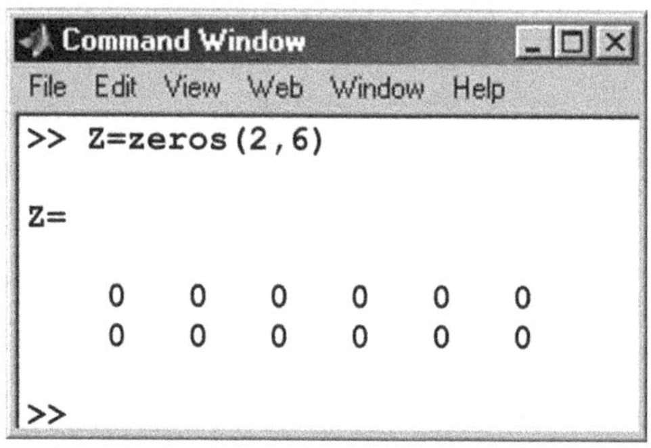

Figure 15.17 Using `zeros`.

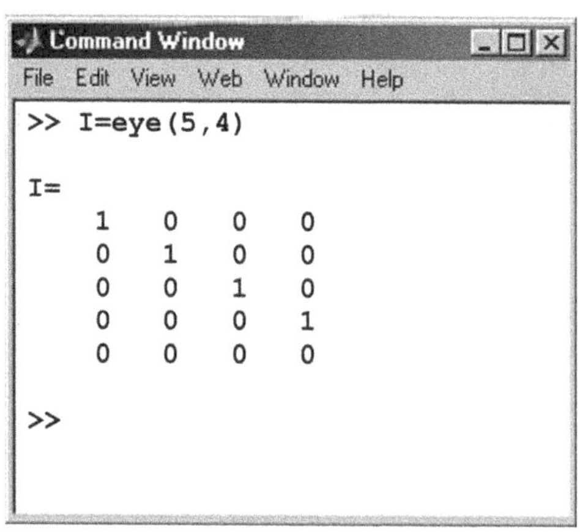

Figure 15.18 Using `eye`.

15.2.5 Matrix Calculations

MATLAB offers many different options for matrix calculations For example, matrix transposition is performed using the apostrophe `'` sign, as shown in Fig. 15.20.

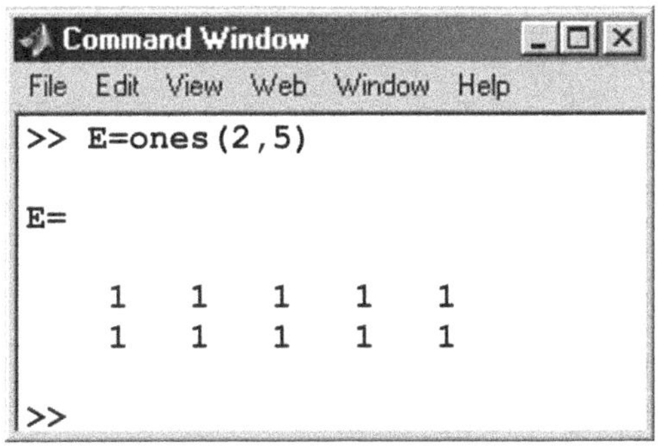
Command Window
File Edit View Web Window Help
>> E=ones(2,5)
E=
1 1 1 1 1
1 1 1 1 1
>>

Figure 15.19 Using **ones**.

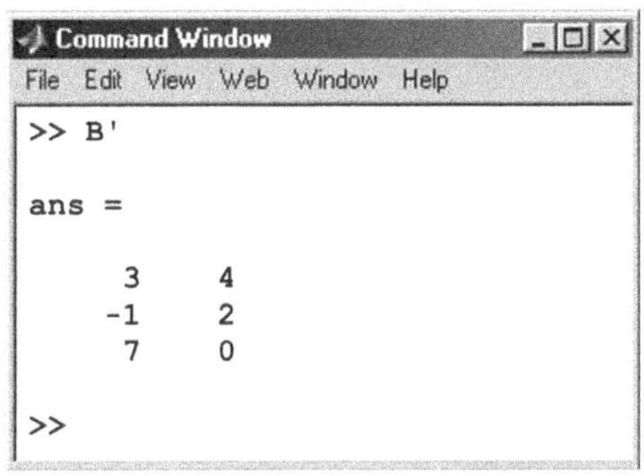
Command Window
File Edit View Web Window Help
>> B'
ans =
3 4
-1 2
7 0
>>

Figure 15.20 Matrix transposition.

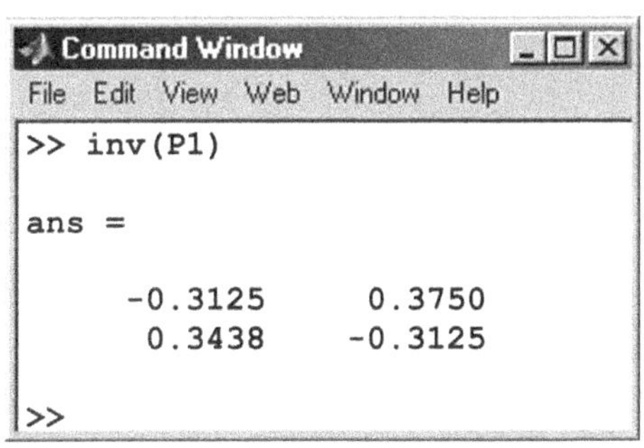
Command Window
File Edit View Web Window Help
>> inv(P1)
ans =
-0.3125 0.3750
0.3438 -0.3125
>>

Figure 15.21 Matrix inversion.

An inverse matrix is obtained by the **inv** function, which is valid for the squared matrices only (see Fig. 15.21).

Pseudoinverse matrices (Section 5.1.15) can be calculated by using the **pinv** function.

More information on matrix calculations can be found in the list of data functions available through the **help datafun** command.

15.3 INTEGRATING EXCEL AND MATLAB®

MATLAB and Excel integration provides a direct access from Excel to numerous MATLAB functions for data processing, computation, and visualization. A standard add-in **excllink.xla** executes this integration.

15.3.1 Configuring Excel

Firstly, make sure that Spreadsheet Link EX software has been installed together with the MATLAB package. If so, the add-in file **excllink.xla** can be found in the folder **exclink**, in folder **toolbox** or in the MATLAB root folder. On starting Excel, click **Add-ins** in the **Tools** menu. A dialog box containing the list of available add-ins will open. Use the **Browse** button to specify the path to **excllink.xla**. A flagged string **Excel Link 2.0 for use with MATLAB** will be displayed in the list of available add-ins. Click **OK** to load the add-in into Excel.

Note a new Excel toolbar **Excel Link**, which contains three buttons: **putmatrix**, **getmatrix**, and **evalstring**. These buttons implement the basic steps required to make a link between Excel and MATLAB, namely the matrix exchange and the MATLAB commands execution from Excel. When Excel is restarted, the add-in **excllink.xla** will be loaded automatically.

Joint Excel/MATLAB work requires a few more adjustments, which are incorporated into Excel by default (but can be changed). Select **Options** from the Excel **Tools** menu, go to the tab **General**, and make sure that the **R1C1 reference style** option is not selected. It means the cells are referenced as A1, A2, etc. Make sure that the **Move selection after Enter** option is selected in the **Edit** tab.

15.3.2 Data Exchange

Start Excel and make sure that all settings are set as described in the section above. MATLAB should be closed. Enter a matrix in the range A1:C3 as it is shown in Fig. 15.22.

Select this range and click the **putmatrix** button. An Excel warning window will appear saying that MATLAB is not running. Click **OK** and wait for MATLAB to start. After that a new Excel dialog box will appear. It will offer to name the MATLAB variable to which the data, exported from Excel, get assigned to. For example, enter `M` and close the window by pressing **OK**. Go to the MATLAB command window and make sure that the workspace has a new variable `M` that contains a (3 × 3) array (see Fig. 15.23).

Let us perform a MATLAB operation with a matrix `M`. For example, let us calculate the inverse of `M`. The MATLAB command **`Inv`** (similar to any other MATLAB command) can

	A	B	C
1	5.5	1.6	-0.8
2	2.3	6.1	0.2
3	0.1	0.4	3.9

Figure 15.22 Matrix to be exported.

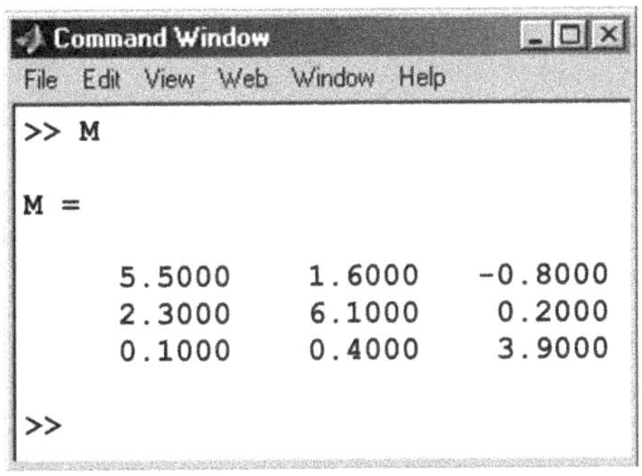

Figure 15.23 Matrix imported to MATLAB.

be called directly from Excel. A click on the **evalstring** button in the **Excel Link** toolbar opens an Excel input box used to type a MATLAB command, for example,

$$\texttt{IM} = \texttt{inv(M)}.$$

Check that the Excel result is identical to that obtained in MATLAB.

Go back to Excel, activate cell A5, and click **getmatrix**. An input dialog box appears where the name of a MATLAB variable imported into Excel should be typed into. In our case, the variable's name is `IM`. Click **OK** and the range A5:A7 will be filled with the inverse matrix.

Thus, to export an Excel matrix into a MATLAB, the whole matrix range should be selected in a spreadsheet. To import a MATLAB variable into Excel, it is enough to select a single cell, which will be the upper left element of the imported array. Other array elements are pasted into adjacent cells in accordance with the array's size. The existing values in the cells will be overwritten, hence be careful when importing arrays.

The approach described above is the simplest way to share information between the applications, that is, the raw data from Excel are exported to MATLAB, where they are processed somehow, and the result is imported back into Excel. The data exchange is performed using the **Excel Link** toolbar. The data can be presented as a matrix, that is, a rectangular area in a worksheet. The values in rows or in columns are converted into MATLAB row and column vectors, respectively. Import to Excel happens in a similar manner.

15.4 PROGRAMMING

15.4.1 M-Files

The command line style of MATLAB becomes cumbersome if there is a need to type many commands and to change them repeatedly. Using the **`diary`** command for logging and saving the workspace does not really resolve the problem. The most convenient way to run groups of the MATLAB commands is to use an M-file. These files store commands, execute them all at once or in parts, and save them for future use. The M-files are created and edited in a special Editor.

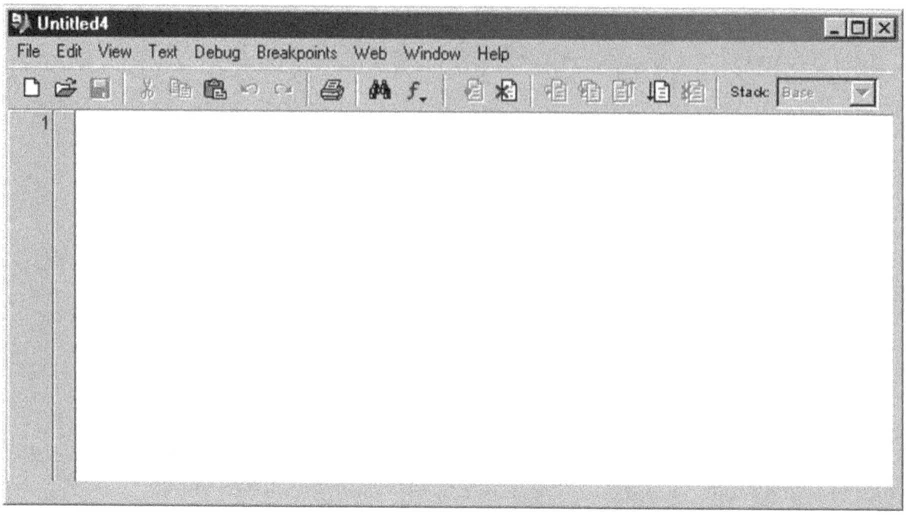

Figure 15.24 MATLAB Editor.

Expand the **File** menu and select the **M-file** item in the submenu **New**. A new file will be opened in the Editor window, the example of which is shown in Fig. 15.24.

Two types of M-files exist in MATLAB. Program files (*Script*), which contain a sequence of commands and function files (*Function*), which present the user-defined functions.

15.4.2 Script File

Open the Editor and type commands that instruct the display of two plots in the same graph window (see Fig. 15.25).

Use the **Save as File** menu item and save the file under the name **mydemo.m** in the subfolder **work** of the root MATLAB folder. To run the script saved in the file, select the

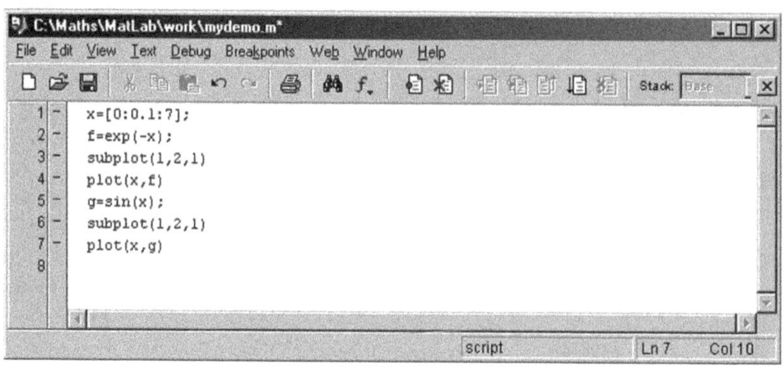

Figure 15.25 Script example.

Run item in the **Debug** menu. This creates a graph window entitled *Figure 1* that contains the plots *Figure 1, 2 ...* are the default names of the graph windows produced by MATLAB. See Fig. 15.30 as an example.

The output of the Script file is shown in the command window. If a command ends with a semicolon, its output is suppressed. If there is a typo in a command and MATLAB cannot recognize it, the execution stops at the corresponding incorrect line and an error message pops up.

A very useful functionality offered by the Editor is the execution of a part of the commands. Close the graph window *Figure 1*. Select the first four commands in the Editor (using the mouse or the arrow keys) and press **Evaluate** using the **Text** menu. Please note that the graphical window contains only one plot now, which is in line with the selected commands.

M-files may be supplemented by comments, which will be ignored during execution. A comment line starts with a percent sign. Comments are automatically highlighted in green in the Editor. An example is shown in Fig. 15.26.

An M-file can be opened via the menu item **Open File** of the workspace or in the Editor.

15.4.3 Function File

The Script file described above is only a sequence of MATLAB commands. It has no input arguments and does not return any values. While developing your own applications in MATLAB, it is necessary to use function files that perform some actions using input arguments and producing results. Let us consider some simple examples that demonstrate function files implementation.

Data centering (see Section 8.1.4) is among the most popular data preprocessing operations. Therefore, there is a good reason to compose a function file, which can be used whenever this action is necessary.

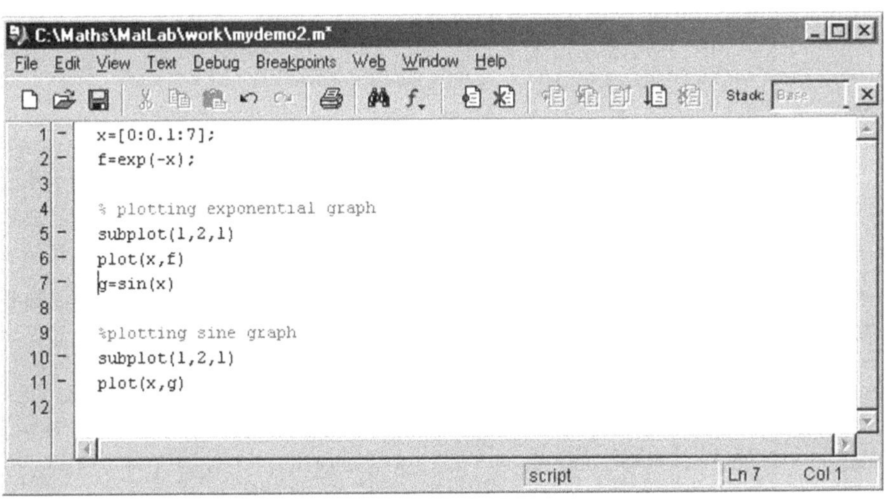

Figure 15.26 Comments example.

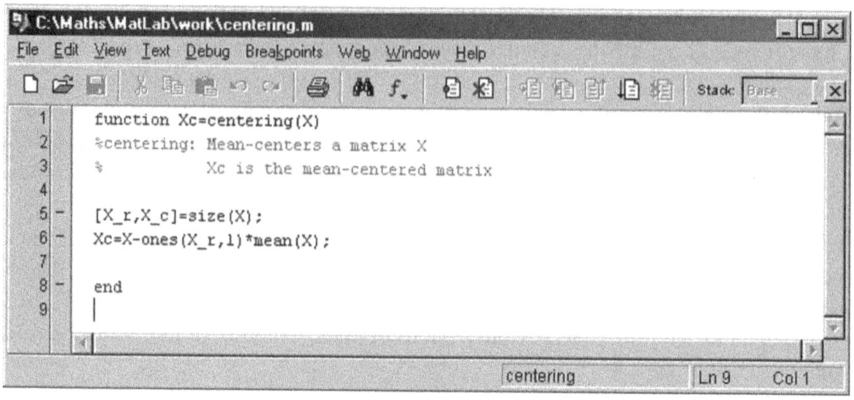

Figure 15.27 Data centering function.

The word `function` in the first line of Fig. 15.27 indicates that this is a function file. This is the header, which includes the function's name and a list of input and output arguments. In this example, the function's name is `centering`. It has one input argument `X` and one output value `Xc`. A few lines below the header contain comments and then the function body follows. In the present example, the body consists of two lines, which calculate the output value. It is important that the result is assigned to variable `Xc`, which is declared in the header line. Note the semicolons that prevent the display of the intermediate screen outputs. Now, save this file in a working folder. Please note that the menu items **Save** and **Save as** open a dialog box in which the **File name** field already contains the name `centering`. Do not change it and save the function file under the name proposed.

This function can be now used in the same way as any standard function, such as `sin` and `cos`. The user-defined functions can be called from a Script file or from another function file. Try writing a function file that scales a matrix, that is, that divides each column by the standard deviation of this column.

A function file can include several input arguments separated by commas. It can also return a number of values. To declare this, the output arguments should be enclosed in brackets and separated by commas. A good example (see Fig. 15.28) is a function that converts the time interval specified in seconds into hours, minutes, and seconds.

When a multiple output function is run, the produced result should be a set of variables or a vector of an appropriate length.

15.4.4 Plotting

MATLAB provides many opportunities for plotting vectors and matrices, including labeling and printing. Some of the important graphical functions are explained below. The **`Plot`** function offers various forms associated with input parameters. For example `plot(y)` creates a piecewise-linear graph of `y` elements plotted against their indices. If two vector arguments are used, such as in `plot(x,y)`, the result is a plot of `y` versus `x`. In the example shown in Fig. 15.29, the following commands are used to plot a graph of `sin` in the range $[0, 2\pi]$.

The program creates a plot, which is shown in Fig. 15.30.

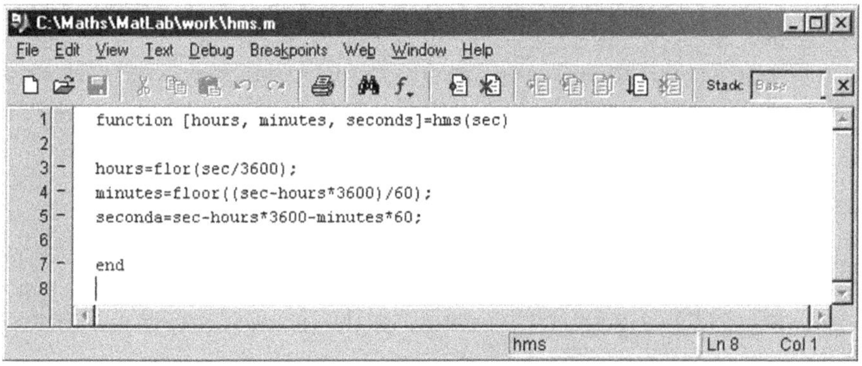

C:\Maths\MatLab\work\hms.m
File Edit View Text Debug Breakpoints Web Window Help
Stack Base
function [hours, minutes, seconds]=hms(sec)
hours=flor(sec/3600);
minutes=floor((sec-hours*3600)/60);
seconda=sec-hours*3600-minutes*60;
end
hms
Ln 8 Col 1

Figure 15.28 Example of a function returning several values.

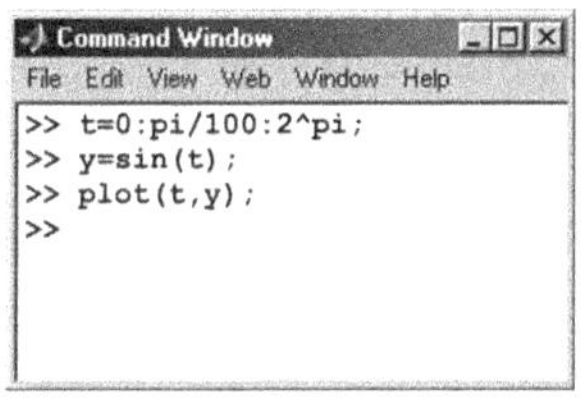

Figure 15.29 Graph example.

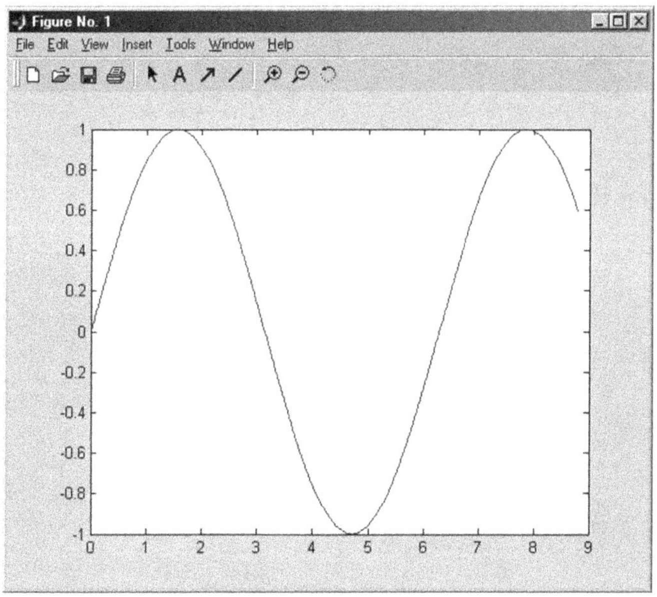

Figure 15.30 Graph window.

MATLAB assigns a specific color to each plotted curve automatically, except for when the user defined it manually. This feature helps to distinguish between different datasets.

The command **hold on** preserves the current plot, and the subsequent graphic commands add new curves to the existing graph. The **Subplot** function (see Fig. 15.31) adds new plots to the existing graphical window providing a possibility to display many plots in one figure as shown in Fig. 15.32.

15.4.5 Plot Printing

Both the **Print** item in the **File** menu and the **print** command print MATLAB plots. The **Print** item opens a dialog box with general printing options. The **print** command provides more flexibility and controls printing from M-files. The output can be sent to the selected printer or saved in a specified file.

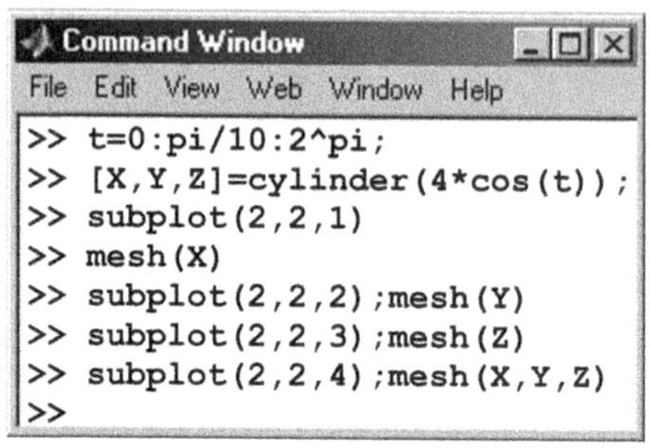

Command Window
File Edit View Web Window Help
>> t=0:pi/10:2^pi;
>> [X,Y,Z]=cylinder(4*cos(t));
>> subplot(2,2,1)
>> mesh(X)
>> subplot(2,2,2);mesh(Y)
>> subplot(2,2,3);mesh(Z)
>> subplot(2,2,4);mesh(X,Y,Z)
>>

Figure 15.31 Example of a multiplot output.

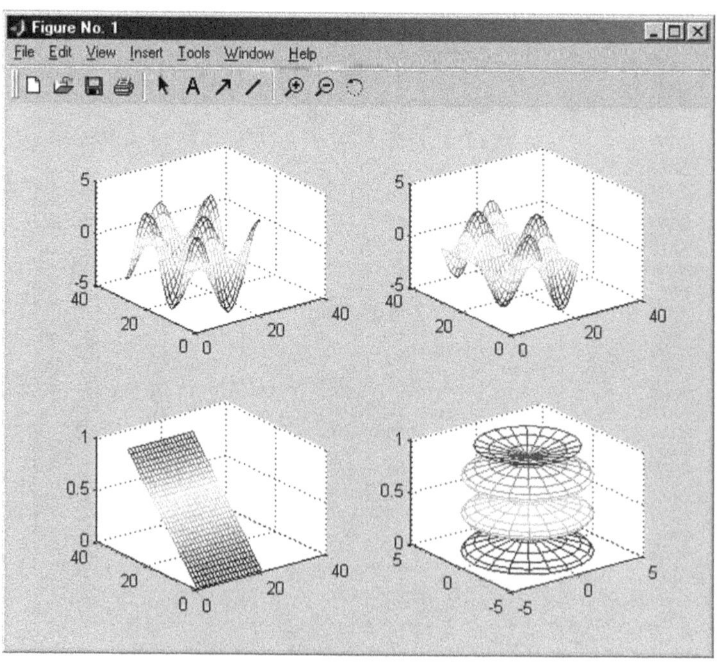

Figure 15.32 Multiplot window.

15.5 SAMPLE PROGRAMS

This section describes the main algorithms used for multivariate data analysis, including the simplest methods of data transformation, such as centering and scaling as well as algorithms for data analysis, such as principal component analysis (PCA) and projection on the latent structures (PLS).

15.5.1 Centering and Scaling

Often, the raw data should be transformed before conducting the analysis. The most popular methods in use are centering and scaling. Section 15.4.3 describes a function applied to matrix centering. The function presented in Fig. 15.33 performs data scaling.

15.5.2 SVD/PCA

The most popular method of the multivariate data compression is PCA described in Chapter 9. Mathematically, PCA can be viewed as a decomposition of **X**, that is, its presentation as a product of two matrices **T** and **P**

$$\mathbf{X} = \mathbf{T}\mathbf{P}^t + \mathbf{E}$$

Matrix **T** is called the *scores matrix*, matrix **P** is called the *loadings matrix*, and **E** is the matrix of residuals. The easiest way to find matrices **T** and **P** is to apply the singular value decomposition (SVD) (Section 5.2.11) using the standard MATLAB function **svd**. Figure 15.34 provides an illustration.

```
function Xs = scaling(X)
% scaling: the output matrix is Xs
% matrix X must be centered

Xs = X * inv(diag(std(X)));

%end of scaling
```

Figure 15.33 Scaling function.

```
function [T, P] = pcasvd(X)
% pcasvd: calculates PCA components.
% The output matrices are T and P.
% T contains scores
% P contains loadings
[U,D,V] = svd(X);
T = U * D;
P = V;
%end of pcasvd
```

Figure 15.34 pcasvd function.

15.5.3 PCA/NIPALS

Developing one's own function for the PCA decomposition can be viewed as a better didactic approach. This can be done by using a recursive non-linear iterative partial least squares (NIPALS) algorithm, which calculates one PCA component at each step. First of all, the initial matrix **X** is transformed (at least centered, see Section 15.4.3) and assigned to matrix $\mathbf{E}_0$, $a = 0$. Subsequently, the following algorithm is applied.

1. Select initial vector **t**
2. $\mathbf{p}^t = \mathbf{t}^t\mathbf{E}_a/\mathbf{t}^t\mathbf{t}$
3. $\mathbf{p} = \mathbf{p}/(\mathbf{p}^t\mathbf{p})^{1/2}$
4. $\mathbf{t} = \mathbf{E}_a\mathbf{p}/\mathbf{p}^t\mathbf{p}$
5. Check convergence, if not, go to step 2

After computing the a-th component, let us set $\mathbf{t}_a = \mathbf{t}$ and $\mathbf{p}_a = \mathbf{p}$. To obtain the next component, it is necessary to calculate the residuals $\mathbf{E}_{a+1} = \mathbf{E}_a - \mathbf{t}\mathbf{p}^t$ and subject them to the same algorithm replacing index a with $a + 1$.

The NIPALS algorithm code can be written by the readers themselves. Our version is presented in Fig. 15.35. Using this function, one can set the number of principal components

```
function [T, P] = pcanipals(X, numberPC)
% pcanipals: calculates PCA components.
% The output matrices are T and P.
% T contains scores
% P contains loadings
% calculation of number of components
[X_r, X_c] = size(X); P=[]; T=[];

if lenfth(numberPC) > 0
    pc = numberPC{1};
elseif (length(numberPC) == 0) & X_r < X_c
    pc = X_r;
else
    pc = X_c;
end;

% calculation of scores and loadings for each component
for k = 1:pc
    P1 = rand(X_c, 1); T1 = X * P1; d0 = T1'*T1;
    P1 = (T1' * X/(T1' * T1))'; P1 = P1/norm(P1); T1 = X
* P1; d = T1' * T1;

    while d - d0 > 0.0001;
        P1 = (T1' * X/(T1' * T1)); P1 = P1/norm(P1); T1
= X * P1; d0 = T1'*T1;
        P1 = (T1' * X/(T1' * T1)); P1 = P1/norm(P1); T1
= X * P1; d = T1'*T1;
    end

    X = X - T1 * P1; P = cat(1, P, P1'); T = [T,T1];
end
```

Figure 15.35 pcanipals function.

(PCs) (`numberPC` argument). If the number of PCs is unknown, one should write [P,T] = `pcanipals(X)` in the command line. In this case, the program sets the number of PCs equal to the smallest dimension of matrix **X**.

Computation of PCA with Chemometrics Add-In is explained in Chapter 8.

15.5.4 PLS1

PLS is the most popular method for multivariate calibration. This method performs a simultaneous decomposition of the predictor matrix **X** and the response matrix **Y**:

$$\mathbf{X} = \mathbf{TP}^t + \mathbf{E}, \quad \mathbf{Y} = \mathbf{UQ}^t + \mathbf{F}, \quad \mathbf{T} = \mathbf{XW}(\mathbf{P}^t\mathbf{W})^{-1}$$

The projections are constructed cojointly in such a way that it leads to a maximum correlation between the **X**-scores vector $\mathbf{t}_a$ and the **Y**-scores vector $\mathbf{u}_a$. If the **Y** block has several responses (i.e., $K > 1$), two projections PLS1 and PLS2 can be obtained. In the former case, each response $\mathbf{y}_k$ generates its own projection subspace. In this approach (called *PLS1*), both scores **T** (**U**) and loadings **P** (**W, Q**) depend on the response in use. The PLS2 method builds only one projection space common for all responses. A more detailed explanation can be found in Section 8.1.3.

The NIPALS algorithm is also used to construct PLS1. Firstly, the initial matrices **X** and **Y** are centered (see Section 15.4.3), and the results are denoted by matrix $\mathbf{E}_0$ and vector $\mathbf{f}_0$, $a = 0$. Afterward, they are subjected to the following algorithm

1. $\mathbf{w}^t = \mathbf{f}_a^t\mathbf{E}_a$
2. $\mathbf{w} = \mathbf{w}/(\mathbf{w}^t\mathbf{w})^{1/2}$
3. $\mathbf{t} = \mathbf{E}_a\mathbf{w}$
4. $q = \mathbf{t}^t\mathbf{f}_a/\mathbf{t}^t\mathbf{t}$
5. $\mathbf{u} = q\mathbf{f}_a/q^2$
6. $\mathbf{p}^t = \mathbf{t}^t\mathbf{E}_a/\mathbf{t}^t\mathbf{t}$

When the a-th component has been computed, $\mathbf{t}_a = \mathbf{t}$ and $\mathbf{p}_a = \mathbf{p}$ are assigned. To obtain the next component, the residuals $\mathbf{E}_{a+1} = \mathbf{E}_a - \mathbf{tp}^t$ should be calculated and then subjected to the same algorithm replacing index a with $a + 1$. Figure 15.36 shows a MATLAB code of the algorithm.

The calculation of scores $\mathbf{T}_{new}$ and $\mathbf{U}_{new}$ for new samples ($\mathbf{X}_{new}$, $\mathbf{Y}_{new}$) is more complicated in the PLS method than in PCA. Using PLS1, matrix $\mathbf{X}_{new}$ and vector $\mathbf{y}_{new}$ should be preprocessed in the same way as the training set (**X**, **y**), and the results are labeled $\mathbf{E}_0$ and $\mathbf{f}_0$, $a = 0$. Then they are subjected to the following algorithm that computes the scores for $a = 1, \ldots, A$.

1. $\mathbf{t} = \mathbf{E}_a\mathbf{w}$
2. $\mathbf{u} = \mathbf{f}_a$

After computing the a-th PLS1 component, it is necessary to calculate new residuals $\mathbf{E}_{a+1} = \mathbf{E}_a - \mathbf{tp}^t$ and $\mathbf{f}_{a+1} = \mathbf{f}_a - q\mathbf{t}$. Afterward the residuals are again subjected to the algorithm above, replacing index a with $a + 1$.

The PLS1 calculation by Chemometrics Add-In is discussed in Section 8.4.

```
function [w, t, u, q, p] = pls(x, y)
%PLS: calculates a PLS component.
%The output vectors are w, t, u, q and p.
%
% Choose a vector from y as starting vector u.

   u = y(:, 1);

% The convergence criterion is set very high.
   kri = 100;

% The commands from here to end are repeated
% until convergence.
   while (kri > 1e - 10)

% Each starting vector u is saved as uold.
      uold = u; w = (u' * x)'; w = w/norm(w);
      t = x * w; q = (t' * y)'/(t' * t);
      u = y * q/(q' * q);

% The convergence criterion is the norm of u-uold
% divided by the norm of u.
      kri = norm(uold - u)/norm(u);
   end;

% After convergence, calculate p.
   p = (t' * x)'/(t' * t);
% End of pls
```

Figure 15.36 pls function.

15.5.5 PLS2

The PLS2 algorithm is as follows. Firstly, the initial matrices **X** and **Y** are preprocessed (at least centered, see Section 15.4.3), and the resulting matrices are denoted by $\mathbf{E}_0$ and $\mathbf{F}_0$, $a = 0$. Then they are subjected to the following algorithm.

1. Select the initial vector **u**
2. $\mathbf{w}^t = \mathbf{u}^t\mathbf{E}_a$
3. $\mathbf{w} = \mathbf{w}/(\mathbf{w}^t\mathbf{w})^{1/2}$
4. $\mathbf{t} = \mathbf{E}_a\mathbf{w}$
5. $\mathbf{q}^t = \mathbf{t}^t\mathbf{F}_a/\mathbf{t}^t\mathbf{t}$
6. $\mathbf{u} = \mathbf{F}_a\mathbf{q}/\mathbf{q}^t\mathbf{q}$
7. Check convergence, if not, go to step 2
8. $\mathbf{p}^t = \mathbf{t}^t\mathbf{E}_a/\mathbf{t}^t\mathbf{t}$

After the a-th PLS2 component is computed, the following values are set: $\mathbf{t}_a = \mathbf{t}$, $\mathbf{p}_a = \mathbf{p}$, $\mathbf{w}_a = \mathbf{w}$, $\mathbf{u}_a = \mathbf{u}$, and $\mathbf{q}_a = \mathbf{q}$. To obtain the next PLS2 component, it is necessary to calculate the residuals $\mathbf{E}_{a+1} = \mathbf{E}_a - \mathbf{t}\mathbf{p}^t$ and $\mathbf{F}_{a+1} = \mathbf{F}_a - \mathbf{t}\mathbf{q}^t$ and subject them to the same algorithm, replacing index a with $a + 1$.

The MATLAB program for PLS2 is shown in Fig. 15.37.

In order to calculate the scores $\mathbf{T}_{new}$ and $\mathbf{U}_{new}$ for new samples ($\mathbf{X}_{new}$, $\mathbf{Y}_{new}$), the matrices $\mathbf{X}_{new}$ and $\mathbf{Y}_{new}$ should be preprocessed in the same way as the training set (**X**, **Y**). The

```
function [W, T, U, Q, P, B, SS] = plsr(x, y, a)
% PLS: calculates a PLS component.
% The output matrices are W, T, U, Q and P.
% B contains the regression coefficients and SS the sums of
% squares for the residuals.
% a is the numbers of components.
%
% For a components: use all commands to end.

   for i=1:a
% Calculate the sum of squares. Use the function ss.
      sx = [sx; ss(x)];
      sy = [sy; ss(y)];

% Use the function pls to calculate one component.
      [w, t, u, q, p] = pls(x, y);

% Calculate the residuals.
      x = x - t * p';
      y = y - t * q';

% Save the vectors in matrices.
      W = [W w];
      T = [T t];
      U = [U u];
      Q = [Q q];
      P = [P p];
   end;

% Calculate the regression coefficients after the loop.
   B=W*inv(P'*W)*Q';

% Add the final residual SS to the sum of squares vectors.
   sx=[sx; ss(x)];
   sy=[sy; ss(y)];

% Make a matrix of the ss vectors for X and Y.
   SS = [sx sy];

%Calculate the fraction of SS used.
   [a, b] = size(SS);
   tt = (SS * diag(SS(1,:).^(-1)) - ones(a, b)) * (-1)

%End of plsr
function [ss] = ss(x)
%SS: calculates the sum of squares of a matrix X.
%
   ss=sum(sum(x. * x));
%End of ss
```

Figure 15.37 plsr function.

results are denoted by $\mathbf{E}_0$ и $\mathbf{F}_0$, $a = 0$. Subsequently the latter are subjected to the following algorithm that computes the scores for $a = 1, \ldots, A$.

1. $\mathbf{t} = \mathbf{E}_a\mathbf{w}$
2. $\mathbf{u} = \mathbf{F}_a\mathbf{q}/\mathbf{q}^t\mathbf{q}$

After computing the a-th PLS2 component, it is necessary to calculate the residuals $\mathbf{E}_{a+1} = \mathbf{E}_a - \mathbf{tp}^t$ and $\mathbf{F}_{a+1} = \mathbf{F}_a - \mathbf{tq}^t$ and subject them to the same algorithm replacing index a with $a + 1$.

Calculation of PLS2 using Chemometrics Add-In is explained in Section 8.5.

CONCLUSION

MATLAB has a special place among the general-purpose tools used in chemometrics. MATLAB's popularity is enormous. This is due to the fact that MATLAB is a powerful and versatile tool to process multivariate data. The structure of the package makes it a convenient tool for matrix computations. The range of problems that can be solved with MATLAB includes matrix analysis, image and signal processing, neural networks, and many others. MATLAB is a high-level language, with an open source. This allows an experienced user to understand the programmed algorithms. After many years of using MATLAB, a huge number of standard functions and Toolboxes (a collection of specialized subroutines) have been created. The most popular in chemometrics is the PLS Toolbox, which is a product of Eigenvector Research, Inc.[2]

MATLAB is a very popular tool for data analysis. According to a survey, about one-third of the research community uses it. The Unscrambler program is employed only by 16% of scientists. The main drawback of MATLAB is its high software price. In addition, MATLAB is effective for routine calculations but lacks interactivity, which makes it unsuitable to perform explorative research of new datasets.

[2] http://www.eigenvector.com/

THE FOURTH PARADIGM

What is science? How does scientific knowledge arise? By what means can an urgent problem be solved? Which problems are urgent? How can a problem be posed? What can and what cannot be achieved by scientific research? All these issues are the eternal questions, the answers of which constitute a system of modern scientific methods, concepts, ideas, etc., briefly called a *paradigm*. Tomas Kuhn defines this term as "universally recognized scientific achievements that, for a time, provide model problems and solutions for a community of researchers."[1]

Looking back at the dark ages, we see that the first paradigm originated in the 5th to 3rd centuries BC. This was, without a doubt, the *geometric* paradigm. *Cognition means drawing*–this concept of scientific knowledge prevailed at the time of Euclid, Pythagoras, and Archimedes. Trisection of an angle, squaring of a circle, duplication of a cube–these great tasks agitated the minds of scientists from the ancient times until the late 18th century. The methods corresponded to the problems, for example, to draw by means of a compass and ruler. Such scientific knowledge was valued above all. Not without reason, Archimede's grave shows a cylinder with an inscribed sphere; he considered this discovery his greatest achievement. Centuries have passed. Apart from the first, geometric paradigm, other concepts had also appeared, but the first one is still prevalent. All scientists fit their papers with schemes, plots, graphics, and diagrams, and this book is not an exception. We believe that it is the true way to explain and illustrate our ideas in a clear and simple way. Not only ordinary scientists, even the great Gauss, who did so much within a differential paradigm, wanted a heptadecagon to be placed on his gravestone. However, the carver refused, saying it would be indistinguishable from a circle. Nevertheless, the pedestal of a statue erected in his honor in his home town of Braunschweig has the shape of a heptadecagon.

[1] T.S. Kuhn. *The Structure of Scientific Revolution*, 3rd Edition, University of Chicago Press, Chicago, 1996.

Chemometrics in Excel, First Edition. Alexey L. Pomerantsev.

The inexorable course of historical progress led to a paradigm shift. As new problems arose, the next *algebraic* paradigm was developed. *Cognition means counting*; this was a new scientific principle that appeared in the 10th to 15th centuries AD through the efforts of Pope Sylvester II, Luca Pacioli, and François Viète. Ruler and compass were replaced with the multiplication table, then with the table of logarithms, and then with the methods of solving the algebraic equations. A point was replaced by a number and a curve was replaced by an equation. Descartes completed the slender building of the algebraic paradigm by plunging geometry into the coordinate system. He showed that the old geometric paradigm is just a special case of a new, algebraic paradigm. We still love numbers, and this book gives numerous examples.

The third, modern paradigm, which can be called *differential*, appeared in the 17th century through the efforts of Newton and Leibniz. They created a new powerful differential calculus method and, since then, *cognition means differentiating*. This paradigm brought fantastic, incredible results. For the first time, scientists were able not only to capture and explain what is happening but also to predict the future. The differential equations were applied everywhere from calculation of the motion of celestial bodies, kinetics of the chemical, biological, and physical processes, modeling of economic and social phenomena, etc. Numerical methods for the solution of the differential equations were developed; the most powerful computers were set up to implement them. Everything seemed fine, but–

The combustion of hydrogen, that is, the reaction $2H_2 + O_2 = 2H_2O$, is probably the simplest chemical process. However, Prof. Vasily Dimitrov, in his wittily titled book *Simple Kinetics*, numbers 30 direct and 30 inverse stages of this process. Each stage is described by a differential equation, which together provides a system of 60 equations. Note that this is the easiest isothermal case not accounting for boundary effects but for heat and mass transfer. The solution of this system is difficult because most of the kinetic parameters are unknown, and they should be identified experimentally. Moreover, many stages are very fast, and they involve short-lived radicals. This is only the beginning of the problems because the system of equations is very stiff, riddled by numerous internal correlations. Finally, all the above should be somehow linked to practice. What can be done with much more complicated problems? For example, how to apply the differential paradigm to the problem of detection of counterfeited medicines? What equations can be used to describe the differences in physiology and preferences of the people of North and South Europe?[2]

All this testifies to the weakness of the differential paradigm, in which every phenomenon should be described by a meaningful (aka white) model that has a self-containing meaning based on fundamental prior knowledge. This approach is now replaced by a new, fourth paradigm on the basis of formal (or black) data-driven modeling. Data, and data only, is the source of new knowledge. This knowledge is hidden in data as a gold scintilla in ore, and the goal of a scientist is to extract it by data analysis. *Cognition means analyzing*. This, apparently, is the slogan of the fourth paradigm, birth of which we are witnessing. In contrast to the third differential paradigm, this approach does not issue a challenge to predict the behavior of a system beyond the experimental region. This new paradigm concentrates on the understanding and interpolation of the existing data on the attainment of the underlying causes in nature around us.

Referencing Karl Marx is now irrelevant. There is a strong opinion that the old chap misunderstood a lot and that he was crazy about the Means of Production concept. In other words, he was a narrow-minded historical materialist. But all the same ... In fact, in addition

[2]V.I. Dimitrov. *Simple Kinetics*, Nauka, Novosibirsk, 1982 (in Russian).

to the science paradigms, there were also production paradigms. And it turns out that those two are matching wonderfully; we can even say correlated–the slavery agronomics with geometric, the feudal manufacture with algebraic, the capitalistic conveyor with the differential. The meaning of this hidden variable is clear because science has never existed apart from production, and they both progressed in parallel. What is happening now, what technological revolution is behind a new, fourth paradigm? Yes, there is such a revolution; it has already begun! This happened in the United States, on September 10, 2004 through the efforts of a group of bureaucrats from the U.S. Food and Drug Administration (FDA).

To understand what is going on, let us consider, for example, the current technology in the automotive industry. Conceptually, it looks like this. Sydney produces the bolts, Taipei makes the nuts, and Salzgitter assembles them. With thanks to standardization and unification (QC does not sleep), the nuts from Taipei fit the bolts from Sydney, and a lot of cars are produced. However, such a concept is not suitable for the biological, chemical, pharmaceutical, and food industries. The standardized burgers, perfectly combined with the standardized buns will always lose to homemade rissoles, lovingly cooked by a thoughtful grandmother. The reason is a control of the cooking process, which is carried out by a grandmother not only at the beginning and at the end of the process, as it is provided by the production schedules, but continuously, at every stage, using the on-line method.

Let us talk about the methods of the process control. There are only four: off-, at-, on- and in-line. They are as follows. Let us say you are cooking a soup and want to know if there is enough salt in it. If you take a portion of the soup and send a sample to the chemical laboratory, this is an off-line method, and the result will be known in 2 weeks. If you drag the lab into the kitchen and check the sample next to the stove, this is an at-line approach, and the result will be obtained in an hour. In case your kitchen is equipped with a specialized high-speed sensor (grandmother), a sample can be placed into the sensor; this is an on-line control. The in-line option enables mounting a sensor into a pan for continuous monitoring. A soup is a simple object in comparison with high-quality medicines. It is clear that such a complex production should be monitored continuously, that is, by an on-line or an in-line method. Attempts to apply the differential paradigm here, that is, to describe all the stages by a system of differential equations and then to control only easily available parameters, such as temperature and pressure, have been undertaken many times. Countless number of papers and theses on this subject have been published. Unfortunately, the example of hydrogen oxidation clearly demonstrates the futility of these attempts. Well, it does not come out at all, no matter how carefully they are endeavoring.

What can and should be done in such a difficult situation? It seems easy to just equip the production line with a variety of sensors and let them continuously monitor all possible indicators during the process. Being collected in a real-time mode, these data contain hidden information on the process evolving. The only thing we need is to learn how to extract this information and how to use it for process control. To do this, first, we should accumulate sufficient historical experience, that is, to collect and organize the data regarding the successful, moderate, and poor production batches. Later, in the course of a new batch, we need to compare the current process data with those obtained before, checking on whether the process is going, as it should without any abnormalities. It is time to recall once again the grandmother, who cooks the soup exactly in this manner, that is, tasting and improving it in accordance with her rich experience collected before.

It was just this "grandmother" technology, based on the real-time continuous monitoring of the quality attributes, the comparing of current data with historical experience, the application of the sensors providing multivariate indirect data about the system state, etc.;

all these have been prescribed to introduce pharmaceutical and similar enterprises. To this end, in September 2004, the FDA issued a document entitled "PAT–A Framework for Innovative Pharmaceutical Development, Manufacturing, and Quality Assurance." It outlines the main principles of this approach, the main motto of which is "building quality into products." We cannot say that nothing has happened before, and we were just waiting for the directive. Over the last 10 years, chemometricians actively cooperated with producers, applying these methods in practice to food, pharmaceutical, iron, and steel industries. Now it has become a law, that is, nobody can now produce or sell a drug in the United States without such a control. Chemometrics has been recognized![3]

Paradoxically, we found the first stirrings of the paradigm in a modest, little-known science, chemometrics, which is mainly regarded as a subsection of analytical chemistry, designed to solve metrological and methodological problems in the interpretation of chemical experiments. Who knows, maybe in 100 years, this paradigm will be called chemometric. Let us wait and see.

Note. This chapter is based on a paper "The Fourth paradigm" published in a popular Russian science journal[4] in the year 2006. Recently I found a book that independently brought forward the same idea on the evolution of science.[5]

[3] http://www.fda.gov/downloads/Drugs/GuidanceComplianceRegulatoryInformation/Guidances/UCM070305.pdf
[4] A.L. Pomerantsev. The Fourth Paradigm, Khimia i zhizn – XXI vek, 2006, issue 6, 22–28 (in Russian).
[5] T. Hey, S. Tansley, K. Tolle (Eds.). *The Fourth Paradigm: Data-Intensive Scientific Discovery*, Microsoft, 2010.

INDEX

Chemometrics in Excel, First Edition. Alexey L. Pomerantsev.